D0204127

DIFFERENTIAL EQUATIONS AND MATHEMATICAL BIOLOGY

CHAPMAN & HALL/CRC
Mathematical Biology and Medicine Series

Aims and scope:
This series aims to capture new developments and summarize what is known over the whole spectrum of mathematical and computational biology and medicine. It seeks to encourage the integration of mathematical, statistical and computational methods into biology by publishing a broad range of textbooks, reference works and handbooks. The titles included in the series are meant to appeal to students, researchers and professionals in the mathematical, statistical and computational sciences, fundamental biology and bioengineering, as well as interdisciplinary researchers involved in the field. The inclusion of concrete examples and applications, and programming techniques and examples, is highly encouraged.

Series Editors

Alison M. Etheridge
Department of Statistics
University of Oxford

Louis J. Gross
Department of Ecology and Evolutionary Biology
University of Tennessee

Suzanne Lenhart
Department of Mathematics
University of Tennessee

Philip K. Maini
Mathematical Institute
University of Oxford

Hershel M. Safer
Informatics Department
Zetiq Technologies, Ltd.

Eberhard O. Voit
Department of Biometry and Epidemiology
Medical University of South Carolina

Proposals for the series should be submitted to one of the series editors above or directly to:
CRC Press UK
23 Blades Court
Deodar Road
London SW15 2NU
UK

Chapman & Hall/ CRC Mathematical Biology and Medicine Series

Differential Equations and Mathematical Biology

D. S. Jones
B. D. Sleeman

CHAPMAN & HALL/CRC

A CRC Press Company
Boca Raton   London   New York   Washington, D.C.

Library of Congress Cataloging-in-Publication Data

Jones, D. S. (Douglas Samuel)
    Differential equations and mathematical biology / D.S. Jones and B.D. Sleeman.
      p.   cm. -- (Chapman & Hall/CRC mathematical biology and medicine series)
    Originally published: London ; Boston : Allen & Unwin, 1983.
    ISBN 1-58488-296-4 (alk. paper)
    1. Biomathematics. 2. Differential equations. I. Sleeman, B. D. II. Title. III. Series.

QH323.5 .J65 2003
570′.15′1535--dc21
                            2002191159

This book contains information obtained from authentic and highly regarded sources. Reprinted material is quoted with permission, and sources are indicated. A wide variety of references are listed. Reasonable efforts have been made to publish reliable data and information, but the author and the publisher cannot assume responsibility for the validity of all materials or for the consequences of their use.

Neither this book nor any part may be reproduced or transmitted in any form or by any means, electronic or mechanical, including photocopying, microfilming, and recording, or by any information storage or retrieval system, without prior permission in writing from the publisher.

The consent of CRC Press LLC does not extend to copying for general distribution, for promotion, for creating new works, or for resale. Specific permission must be obtained in writing from CRC Press LLC for such copying.

Direct all inquiries to CRC Press LLC, 2000 N.W. Corporate Blvd., Boca Raton, Florida 33431.

Trademark Notice: Product or corporate names may be trademarks or registered trademarks, and are used only for identification and explanation, without intent to infringe.

Visit the CRC Press Web site at www.crcpress.com

© 2003 by Chapman & Hall/CRC

No claim to original U.S. Government works
International Standard Book Number 1-58488-296-4
Library of Congress Card Number 2002191159
Printed in the United States of America   2 3 4 5 6 7 8 9 0
Printed on acid-free paper

This book is dedicated with deep gratitude to our wives, Ivy and Julie, for their constant encouragement, support and understanding during the long period of its preparation

Preface

In recent years, mathematics has made a considerable impact as a tool with which to model and understand biological phenomena. In return biology has confronted the mathematician with a variety of challenging problems which have stimulated developments in the theory of non-linear differential equations. This book is the outcome of the need to introduce undergraduates of mathematics, the physical and biological sciences to some of these developments. It is primarily directed to university students who are interested in modelling and the application of mathematics to biological and physical situations.

Chapter 1 is introductory, showing how the study of first-order ordinary differential equations may be used to model the growth of a population, monitoring the administration of drugs and the mechanism by which living cells divide. In Chapter 2, a fairly comprehensive account of a linear ordinary differential equation with constant coefficients is given while Chapter 3 extends the theory to systems of equations. Such equations arise frequently in the discussion of the biological models encountered throughout the text. Chapter 4 is devoted to modelling biological phenomena and in particular includes (i) physiology of the heart beat cycle, (ii) blood flow, (iii) the transmission of electrochemical pulses in the nerve, (iv) the Belousov-Zhabotinskii chemical reaction and (v) predator-prey models.

Nearly all the biological models described in Chapter 4 have special solutions which arise as solutions to first-order autonomous systems of non-linear differential equations. Chapter 5 gives an account of such systems through the use of the Poincaré phase plane.

With the knowledge of differential equations developed thus far, we are in a position to begin an analysis of the heart beat, nerve impulse transmission, chemical reactions and predator-prey problems. These are the subjects of Chapters 6–9.

In order to gain a deeper insight into biological models, it is necessary to have a knowledge of partial differential equations. These are the subject of Chapters 10 and 11. In particular, a number of the models discussed in Chapter 4 involve processes of diffusion (Chapter 12), and the evolutionary equations considered in Chapter 11 are basic for an understanding of these processes. A special feature of Chapter 12 is a treatment of pattern formation in developmental biology based on Turing's famous idea of diffusion driven instabilities. The theory of bifurcation and chaotic behaviour is playing an increasing role in fundamental problems of biological modelling. An introduction to these topics is contained in Chapter 13. Chapter 14 models and studies problems of growth of solid avascular tumours. Again differential equations play a fundamental part. However, a new feature here is that we encounter moving boundary problems. The book concludes in Chapter 15 with a discussion of epidemics and the spread of infectious diseases, modelled via various differential equations.

As an encouragement to further study, some of the chapters have notes indicating sources of material as well as references to additional literature. Each chapter has a set of exercises which either illustrate some of the ideas discussed or require readers to develop and test models of their own.

In writing the book, the authors have endeavoured to give it a multipurpose role. For example, it can be used (i) as a course in differential equations based on Chapters 1, 2, 3, 5, 10 and 11, (ii) as a course in biological modelling for students of mathematics and the physical sciences or (iii) as a course in differential equation models of biology for life science students based on Chapters 1, 2, 3, 5, 10, 11, 12 and 13, together with a selection of the remaining chapters depending on the students' interests. Throughout the stages of writing this book, the authors have benefited from discussions and advice from colleagues in the Department of Mathematics, University of Dundee and the School of Mathematics, University of Leeds. To them all, we express our appreciation.

Finally, it is a pleasure to thank Nick Hill, Mel Holmes, Michael Plank, Mrs. Doreen Ross and David Sleeman for all their efforts in connection with the preparation of this book.

D. S. JONES and B. D. SLEEMAN
Dundee and Leeds

Contents

Chapter 1

Introduction

1.1 Population growth

To indicate why study of differential equations can be useful some simple examples will be considered.

The way that the size of a population varies in time is a matter of interest in several contexts. Let the number of individuals in a given area at time t be $p(t)$. At time $t+T$ the number of individuals is $p(t+T)$ so that $p(t+T)-p(t)$ must be the number of individuals that have been added to the population during the time interval T. The longer the interval the more individuals can be expected to arrive and the shorter the time the fewer additions can be expected. So write the change in the time interval T as NT and then

$$p(t+T) - p(t) = NT$$

or

$$\frac{p(t+T) - p(t)}{T} = N.$$

Letting $T \to 0$ we see that the left-hand side becomes the derivative of p with respect to t. Consequently, we have

$$\frac{dp(t)}{dt} = N.$$

To gain an idea of the properties of N suppose that the change in size of the population is due entirely to individuals being born. As time progresses the fertility of parents may alter so that more or less offspring are born. Then the number born in the interval T may vary as time proceeds. In other words N may alter as t does. Another effect is introduced by the simple hypothesis that the more individuals there are at time t the more births are likely to occur. Then N will depend on $p(t)$ also. Both possibilities can be allowed for by rewriting our equation as

$$\frac{dp(t)}{dt} = N\{t, p(t)\} \tag{1.1.1}$$

to show explicitly quantities on which N depends.

Sometimes

$$\frac{1}{p}\frac{dp}{dt}$$

is known as the **specific growth rate**. So another way of describing (1.1.1) is to say that the specific growth rate is $N(t,p)/p$.

It is plausible to assume that, in a short time interval, there will be about twice as many births as in a time interval of half its length. Thus, one could expect that the number of births would be proportional to $p(t)T$ when T is small. If the fertility of the parents does not change the actual number of births in the time interval, T can be expressed as $N_0 p(t)T$ with N_0 a suitable constant. Then (1.1.1) becomes

$$\frac{dp(t)}{dt} = N_0 p(t) \tag{1.1.2}$$

which states that the specific growth rate is N_0, the same for all times and all sizes of population.

The solution of (1.1.2) is

$$p(t) = p_0 e^{N_0 t} \tag{1.1.3}$$

where p_0 is any constant. This may be confirmed by taking a derivative of (1.1.3) with respect to t. The value of p_0 can be fixed by putting $t = 0$ in (1.1.3); evidently p_0 is the size of the population at $t = 0$.

Another way of verifying the result in (1.1.3) is to integrate (1.1.2). Thus

$$N_0 \int_0^t dt = \int_0^t \frac{1}{p}\frac{dp}{dt}dt$$
$$= \int_{p(0)}^{p(t)} \frac{dp}{p}$$

on changing the variable of integration from t to p. Hence

$$N_0 t = \ln\{p(t)/p(0)\} \tag{1.1.4}$$

which agrees with (1.1.3).

The behaviour of the population as time increases according to (1.1.3) is displayed in Figure 1.1.1. The size grows steadily, and the increase becomes dramatic as time goes on. Of course, in any real situation, there will be a limit to the growth because of a shortage of essential supplies, insufficient food and self-pollution of the environment. Nevertheless, many organisms exhibit exponential growth in their initial stages. It is always easy to check whether a population is growing exponentially by plotting $\ln p$ against time; a straight line should be obtained. The slope of the line is the specific growth rate N_0 as is clear from (1.1.4).

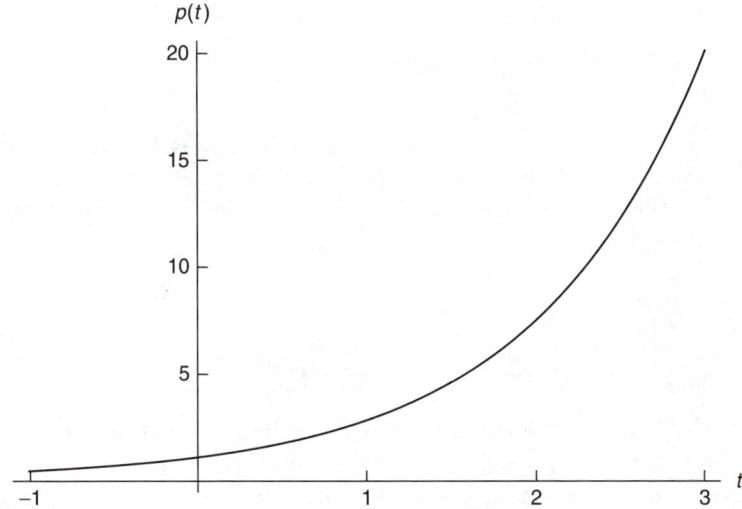

FIGURE 1.1.1: Graph of exponential growth.

Equation (1.1.2) has been derived on the assumption that only births occur. In the event that there are deaths but no births the same equation can be reached. However, N_0 is now a negative number since the population decreases in the time interval T. It follows from (1.1.3) that the population decays exponentially with time from its size at $t = 0$.

More facets of the population problem can be incorporated. For instance, we may postulate that the number of deaths in the short time interval T is $D_0 p(t)T$. Similarly, individuals may enter the given area from outside, say $I(t)T$ immigrants in the interval T. Likewise, some may depart from the area giving rise to $E(t)T$ emigrants. Then

$$p(t+T) - p(t) = N_0 p(t)T - D_0 p(t)T + I(t)T - E(t)T$$

leading to

$$\frac{dp(t)}{dt} = (N_0 - D_0)p(t) + I(t) - E(t) \tag{1.1.5}$$

when $T \to 0$.

More generally, I and E could be made to depend on p so that (1.1.5), which is often called **Verhulst's differential equation**, can be difficult to solve. In fact, (1.1.5) could be regarded as a particular case of (1.1.1) with an appropriate definition of N but that does not make it any easier to solve. Notwithstanding, it is transparent that, if we hope to predict the size of a population at a given time, finding the solution of a differential equation will be an essential requirement.

1.2 Administration of drugs

When a drug is administered, it forms a concentration in the body fluids. This concentration diminishes in time through elimination, destruction or inactivation. The rate of reduction of the concentration is found, in most cases, to be proportional to the concentration. Therefore, if $c(t)$ is the concentration at time t, we have

$$\frac{d}{dt}c(t) = -\frac{c(t)}{\tau}, \tag{1.2.1}$$

where τ is a constant which measures the rapidity at which the concentration falls.

Exactly the same differential equation can be derived from (1.1.5) for a population which changes only by deaths which happen at a constant specific rate. So conclusions about drugs can be transferred easily to populations which alter by death alone.

Analogous to (1.1.3) the solution of (1.2.1) is

$$c(t) = c_0 e^{-t/\tau}, \tag{1.2.2}$$

where c_0 is the concentration at time $t = 0$. Notice that when $t = \tau$ the concentration has dropped to c_0/e, so that in time τ the concentration has been reduced to $1/e$ of its initial value. This explains the significance of the time τ; the larger it is, the more slowly the drug disperses.

According to (1.2.2) the drug never disappears completely from the body except after infinite time. However, the residue will usually be negligible when $t \gg \tau$. Notwithstanding this, the fact that some of the drug is always left is relevant when repeated doses are made, as is common practice. The level to which the drug accumulates is then of particular importance.

Imagine that the dose c_0 is administered regularly at the times $t = 0, t_0, 2t_0, 3t_0, \ldots$. At time t_0 the residue r_1 **just before the second dose** is

$$r_1 = c_0 e^{-t_0/\tau},$$

and then the second dose is given so that the total concentration c_1 is given by

$$c_1 = c_0 + c_0 e^{-t_0/\tau}.$$

From $t = t_0$ to $t = 2t_0$ the concentration will fall exponentially so that the residue r_2 **just before the third dose at** $t = 2t_0$ is

$$r_2 = c_1 e^{-t_0/\tau} = c_0 e^{-t_0/\tau}(1 + e^{-t_0/\tau}).$$

After the third dose the concentration c_2 is

$$c_2 = c_0 + r_2 = c_0(1 + e^{-t_0/\tau} + e^{-2t_0/\tau}).$$

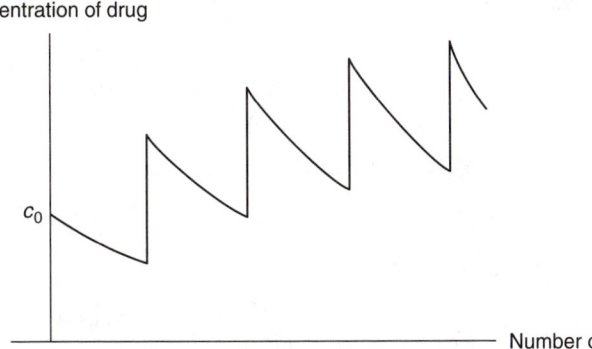

FIGURE 1.2.1: Concentration of drug initially.

Clearly, if we keep doing this, we shall discover that at $t = (n-1)t_0$, with n a positive integer,

$$\begin{aligned} c_{n-1} &= c_0(1 + e^{-t_0/\tau} + e^{-2t_0/\tau} + \cdots + e^{-(n-1)t_0/\tau}) \\ &= c_0 \frac{1 - e^{-nt_0/\tau}}{1 - e^{-t_0/\tau}} \end{aligned} \qquad (1.2.3)$$

on summing the geometric series. The residue r_n at $t = nt_0$ will be

$$r_n = c_{n-1}e^{-t_0/\tau} = c_0 e^{-t_0/\tau}\frac{1 - e^{-nt_0/\tau}}{1 - e^{-t_0/\tau}} \qquad (1.2.4)$$

from (1.2.3).

The manner in which the drug builds up as the number of doses increases is illustrated in Figure 1.2.1. The level of concentration grows in an oscillatory fashion. At first sight it looks as though there is no limit to the concentration. To check whether this is true examine (1.2.3). The only term involving n is $e^{-nt_0/\tau}$ in the numerator. Since this term decreases as n increases c_n does increase with n. However, the growth is not unlimited because the exponential decays to zero. Thus the concentration never exceeds c_M where

$$c_M = \frac{c_0}{1 - e^{-t_0/\tau}}. \qquad (1.2.5)$$

Although the concentration becomes c_M only in the limit as $n \to \infty$, yet c_M is a good estimate of the concentration immediately after a dose when $nt_0/\tau \gg 1$. Indeed, if $nt_0 > 5\tau$, c_{n-1} differs from c_M by less than 1% so that, unless t_0/τ is small, the level of concentration will not be far from c_M after a few doses. To put it another way, if you want to reach the maximum concentration in about 5 doses, you should make the interval between doses larger than τ. Naturally, the larger t_0/τ is made the closer the maximum becomes to c_0, the concentration of a single dose.

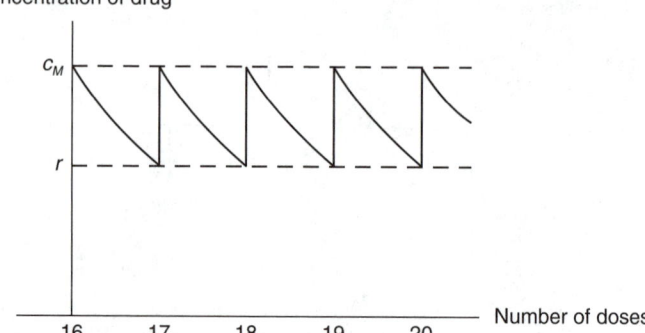

FIGURE 1.2.2:   Behaviour of concentration eventually.

Similarly, just before a dose the residue approaches r as n increases where r is given by

$$r = c_M e^{-t_0/\tau} = \frac{c_0}{e^{t_0/\tau} - 1}. \tag{1.2.6}$$

Observe that $c_M = c_0 + r$. Notice also that (1.2.6) implies that r becomes small when t_0/τ is large. The larger t_0/τ the more the level of concentration varies between doses. Thus, there is a trade-off between keeping the residue above a certain level and reaching c_M in a few doses.

When sufficient doses have been administered for (1.2.5) and (1.2.6) to be good approximations the concentration behaves as in Figure 1.2.2. It swings between c_M and r, never exceeding c_M nor falling below r.

The oscillatory build-up of Figure 1.2.1 may be undesirable. Several antibiotics can have harmful effects until their concentration has surpassed a certain threshold, since sub-optimal concentrations may induce resistance to the drug by the micro-organisms occurring. The oscillatory growth can be avoided by taking advantage of the behaviour in Figure 1.2.2. An initial large dose of $c_0 + r$ or c_M is given and thereafter doses of c_0 are supplied at intervals of t_0. The first dose takes the concentration to c_M and from then on the level follows the curve of Figure 1.2.2. So long as r is above any threshold imposed the difficulty referred to has been surmounted.

1.3   Cell division

When cells divide, their numbers grow by a process akin to that of Section 1.1. A new feature is that the multiplication in numbers is restricted by crowding effects. Biochemically, these may be due to lack of nutrient, shortage of oxygen, change in pH or the production of inhibitors, for example. Whatever the cause, the cells are interacting with one another. Since each cell can interact with p others there are p^2 possibilities. This suggests that, in (1.1.1), we should put

$$N\{t, p(t)\} = N_0 p(t) - ap(t)^2 \qquad (1.3.1)$$

where N_0 and a are positive constants. The term involving N_0 is the same as before and accounts for the increase due to division. The term containing a represents the inhibition on growth caused by crowding. With the substitution (1.3.1), (1.1.1) gives

$$\frac{dp}{dt} = N_0 p - ap^2 \qquad (1.3.2)$$

which is called the **differential equation of logistics**.

If we integrate (1.3.2) from 0 to t, as in Section 1.1, we obtain

$$\int_0^t dt = \int_0^t \frac{1}{N_0 p - ap^2} \frac{dp}{dt} dt$$
$$= \int_{p(0)}^{p(t)} \frac{dp}{N_0 p - ap^2},$$

when p is employed as the variable of integration. Now,

$$\int \frac{dp}{N_0 p - ap^2} = \frac{1}{N_0} \int \left(\frac{1}{p} - \frac{a}{ap - N_0} \right) dp = \frac{1}{N_0} \ln \left(\frac{p}{ap - N_0} \right).$$

Hence we have

$$t = \frac{1}{N_0} \ln \left(\frac{p(t)\{ap(0) - N_0\}}{\{ap(t) - N_0\}p(0)} \right)$$

whence

$$p(t)\{ap(0) - N_0\} = \{ap(t) - N_0\}p(0)e^{N_0 t}.$$

Consequently,

$$p(t) = \frac{N_0 p(0)}{ap(0) + \{N_0 - ap(0)\}e^{-N_0 t}} \qquad (1.3.3)$$

which is known as the **logistic law of growth**.

The curve of logistic growth is shown in Figure 1.3.1, assuming that $N_0 > ap(0)$. The curve rises steadily from the value $p(0)$ at $t = 0$ to an eventual value of N_0/a, there being no maxima or minima in between. There is, however, a point of inflexion where the curve crosses its tangent at $t = t_0$ where

$$t_0 = \frac{1}{N_0} \ln \left(\frac{N_0}{ap(0)} - 1 \right)$$

and $p(t_0) = N_0/2a$.

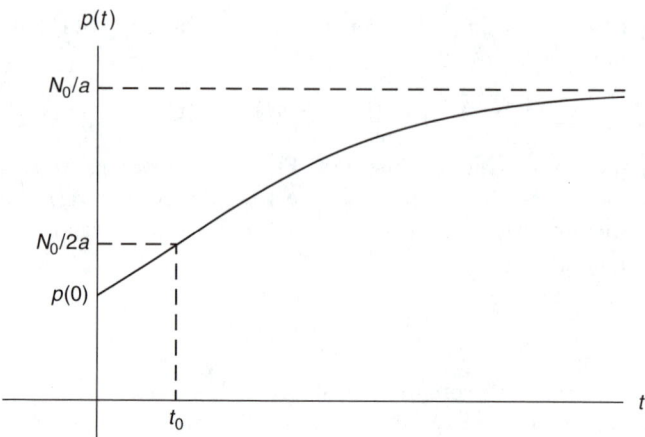

FIGURE 1.3.1:   The curve of logistic growth.

Observe that the final value N_0/a of p does not involve $p(0)$, so that, no matter what the initial size of the population, its final size is always the same and does not depend on the starting size of the population.

The logistic law assumes that all the cells divide at the same rate. This is not always true. There are types in which some cells divide faster than others. Whether the logistic law can be applied still depends upon the differences between the various rates of division present. If the rates are not too far apart it is probably feasible to take N_0 as their average. For greater deviations it may be necessary to adopt a model in which the statistics of the number of cells of a given age and type at a given time play a part.

1.4   Differential equations with separable variables

Having seen from the preceding sections that models lead naturally to differential equations, we shall devote the rest of this chapter to investigating some of the methods of solution for differential equations. In general, a differential equation picked out of a hat will be insoluble, at any rate in terms of elementary functions. However, progress can be made with particular types and attention will be concentrated on those which yield to analytical attack.

The first type to be considered is that which can be written

$$f(y)\frac{dy}{dt} = g(t). \tag{1.4.1}$$

It is called a **differential equation of the first order with separable variables**. The phrase "of the first order" signifies that the only derivative which is present, namely dy/dt, is of the first order. Integrate both sides of (1.4.1) to obtain

$$\int_{t_0}^{t} f(y)\frac{dy}{dt}dt = \int_{t_0}^{t} g(t)dt$$

where t_0 is some constant. If t_1 is another constant

$$\int_{t_0}^{t} g(t)dt = \int_{t_1}^{t} g(t)dt + \int_{t_0}^{t_1} g(t)dt$$

and the second term on the right-hand side is just a constant. Therefore we can write

$$\int^{t} f(y)\frac{dy}{dt}dt = \int^{t} g(t)dt + C \qquad (1.4.2)$$

without indicating the lower limit of integration (other than that it is a constant) providing that the constant C is left at our disposal. On the left-hand side of (1.4.2) change the variable of integration to y and then

$$\int^{y} f(y)dy = \int^{t} g(t)dt + C. \qquad (1.4.3)$$

If (1.4.1) were expressed formally as

$$f(y)dy = g(t)dt,$$

(1.4.3) could be obtained by placing an integral sign on each term and adding the constant C.

A value cannot be assigned to C unless y is prescribed for some value of t. Thus a general solution such as (1.4.3) is bound to involve an arbitrary constant.

Example 1.4.1
Find the general solution of

$$(T^2 - t^2)\frac{dy}{dt} + ty = 0,$$

T being a constant.

Using the formal approach described above, we rewrite the differential equation as

$$\frac{dy}{y} + \frac{t dt}{T^2 - t^2} = 0,$$

so that the general solution is

$$\int^{y} \frac{dy}{y} + \int^{t} \frac{t dt}{T^2 - t^2} = C.$$

By carrying out the integrations, we obtain

$$\ln y - \tfrac{1}{2} \ln |T^2 - t^2| = C$$

whence

$$y = e^C |T^2 - t^2|^{1/2}.$$

Put $e^C = D$ and then the general solution is

$$y = D|T^2 - t^2|^{1/2}, \tag{1.4.4}$$

where now D is the arbitrary constant.

If it is known that $y = 1$ when $t = T/2$, substitution in (1.4.4) gives

$$1 = DT\sqrt{3}/2$$

from which is deduced that $D = 2/T\sqrt{3}$. Thus

$$y = 2|T^2 - t^2|^{1/2}/T\sqrt{3} \tag{1.4.5}$$

is the solution of the differential equation which takes the value 1 when $t = T/2$. ▯

All the solutions derived in the first three sections were based on separable equations but the existence of Verhulst's equation (1.1.5) indicates that other types need to be discussed.

1.5   General properties

A typical differential equation of the first order can be expressed as

$$\frac{dy}{dt} = f(t, y). \tag{1.5.1}$$

Precise attributes of the function f that will guarantee that the differential equation does possess a solution are not of concern here. It will be sufficient to suppose that corresponding to each pair of values (t, y) there is a definite value of f and that small changes of t, y will be accompanied by only a small variation in f. Then, if a point (t, y) in the (t, y)-plane is chosen, (1.5.1) asserts that at (t, y) the derivative dy/dt has the value $f(t, y)$. In other words, (1.5.1) assigns a direction to a point of the (t, y)-plane. If a curve can be drawn so that, at each of its points, its gradient satisfies (1.5.1), then this curve will have an equation which is a solution of the differential equation. Any such curve may be called a **solution curve** of the differential equation.

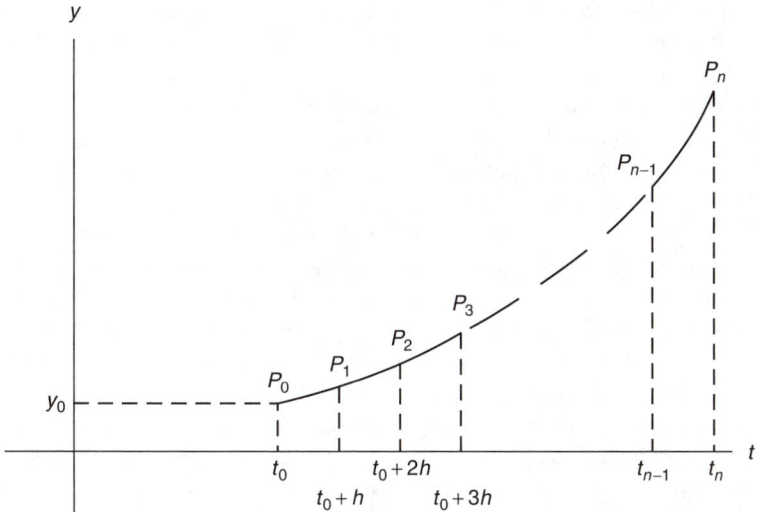

FIGURE 1.5.1: Polygonal approximation to solution curve.

An approximate solution curve can be constructed graphically. Let it be required that $y = y_0$ when $t = t_0$. Take the point (t_0, y_0) in the (t, y)-plane as P_0 (Figure 1.5.1). Draw the straight line through P_0 with slope $f(t_0, y_0)$ and let it intersect $t = t_0 + h$ at P_1. Let P_1 be the point $(t_0 + h, y_1)$. At P_1 draw the straight line of slope $f(t_0 + h, y_1)$ and let it intersect $t = t_0 + 2h$ at P_2, with ordinate y_2. Draw the straight line $P_2 P_3$ with slope $f(t_0 + 2h, y_2)$. Continuing in this way, we construct a polygon $P_0 P_1 P_2 P_3 \ldots$ which has the slope prescribed by the differential equation at the points $t = t_0, t_0 + h$, $t_0 + 2h, \ldots$. If h is kept small, the slope of the polygonal curve will never deviate by much from that demanded by the differential equation. It would therefore seem that as $h \to 0$ the polygonal curve would tend to the solution curve of the differential equation which passes through (t_0, y_0), i.e., provide the solution of (1.5.1) such that $y(t_0) = y_0$. Since only one slope can be drawn at each point, there is only one solution curve.

The procedure can be put in analytical terms as follows. Let $t_1 = t_0 + h$; then P_1 is the point (t_1, y_1) where

$$t_1 = t_0 + h, \qquad y_1 = y_0 + hf(t_0, y_0).$$

Similarly, P_2 is (t_2, y_2) where

$$t_2 = t_1 + h, \qquad y_2 = y_1 + hf(t_1, y_1)$$

and, in general, P_n is defined by

$$t_n = t_{n-1} + h, \qquad y_n = y_{n-1} + hf(t_{n-1}, y_{n-1}).$$

We anticipate that $y_1, y_2, \ldots$ will be reasonable approximations to the values of the solution curve at $t_1, t_2, \ldots$ though it must be stressed that we have not *proved* that this is so.

This approximate technique is known as **Euler's method** and is the prototype for all numerical methods of solving a differential equation by a step-by-step process. Many sophisticated methods are now available for solving differential equations numerically and are capable of providing solutions to any required degree of accuracy. Indeed, there are programs which automatically adjust themselves so as to achieve the accuracy prescribed by the user at the outset. However, the study of numerical techniques would take us too far afield and we refer the reader to specialist texts such as Lambert (1973, 1991).

The solution curve has an equation which expresses y as a function of t in such a way that $y = y_0$ at $t = t_0$. If the starting point is altered, another solution curve is generated. So, by varying the initial values, a family of solution curves is produced. The family must be arbitrary to the extent of covering all initial values that can be contemplated. In fact, if the equation of the family is

$$F(t, y) = 0,$$

the arbitrary element is determined completely by

$$F(t_0, y_0) = 0,$$

because only one solution curve passes through (t_0, y_0). Hence the equation $F(t, y) = 0$ can be written as

$$g(t, y) = C, \tag{1.5.2}$$

where C is an arbitrary constant and g is completely specified.

The strict proof of the validity of the above statements is beyond the scope of this book. Nevertheless, the arguments set out are sufficiently plausible for the reader to understand that, when a solution of (1.5.1) has been found in the form (1.5.2) involving a single arbitrary constant, that solution constitutes a **general solution** of the differential equation. Such general solutions have already been encountered when the variables are separable.

1.6 Equations of homogeneous type

A differential equation that has the form

$$\frac{dy}{dt} = f\left(\frac{y}{t}\right), \tag{1.6.1}$$

where $f(y/t)$ is a function of the single variable y/t, is said to be of **homogeneous type**.

To solve (1.6.1), make the substitution

$$y = tz. \tag{1.6.2}$$

Then, by the product rule for taking a derivative,

$$\frac{dy}{dt} = z + t\frac{dz}{dt}$$

and so (1.6.1) becomes

$$z + t\frac{dz}{dt} = f(z)$$

or

$$t\frac{dz}{dt} = f(z) - z. \tag{1.6.3}$$

In (1.6.3) the variables are separable, and so it may be solved as in Section 1.4. All that the reader needs to remember, therefore, is to make the change of variable (1.6.2) when an equation of homogeneous type is met.

Example 1.6.1

Find the general solution of

$$(3t - y)\frac{dy}{dt} + t = 3y.$$

Since the differential equation can be written as

$$\frac{dy}{dt} = \frac{3y - t}{3t - y} = \frac{3(y/t) - 1}{3 - (y/t)},$$

it is of homogeneous type. With the substitution (1.6.2), we have

$$z + t\frac{dz}{dt} = \frac{3z - 1}{3 - z},$$

which implies that

$$t\frac{dz}{dt} = \frac{z^2 - 1}{3 - z}.$$

Hence the general solution is

$$\int \frac{3 - z}{z^2 - 1}\,dz = \int \frac{dt}{t} + C.$$

Since

$$\frac{3}{z^2 - 1} = \frac{3}{2}\left(\frac{1}{z-1} - \frac{1}{z+1}\right),$$

we obtain

$$\frac{3}{2}\ln\left|\frac{z-1}{z+1}\right| - \frac{1}{2}\ln|z^2 - 1| = \ln t + C.$$

Consequently

$$\frac{|z-1|}{(z+1)^2} = Dt$$

where $D = e^C$. Substituting for z from (1.6.2), we obtain

$$y - t = D(y + t)^2$$

as the general solution in the original variables.   ⬜

1.7   Linear differential equations of the first order

A differential equation that can be expressed as

$$\frac{dy}{dt} + f(t)y = g(t), \tag{1.7.1}$$

where y does not occur in either $f(t)$ or $g(t)$, is called a **linear differential equation of the first order**.

Why the nomenclature "linear" is employed can be understood from consideration of (1.7.1) when g is replaced by zero so that the differential equation becomes

$$\frac{dy}{dt} + f(t)y = 0.$$

Suppose that y_1 and y_2 are solutions of this differential equation, i.e.,

$$\frac{dy_1}{dt} + f(t)y_1 = 0 \tag{1.7.2}$$

$$\frac{dy_2}{dt} + f(t)y_2 = 0. \tag{1.7.3}$$

The addition of (1.7.2) and (1.7.3) shows that $y_1 + y_2$ is also a solution. Actually, if A and B are constants, $Ay_1 + By_2$ is a solution. This construction of solutions by adding together constant multiples of solutions is the property of linearity. In non-linear differential equations, the sum of two solutions cannot be asserted to be a solution without independent verification. Another linear property is that, if $y_1 = 1$ when $t = t_0$ and $y_2 = C$ when $t = t_0$, then $y_2(t) = Cy_1(t)$. Thus, if the initial value of a solution is increased, all subsequent values are increased in the same proportion. The problem of cell division in Section 1.3 led to a solution which after a long time was the same for all initial populations and so the differential equation of logistics is non-linear, as is also evident from the structure of (1.3.2).

The technique for solving (1.7.1) is to multiply by $h(t)$ to form

$$h(t)\frac{dy}{dt} + h(t)f(t)y = h(t)g(t).$$

The function h is to be chosen so that the left-hand side is the same as $d(hy)/dt$; it is then known as an *integrating factor*.

Now

$$\frac{d}{dt}(hy) = h\frac{dy}{dt} + \frac{dh}{dt}y$$

which is the same as the left-hand side provided that

$$\frac{dh}{dt} = hf.$$

In this differential equation for h the variables are separable and so

$$\ln h(t) = \int^t f(u)du + C_1$$

where C_1 is a constant, the lower limit of integration being omitted for the reasons given in deriving (1.4.2). Hence

$$h(t) = \exp\left(C_1 + \int^t f(u)du\right).$$

However, if $h(t)$ is an integrating factor so is $C_2 h(t)$ when C_2 is a constant. Therefore we can take for our integrating factor

$$h(t) = \exp\left(\int^t f(u)du\right). \tag{1.7.4}$$

The differential equation, after multiplication by the integrating factor, can be written as

$$\frac{d}{dt}(hy) = hg$$

with the general solution

$$h(t)y(t) = \int^t h(v)g(v)dv + C.$$

Substitution from (1.7.4) gives

$$y(t) = \left[\int^t g(v)\exp\left(\int^v f(u)du\right)dv + C\right]\exp\left(-\int^t f(u)du\right). \qquad (1.7.5)$$

The rule for the reader to remember for a linear equation of the first order is to multiply by the integrating factor defined by (1.7.4), making sure first that the differential equation is in the form (1.7.1). It is not worth committing (1.7.5) to memory.

Example 1.7.1

Find the general solution of

$$(t^2 + 1)\frac{dy}{dt} + ty = \frac{1}{2}. \qquad (1.7.6)$$

The differential equation is linear with

$$f(t) = \frac{t}{t^2 + 1}, \qquad g(t) = \frac{1}{2}\frac{1}{t^2 + 1}.$$

Therefore

$$\int f(t)dt = \int \frac{t}{t^2 + 1}dt = \tfrac{1}{2}\ln(t^2 + 1)$$

and, according to (1.7.4), the integrating factor is

$$h(t) = \exp\left\{\tfrac{1}{2}\ln(t^2 + 1)\right\} = (t^2 + 1)^{1/2}.$$

Hence, multiply

$$\frac{dy}{dt} + \frac{t}{t^2 + 1}y = \frac{1}{2}\frac{1}{t^2 + 1}$$

by $(t^2 + 1)^{1/2}$ with the result

$$(t^2 + 1)^{1/2}\frac{dy}{dt} + \frac{t}{(t^2 + 1)^{1/2}}y = \frac{1}{2}\frac{1}{(t^2 + 1)^{1/2}}.$$

Consequently

$$\frac{d}{dt}\{(t^2 + 1)^{1/2}y\} = \frac{1}{2}\frac{1}{(t^2 + 1)^{1/2}},$$

which gives on integration

$$(t^2 + 1)^{1/2}y = \tfrac{1}{2}\ln\{t + (t^2 + 1)^{1/2}\} + C \qquad (1.7.7)$$

where C is an arbitrary constant. Equation (1.7.7) provides the general solution of (1.7.6). $\square$

Observe that, if $f(t)$ has the constant value f_0, the integrating factor is $e^{f_0 t}$ so that the general solution of (1.7.1) is then

$$y(t) = e^{-f_0 t} \int^t g(x)e^{f_0 x}dx + Ce^{-f_0 t}, \qquad (1.7.8)$$

C being an arbitrary constant.

1.8 Notes

For a specialist text on numerical techniques, the reader is referred to J. D. Lambert, *Computational Methods in Ordinary Differential Equations*, John Wiley & Sons, New York, 1973 and *Numerical Methods for Ordinary Differential Systems*, John Wiley & Sons, New York, 1991.

Exercises

1.1 A spherical water drop loses volume by evaporation at a rate proportional to its surface area. Express its radius at time t in terms of the constant of proportionality and its radius r_0 at $t = 0$.

1.2 The rate of increase of bacteria in a culture is proportional to the number present. The population multiplies by the factor n in the time interval T. Find the number of bacteria at time t when the initial population is p_0.

1.3 In Exercise 1.2 the population is found to increase by 2455 bacteria from $t = 2$ to $t = 3$ and by 4314 bacteria from $t = 4$ to $t = 5$. Show that $p_0 = 4291$ approximately and that, when $T = 3, n$ is about 2.33.

1.4 What changes take place in the curve of logistic growth in Figure 1.3.1 if $N_0 < ap(0)$?

1.5 All observations on animal tumours indicate that their sizes obey the **Gompertz growth law**

$$\frac{ds}{dt} = ks\ln\left(\frac{S}{s}\right)$$

rather than the logistic law. Here k and S are positive constants. By putting $y = \ln s$ prove that

$$s(t) = S\exp(-Ae^{-kt})$$

where $A = \ln(S/s_0)$, s_0 being the size at $t = 0$. Deduce that, in Gompertz growth, the size moves steadily from its initial value to an eventual value of S without passing through maxima or minima, though there is a point of inflexion if $s_0 < Se^{-1}$.

1.6 Find the general solutions of

(a) $t\dfrac{dy}{dt} = (1+t)y$,

(b) $t(2y+3)\dfrac{dy}{dt} = y(3+y)$,

(c) $2ty(1+t)\dfrac{dy}{dt} = 1+y^2$.

1.7 Find the general solutions of

(a) $\dfrac{dy}{dt} = \dfrac{t}{y} + \dfrac{y}{t}$,

(b) $\dfrac{dy}{dt} = \left(\dfrac{t}{y}\right)e^{-y/t} + \dfrac{y}{t}$,

(c) $t^2\dfrac{dy}{dt} = ty - y^2$.

1.8 (a) By means of the substitution $w - 1 = y, u + 2 = t$ show that

$$\frac{dw}{du} = \frac{u+w+1}{u-w+3}$$

can be solved.

(b) Generalise this result to

$$\frac{dw}{du} = \frac{au+bw+c}{a'u+b'w+c'}$$

when $ab' \neq a'b$ by substituting $w = y + h, u = t + k$ where

$$ak + bh + c = 0,$$
$$a'k + b'h + c' = 0.$$

1.9 Find the general solutions of

(a) $(1 - t^2)\dfrac{dy}{dt} - ty = (1 - t^2)^{1/2}$,

(b) $\dfrac{dy}{dt} = (t - 4)e^{4t} + ty$,

(c) $tw\dfrac{dw}{dt} = t^4 + w^2$ by putting $w^2 = y$.

1.10 **Bernoulli's differential equation** has the form

$$\frac{dy}{dt} + f(t)y = g(t)y^\nu.$$

Show that it can be made linear by the substitution $w = y^{1-\nu}$. Hence find the general solution of

$$\frac{dy}{dt} + ty = ty^2.$$

1.11 Use Exercise 1.10 to solve

(a) $(t + 1)\left(y\dfrac{dy}{dt} - 1\right) = y^2$,

(b) $(t^2 - 2y + 1)\dfrac{dy}{dt} = t$.

1.12 The function $y_1(t)$ is known to be a solution of the **Riccati equation**

$$\frac{dy}{dt} + a(t)y + b(t)y^2 = c(t).$$

Show that the general solution can be found by putting $y = y_1 + w$ and using Exercise 1.10. Hence solve

$$\frac{dy}{dt} + 2y^2 = \frac{6}{t^2}$$

given that $y = 2/t$ is a solution.

1.13 Use Exercise 1.12 to find the general solutions of

(a) $t^2\dfrac{dy}{dt} + ty + t^2y^2 = 4$ given that $y = 2/t$ is a solution,

(b) $t\dfrac{dy}{dt} - (2t + 1)y + y^2 = -t^2$ given that $y = t$ is a solution.

1.14 Find the general solutions of

(a) $(t^2 - 1)\dfrac{dy}{dt} + y = \frac{1}{2}(t^2 - 1)^{1/2}$,

(b) $t\dfrac{dy}{dt} = y + \left(\tfrac{1}{4}t^2 + y^2\right)^{1/2}$,

(c) $2\dfrac{dy}{dt} = e^{t+2y}$,

(d) $(2t + 2y + 5)\dfrac{dy}{dt} = 2y - 2t + 1$,

(e) $\dfrac{dy}{dt} = 2(2t - y)^2$,

(f) $(t + a)\dfrac{dy}{dt} - 3y = 2(t + a)^5$,

(g) $t\dfrac{dy}{dt} = y - \tfrac{1}{2}t\cos^2(2y/t)$,

(h) $(2y - 2t + 5)\dfrac{dy}{dt} = 2y - 2t + 1$,

(i) $\dfrac{dy}{dt} + (1 - 4y^2)\tan 2t = 0$,

(j) $(1 - t^2)\dfrac{dy}{dt} - \tfrac{1}{2}(1 + t)y = (1 - t^2)^{1/2}$,

(k) $(t + y)^2\dfrac{dy}{dt} = (t + y + 1)^2$,

(l) $1 + t\dfrac{dy}{dt} = e^{-y}\dfrac{dy}{dt}\sec^2 y$.

1.15 According to Newton's law of cooling, the rate of decrease of temperature of a body is proportional to the difference between its temperature and that of its environment. If the temperature of the environment is $20°C$ and the body cools from $80°C$ to $60°C$ in 1 h, show that it will take somewhat over 4 h to cool to $30°C$.

1.16 A body cools in 10 min from $100°C$ to $60°C$ when the environment is at $20°C$. How long does it take to cool to $25°C$?

1.17 After administration of a dose, the concentration of a drug decreases by 50% in 30 h. How long does it take to fall to 1% of its initial value?

1.18 The amount of light absorbed by a layer of material is proportional to the incident light and to the thickness of the layer. If a layer 35 cm thick absorbs half the light incident on its surface, what percentage of the incident light will be absorbed by a layer 200 cm thick?

1.19 When the drug theophylline is administered for asthma, a concentration below 5 mg l^{-1} has little effect and undesirable side-effects appear if the concentration exceeds 20 mg l^{-1}. For a body that weights W kg, the concentration when M mg is present is $2M/W$ mg l^{-1}. If the constant that measures the rapidity at which the concentration falls is $\tau = 6$ h, find the concentration at time t h after an initial dose of D mg.

If $D = 500$ and $W = 70$, show that a second dose is necessary after about 6 h to prevent the concentration from becoming ineffective. What further time can elapse before a third dose is necessary?

What is the shortest safe time interval t_0 at which doses of 500 mg can be given regularly?

1.20 If the drug in Exercise 1.19 is fed into the blood stream continuously by infusion at a rate of D_1 mg h^{-1}, instead of being given by separate doses, show that the concentration approaches a steady level of $12D_1/W$ mg l^{-1}. What permissible range of D_1 does this imply for a 60-kg patient?

1.21 In the reservoir model of the heart, it is imagined as a balloon. The balloon or heart is blown up by the influx of blood during the systole and, when the heart valve closes, the reservoir forces blood out during the diastole. At time t, the volume is $v(t)$, the inflow per unit time is $I(t)$ and the outflow per unit time is $F(t)$. Interpret

$$dv/dt = I(t) - F(t).$$

If the pressure $p(t)$ is such that

$$p(t) = K\{v(t) - v_0\}, \qquad F(t) = p(t)/R,$$

where K, v_0 and R are constants, find the differential equation satisfied by p.

In the diastole which lasts from $t = t_0$ to $t = T, I(t) = 0$. Find $p(t)$ in terms of p_0, the value of $p(t_0)$, during the time interval. In the systole, from $t = 0$ to $t = t_0, I(t) = I_0$ (a constant). Find p.

Since the heart is cyclic, $p(0) = p(T)$. Deduce that

$$R = \frac{p_0}{I_0} \frac{1 - \exp(-KT/R)}{1 - \exp(-Kt_0/R)}.$$

Chapter 2

Linear Ordinary Differential Equations with Constant Coefficients

2.1 Introduction

In Chapter 1, the solution of a first-order differential equation was considered. Higher derivatives can occur in some problems, and so we are led to the general ordinary differential equation

$$F\left(t, y, \frac{dy}{dt}, \frac{d^2y}{dt^2}, \dots, \frac{d^ny}{dt^n}\right) = 0, \tag{2.1.1}$$

where F is some function with $n + 2$ arguments. It is called **ordinary** because it involves only the ordinary derivatives of y with respect to the single variable t. Later on we will study cases in which y is a function of more than one variable. In that case, partial derivatives of y can arise and a partial differential equation has to be solved.

The **order** of an ordinary differential equation is the order of the highest derivative appearing. Thus

$$\frac{d^4y}{dt^4} = \frac{dy}{dt}\left(\frac{d^3y}{dt^3}\right)^4 + y^2$$

is of order 4, whereas

$$\left(\frac{dy}{dt}\right)^2 = t^2 + y^2$$

is of order 1.

The two main categories into which ordinary differential equations are classified are **linear** and **non-linear**. The form of the general linear ordinary differential equation of order n is

$$a_n(t)\frac{d^ny}{dt^n} + a_{n-1}(t)\frac{d^{n-1}y}{dt^{n-1}} + \cdots + a_1(t)\frac{dy}{dt} + a_0(t)y = f(t), \tag{2.1.2}$$

where $a_0(t), \dots, a_n(t)$ are known functions of t. If all of $a_0(t), \dots, a_n(t)$ are constants, (2.1.2) is known as a **linear ordinary differential equation with**

constant coefficients. Any ordinary differential equation which does not have the structure of (2.1.2) is called non-linear; it will contain products such as

$$y^2, \qquad \frac{dy}{dt}\frac{d^2y}{dt^2}$$

or functions such as e^y. For example,

$$t\frac{d^2y}{dt^2} + e^t\frac{dy}{dt} + y\cos t = t^3 \tan t$$

is linear and of order 2,

$$5\frac{d^2y}{dt^2} + 4\frac{dy}{dt} + 3y = \ln^2 t$$

is linear with constant coefficients and of order 2, while

$$y^2\frac{dy}{dt} = t$$

is non-linear.

The solution of an ordinary differential equation is always sought on an interval (a, b) $(a < b)$ of t. It is a relation between y and t which satisfies the ordinary differential equation when t is any point of the interval and does not contain any derivatives or integrals of y. Integrals of functions of t may be involved but these should be evaluated when it is reasonable to do so.

The **general solution** (sometimes called the **complete primitive**) of an ordinary differential equation of order n must contain n arbitrary constants. Any solution that does not have n arbitrary constants is not the general solution. For instance, you can check that

$$y = e^t - 1/t \tag{2.1.3}$$

satisfies the ordinary differential equation

$$t^3\left(\frac{d^2y}{dt^2} - y\right) = t^2 - 2, \tag{2.1.4}$$

but it is not the general solution because the general solution must contain two arbitrary constants whereas (2.1.3) has none. Similarly

$$y = Ce^{-t} - 1/t, \tag{2.1.5}$$

with C an arbitrary constant, is a solution of (2.1.4) but is not the general solution. On the other hand,

$$y = C_1e^t + C_2e^{-t} - 1/t, \tag{2.1.6}$$

with C_1 and C_2 arbitrary constants, is the general solution of (2.1.4).

When additional information is available, it may be possible to assign particular values to the arbitrary constants in the general solution. For example, suppose that a solution of (2.1.4) is desired such that $y = 1$ and $dy/dt = 1$ when $t = 1$. From the general solution (2.1.6)

$$dy/dt = C_1 e^t - C_2 e^{-t} + 1/t^2,$$

and so the conditions at $t = 1$ can be satisfied if

$$C_1 e + C_2 e^{-1} - 1 = 1,$$
$$C_1 e - C_2 e^{-1} + 1 = 1.$$

These require that $C_1 = e^{-1}$ and $C_2 = e$; hence

$$y = e^{t-1} + e^{1-t} - 1/t$$

is the solution of (2.1.4) which takes the correct values at $t = 1$.

An ordinary differential equation may not possess a solution. There are theorems, called **existence theorems**, which tell you that certain types of differential equations have a solution. If a solution exists and you can prove that it is the only one that satisfied any conditions imposed, then you have demonstrated a **uniqueness theorem**. Existence and uniqueness are beyond the scope of this chapter (for some information, see the Appendix to Chapter 5).

Even when existence and uniqueness theory is available, the actual finding of a solution may be a difficult task. For instance, there is an existence theorem for linear ordinary differential equations with variable coefficients, but the solution cannot always be written down easily. Again, the solution of

$$\frac{dy}{dt} = t^2 + y^2$$

cannot be expressed in terms of elementary functions although the solution is known to exist. There is, therefore, a need for approximate and numerical techniques for tackling differential equations, but we shall not mention them at this stage.

In fact, this chapter will be confined to discussing linear ordinary differential equations with constant coefficients. For these, not only is existence theory available but also the general solution can be determined explicitly.

2.2 First-order linear differential equations

The linear ordinary differential equation with constant coefficients of the first order is

$$a_1 \frac{dy}{dt} + a_0 y = f(t)$$

where a_1 and a_0 are constants. However, $a_1 \neq 0$ otherwise there would not be a differential equation to solve. Therefore we can divide by a_1 or, equivalently, put $a_1 = 1$ and take as the standard form

$$\frac{dy}{dt} + a_0 y = f(t). \tag{2.2.1}$$

This has been solved already in Section 1.7 and we may quote the general solution

$$y = e^{-a_0 t} \int^t e^{a_0 u} f(u) du + C e^{-a_0 t}, \tag{2.2.2}$$

where C is an arbitrary constant.

The term $C e^{-a_0 t}$ satisfies

$$\left(\frac{d}{dt} + a_0 \right) C e^{-a_0 t} = 0.$$

Since $C e^{-a_0 t}$ is the general solution of (2.2.1) with zero right-hand side it is called the **complementary function**. The integral term in (2.2.2) satisfies (2.2.1) and any solution of (2.2.1) is called a **particular integral**. Consequently,

the general solution of (2.2.1)

$= $ particular integral $+$ complementary function.

It does not matter which particular integral is chosen as we shall show now. Suppose $h(t)$ is any function such that

$$\frac{dh}{dt} + a_0 h = f(t).$$

Put $y = h + z$; then

$$\frac{dy}{dt} + a_0 y = \frac{dh}{dt} + \frac{dz}{dt} + a_0 h + a_0 z = f(t) + \frac{dz}{dt} + a_0 z$$

so that, if y satisfies (2.2.1), we must have

$$\frac{dz}{dt} + a_0 z = 0,$$

i.e., z is the complementary function.

2.3 Linear equations of the second order

The linear ordinary differential equation of the second order with constant coefficients can be expressed as

$$\frac{d^2 y}{dt^2} + a_1 \frac{dy}{dt} + a_0 y = f(t), \tag{2.3.1}$$

after division by the coefficient of the second derivative (which must be non-zero if the differential equation is to be of the second order). With zero right-hand side, (2.3.1) is

$$\frac{d^2y}{dt^2} + a_1 \frac{dy}{dt} + a_0 y = 0 \tag{2.3.2}$$

and is known as the **associated homogeneous differential equation**. Let the general solution of (2.3.2) be denoted by y_c and called the **complementary function**. Let y_p be any solution of (2.3.1) and designate it as a **particular integral**. Put $y = y_p + z$. Then

$$\frac{d^2y}{dt^2} + a_1 \frac{dy}{dt} + a_0 y = \frac{d^2y_p}{dt^2} + a_1 \frac{dy_p}{dt} + a_0 y_p + \frac{d^2z}{dt^2} + a_1 \frac{dz}{dt} + a_0 z$$

$$= f(t) + \frac{d^2z}{dt^2} + a_1 \frac{dz}{dt} + a_0 z.$$

Therefore y satisfies (2.3.1) provided that

$$\frac{d^2z}{dt^2} + a_1 \frac{dz}{dt} + a_0 z = 0,$$

i.e., $z = y_c$. Hence $y = y_p + y_c$ and

$$\text{general solution} = \text{particular integral} + \text{complementary function}$$

as before.

Thus the general solution of both first- and second-order equations has the same structure, and we shall find that this is true for the linear ordinary differential equation with constant coefficients of order n.

2.4 Finding the complementary function

The determination of the complementary function requires the general solution of (2.3.2), i.e., of

$$\ddot{y} + a_1 \dot{y} + a_0 y = 0 \tag{2.4.1}$$

if we use the notation $\dot{y} \equiv dy/dt, \ddot{y} \equiv d^2y/dt^2$.

Consider the equation for λ

$$\lambda^2 + a_1 \lambda + a_0 = 0, \tag{2.4.2}$$

which will be called the **characteristic equation**. It has two roots λ_1 and λ_2 such that

$$\lambda_1 + \lambda_2 = -a_1$$

and

$$\lambda_1 \lambda_2 = a_0.$$

Now

$$\frac{d}{dt}\left(\frac{dy}{dt} - \lambda_1 y\right) = \frac{d^2 y}{dt^2} - \lambda_1 \frac{dy}{dt},$$

so that

$$\left(\frac{d}{dt} - \lambda_2\right)\left(\frac{dy}{dt} - \lambda_1 y\right) = \frac{d^2 y}{dt^2} - (\lambda_1 + \lambda_2)\frac{dy}{dt} + \lambda_1 \lambda_2 y$$
$$= \ddot{y} + a_1 \dot{y} + a_0 y$$

because of the properties of λ_1 and λ_2. Consequently (2.4.1) can be written as

$$\left(\frac{d}{dt} - \lambda_2\right)\left(\frac{dy}{dt} - \lambda_1 y\right) = 0.$$

Write

$$w = \frac{dy}{dt} - \lambda_1 y.$$

Then

$$\frac{dw}{dt} - \lambda_2 w = 0.$$

This is of the first order and has general solution

$$w = Ce^{\lambda_2 t}.$$

Therefore

$$\frac{dy}{dt} - \lambda_1 y = Ce^{\lambda_2 t}.$$

Since this is linear and of the first order, it can be solved by multiplying by the integrating factor $e^{-\lambda_1 t}$, which gives

$$\frac{d}{dt}(ye^{-\lambda_1 t}) = Ce^{(\lambda_2 - \lambda_1)t}$$

whence

$$ye^{-\lambda_1 t} = C\int^t e^{(\lambda_2 - \lambda_1)u}du + D.$$

If $\lambda_1 \neq \lambda_2$,

$$\int^t e^{(\lambda_2-\lambda_1)u}\,du = \frac{e^{(\lambda_2-\lambda_1)t}}{\lambda_2 - \lambda_1}$$

and

$$ye^{-\lambda_1 t} = \frac{Ce^{(\lambda_2-\lambda_1)t}}{\lambda_2 - \lambda_1} + D$$

or

$$y = C_2 e^{\lambda_2 t} + C_1 e^{\lambda_1 t},$$

where C_1 and C_2 are arbitrary constants.

If $\lambda_1 = \lambda_2$,

$$\int^t e^{(\lambda_2-\lambda_1)u}\,du = \int^t du = t$$

and

$$ye^{-\lambda_1 t} = Ct + D$$

or

$$y = (C_1 + C_2 t)e^{\lambda_1 t}.$$

The rule for finding the complementary function can be stated now as: *solve the characteristic equation (2.4.2):*

(a) *if the roots are different*

$$y_c = C_1 e^{\lambda_1 t} + C_2 e^{\lambda_2 t};$$

(b) *if the roots are the same*

$$y_c = (C_1 + C_2 t)e^{\lambda_1 t}.$$

The characteristic equation may be arrived at in the following way. Try $y = e^{\lambda t}$ so that $\dot{y} = \lambda e^{\lambda t}, \ddot{y} = \lambda^2 e^{\lambda t}$. Then (2.4.1) is satisfied if

$$(\lambda^2 + a_1\lambda + a_0)e^{\lambda t} = 0$$

or

$$\lambda^2 + a_1\lambda + a_0 = 0,$$

which is the characteristic equation.

Example 2.4.1

Find the general solution of

$$16\ddot{y} - 8\dot{y} + y = 0.$$

The characteristic equation is

$$16\lambda^2 - 8\lambda + 1 = 0$$

so that $\lambda_1 = \frac{1}{4} = \lambda_2$. The rule now gives

$$y = (C_1 + C_2 t)e^{t/4}$$

as the general solution. ◻

It may happen that λ_1 or λ_2 is complex. No problem occurs because, if $\lambda_1 = a + ib$ where a and b are real, it is known that (compare De Moivre's theorem)

$$e^{\lambda_1 t} = e^{at} e^{ibt} = e^{at}(\cos bt + i \sin bt),$$

so that the solution can be expressed in terms of trigonometric functions if desired.

If a_0 and a_1 are real, we may be seeking a real solution of (2.4.1). In this case when $\lambda_1 = a + ib$ the characteristic equation forces λ_2 to be the complex conjugate, i.e., $\lambda_2 = a - ib$. Since $b \neq 0$, because we are assuming a complex root of (2.4.2), the rule gives

$$y_c = C_1 e^{(a+ib)t} + C_2 e^{(a-ib)t}$$

where C_1 and C_2 are arbitrary (complex) constants. To make y_c real, we must make $y_c^* = y_c$, the asterisk indicating a complex conjugate. But, when a_0 and a_1 are real, y_c^* also satisfies (2.4.1) and so

$$y_c^* = C_1^* e^{(a-ib)t} + C_2^* e^{(a+ib)t}.$$

Thus $y_c^* = y_c$ demands that $C_2^* = C_1$. This means that, if $C_1 = A + iB$ with A and B real, $C_2 = A - iB$. It follows that

$$y_c = (A + iB)e^{(a+ib)t} + (A - iB)e^{(a-ib)t}$$
$$= 2e^{at}(A \cos bt - B \sin bt),$$

which gives a real complementary function with real arbitrary constants A and B.

2.5 Determining a particular integral

There are various devices for finding a particular integral, each of which has advantages and disadvantages. Three methods will be described here.

(a)   Undetermined coefficients

The method consists essentially of guessing an appropriate form for the answer and substituting it in the differential equation. It is suitable if the right-hand side is a polynomial, exponential, sine, cosine or a product of these. The technique will be illustrated by means of examples.

Example 2.5.1
Find a particular integral of

$$\ddot{y} - 2\dot{y} - 3y = 3t^2.$$

The idea is to try to find a polynomial solution, and we show how this can be done by starting from the simplest and gradually making it more complicated.

Try $y = C$. Then $\ddot{y} - 2\dot{y} - 3y = -3C$. Since this can never agree with $3t^2$, it must be rejected.

Try $y = Bt$. Then $\ddot{y} - 2\dot{y} - 3y = -2B - 3Bt$, and again the attempt is unsatisfactory.

Try $y = At^2$. Then $\ddot{y} - 2\dot{y} - 3y = 2A - 4At - 3At^2$, which again is not suitable but does at least contain a term involving t^2.

This suggests that we should try $y = At^2 + Bt + C$. Then

$$\ddot{y} - 2\dot{y} - 3y = 2A - 2B - 3C - (4A + 3B)t - 3At^2.$$

We can make this the same as $3t^2$ if

$$-3A = 3,$$
$$4A + 3B = 0,$$
$$2A - 2B - 3C = 0.$$

These equations are satisfied if $A = -1, B = \frac{4}{3}$ and $C = -\frac{14}{9}$. Therefore

$$y_p = -t^2 + \tfrac{4}{3}t - \tfrac{14}{9}$$

supplies a particular integral.   ☐

In general this suggests, since the derivative of a polynomial of degree n is a polynomial of degree $n - 1$, that *when the right-hand side of (2.3.1) is a polynomial of degree n and $a_0 \neq 0$ try a polynomial of degree n with arbitrary coefficients. If $a_0 = 0$ but $a_1 \neq 0$ try a polynomial of degree $n + 1$.*

Example 2.5.2
Find a particular integral of

$$\ddot{y} - 2\dot{y} - 3y = e^{2t}.$$

In this case we try $y = Ae^{2t}$. Then the left-hand side is

$$4Ae^{2t} - 4Ae^{2t} - 3Ae^{2t} = -3Ae^{2t}$$

which is the same as e^{2t} if $A = -\frac{1}{3}$. Therefore

$$y_p = -\frac{1}{3}e^{2t}.$$

☐

Example 2.5.3

Find a particular integral of

$$\ddot{y} - 2\dot{y} - 3y = e^{-t}.$$

Again we try $y = Ae^{-t}$. The left-hand side becomes

$$Ae^{-t} + 2Ae^{-t} - 3Ae^{-t} = 0$$

so that A cannot be found. The try fails in this case because e^{-t} satisfies the associated homogeneous differential equation and so is part of the complementary function.

The lesson to be learned is that the complementary function should always be found first. The characteristic equation is

$$\lambda^2 - 2\lambda - 3 = 0$$

so that $\lambda_1 = 3, \lambda_2 = -1$ and

$$y_c = C_1 e^{3t} + C_2 e^{-t}.$$

If the right-hand side is part of the complementary function, it cannot be a suitable particular integral.

Try instead $y = Ate^{-t}$ so that $\dot{y} = A(1 - t)e^{-t}, \ddot{y} = A(t - 2)e^{-t}$. Then

$$\ddot{y} - 2\dot{y} - 3y = \{t - 2 - 2(1 - t) - 3t\}Ae^{-t}$$
$$= -4Ae^{-t}$$

which agrees with e^{-t} if $A = -\frac{1}{4}$. Therefore

$$y_p = -\frac{1}{4}te^{-t}.$$

☐

Example 2.5.4

Find a particular integral of

$$\ddot{y} - 4\dot{y} + 4y = e^{2t}.$$

To determine the complementary function, solve the characteristic equation

$$\lambda^2 - 4\lambda + 4 = 0,$$

which gives $\lambda_1 = 2 = \lambda_2$. Therefore

$$y_c = (C_1 + C_2 t)e^{2t}.$$

In this case, both e^{2t} and te^{2t} will fail as particular integrals. So try $At^2 e^{2t}$, which will be found to work.

Example 2.5.5

Find a particular integral of

$$\ddot{y} - 2\dot{y} - 3y = -9te^{2t}.$$

The right-hand side is not part of the complementary function and is a product of a polynomial and an exponential. This suggests that we try $(At + B)e^{2t}$. The left-hand side becomes then

$$(2A - 3B - 3At)e^{2t},$$

which agrees with the right-hand side if $A = 3$ and $B = 2$. Therefore

$$y_p = (3t + 2)e^{2t}.$$

If e^{2t} had been part of the complementary function but not te^{2t}, we would have tried $(At + B)te^{2t}$; if te^{2t} had been part of the complementary function also, the trial function would have been $(At + B)t^2 e^{2t}$.

From these examples we can construct the following prescription for a particular integral. *If*

$$\ddot{y} + a_1\dot{y} + a_0 y = t^m e^{at} \tag{2.5.1}$$

where $a_0 \neq 0$ and m is a non-negative integer,

(I) *when a is not a root of the characteristic equation $\lambda^2 + a_1\lambda + a_0 = 0$, try*

$$y = e^{at}(A_m t^m + A_{m-1}t^{m-1} + \cdots + A_0);$$

(II) *when a is a single root of the characteristic equation, multiply the expansion in (I) by t;*

(III) *when a is a double root of the characteristic equation, multiply the expansion in (I) by t^2.*

Substitution of the proposed expansion in the differential equation will confirm that the right-hand side can be obtained by an appropriate choice of $A_0, \ldots, A_m$.

A right-hand side that is composed of a sum of the type in (2.5.1) can be handled because, if

$$\ddot{y} + a_1\dot{y} + a_0 y = f_1(t) + f_2(t) + \cdots + f_n(t) \qquad (2.5.2)$$

and y_i is a particular integral of

$$\ddot{y} + a_1\dot{y} + a_0 y = f_i(t),$$

then

$$y_p = y_1 + y_2 + \cdots + y_n$$

is a particular integral of (2.5.2) as may be confirmed by substitution in the differential equation.

Trigonometric functions are also covered by the rule given because we can write

$$\cos t = \tfrac{1}{2}(e^{it} + e^{-it}), \quad \sin t = \tfrac{1}{2i}(e^{it} - e^{-it}) \qquad (2.5.3)$$

and use the fact that a can be complex in (2.5.1).

Example 2.5.6
Find a particular integral of

$$\ddot{y} - 2\dot{y} - 3y = 2\sin t.$$

First consider the right-hand side of e^{it}. This is not part of the complementary function as can be seen from Example 2.5.3. Therefore try $y = Ae^{it}$. The left-hand side is $(-4 - 2i)Ae^{it}$, which agrees with e^{it} if $A = -1/(4+2i)$. Similarly, the particular integral corresponding to e^{-it} is $-e^{-it}/(4 - 2i)$. Hence, from (2.5.3),

$$yp = \frac{1}{i}\left(-\frac{e^{it}}{4 + 2i} + \frac{e^{-it}}{4 - 2i}\right)$$

$$= \tfrac{1}{5}(\cos t - 2\sin t). \qquad \square$$

Example 2.5.7
Find a particular integral of

$$\ddot{y} + 4y = 32t\cos 2t - 8\sin 2t.$$

The characteristic equation is $\lambda^2 + 4 = 0$, which implies that $\lambda_1 = 2i, \lambda_2 = -2i$. The complementary function may be expressed in terms of e^{2it} and e^{-2it} or, in real form, by

$$y_c = C_1\cos 2t + C_2\sin 2t.$$

For the right-hand side we consider first e^{2it}. Since $2i$ is a single root of the characteristic equation, case (II) above applies and we try Ate^{2it}. Similarly for te^{2it} we would try $(Bt + C)te^{2it}$. However, it is more economical to combine the two and make $(Bt + C)te^{2it}$ reproduce the term $(16t + 4i)e^{2it}$ required on the right-hand side. We find $B = -2i$ and $C = 2$, so that

$$y_p = (-2it + 2)te^{2it} + (2it + 2)te^{-2it}$$
$$= 4t\cos 2t + 4t^2 \sin 2t.$$

The general solution is

$$y = C_1 \cos 2t + C_2 \sin 2t + 4t(\cos 2t + t\sin 2t). \qquad \square$$

Example 2.5.8

Find a particular integral of

$$\ddot{y} - \dot{y} - 2y = \cosh t.$$

The characteristic equation is $\lambda^2 - \lambda - 2 = 0$, so that $\lambda_1 = 2, \lambda_2 = -1$ and the complementary function is

$$y_c = C_1 e^{2t} + C_2 e^{-t}.$$

The right-hand side can be converted to standard form by using $\cosh t = \frac{1}{2}(e^t + e^{-t})$. For e^t case (I) applies because 1 is not a root of the characteristic equation, but for e^{-t} case (II) is relevant because -1 is a single root of the characteristic equation. Therefore, try $Ae^t + Bte^{-t}$; it is found that $A = -\frac{1}{4}$, $B = -\frac{1}{6}$ and

$$y_p = -\tfrac{1}{4}e^t - \tfrac{1}{6}te^{-t}$$

is a particular integral. $\qquad \square$

The differential equation

$$\ddot{y} + 2\dot{y} + y = te^{-t}$$

provides an illustration of case (III). A particular integral is

$$y_p = \tfrac{1}{6}t^3 e^{-t}.$$

(b)   Factorisation of the operator

In this method the differential equation is split into two first-order differential equations in the same way as was employed for discovering the complementary function in Section 2.4. An example will demonstrate the technique.

Example 2.5.9

Reconsider

$$\ddot{y} - \dot{y} - 2y = \cosh t$$

which was discussed in Example 2.5.8.

The differential equation can be expressed as

$$\left(\frac{d}{dt} + 1\right)\left(\frac{dy}{dt} - 2y\right) = \cosh t.$$

Put $u = \dot{y} - 2y$. Then

$$\frac{du}{dt} + u = \cosh t = \tfrac{1}{2}(e^t + e^{-t}).$$

The integrating factor is e^t and

$$\frac{d}{dt}(ue^t) = \tfrac{1}{2}(e^{2t} + 1)$$

so that

$$ue^t = \tfrac{1}{4}e^{2t} + \tfrac{1}{2}t + C.$$

Therefore

$$\dot{y} - 2y = \tfrac{1}{4}e^t + \tfrac{1}{2}te^{-t} + Ce^{-t}.$$

The integrating factor is e^{-2t} and

$$\frac{d}{dt}(ye^{-2t}) = \tfrac{1}{4}e^{-t} + \tfrac{1}{2}te^{-3t} + Ce^{-3t}.$$

Since

$$\int^t xe^{-3x}\,dx = -\tfrac{1}{3}te^{-3t} + \tfrac{1}{3}\int^t e^{-3x}\,dx = -\tfrac{1}{3}te^{-3t} - \tfrac{1}{9}e^{-3t},$$

$$ye^{-2t} = -\tfrac{1}{4}e^{-t} - \tfrac{1}{6}te^{-3t} - \tfrac{1}{18}e^{-3t} - \tfrac{1}{3}Ce^{-3t} + D$$

whence

$$y = -\tfrac{1}{4}e^t - \tfrac{1}{6}te^{-t} - \tfrac{1}{18}e^{-t} - \tfrac{1}{3}Ce^{-t} + De^{2t}.$$

Actually, this analysis has given the general solution, but we could have obtained a particular integral by leaving C and D out when they arose. The resulting particular integral differs from that of Example 2.5.8 by $-\tfrac{1}{18}e^{-t}$ but, since this is part of the complementary function, there is no change to the general solution. □

The advantages of this method are that it can give the general solution directly and always works even when the right-hand side is not a polynomial, exponential, sine or cosine. On the other hand, it often involves more labour than undetermined coefficients when the right-hand side is such that either method is applicable. It can also be awkward to implement when the roots of the characteristic equation are complex.

(c) Variation of parameters

Another method that can be adopted for any right-hand side is variation of parameters.

Suppose that the complementary function of

$$\ddot{y} + a_1\dot{y} + a_0 y = f(t) \tag{2.5.4}$$

is

$$y_c = C_1 y_1(t) + C_2 y_2(t).$$

We seek a solution of (2.5.4) in the form

$$y = u(t)y_1(t) + v(t)y_2(t), \tag{2.5.5}$$

i.e., we allow the parameters C_1 and C_2 to vary—which explains the nomenclature. There are two unknown functions u and v so that two conditions are needed to determine them. One condition is obtained by substituting (2.5.5) in (2.5.4). The other we can pick for ourselves and we want to do it so as to avoid second derivatives of u and v if possible; otherwise we are no better off. Now, from (2.5.5),

$$\dot{y} = u\dot{y}_1 + v\dot{y}_2 + \dot{u}y_1 + \dot{v}y_2$$

and, if second derivatives of u and v are not to occur in $\ddot{y}$, we must insist that

$$\dot{u}y_1 + \dot{v}y_2 = 0. \tag{2.5.6}$$

Then

$$\dot{y} = u\dot{y}_1 + v\dot{y}_2$$

and

$$\ddot{y} = u\ddot{y}_1 + v\ddot{y}_2 + \dot{u}\dot{y}_1 + \dot{v}\dot{y}_2.$$

Now (2.5.4) can be satisfied if

$$u\ddot{y}_1 + v\ddot{y}_2 + \dot{u}\dot{y}_1 + \dot{v}\dot{y}_2 + a_1(u\dot{y}_1 + v\dot{y}_2) + a_0(uy_1 + vy_2) = f(t)$$

or

$$(\ddot{y}_1 + a_1\dot{y}_1 + a_0 y_1)u + (\ddot{y}_2 + a_1\dot{y}_2 + a_0 y_2)v + \dot{u}\dot{y}_1 + \dot{v}\dot{y}_2 = f(t).$$

But y_1 and y_2 both satisfy the associated homogeneous differential equation, and so the terms in the two sets of parentheses vanish. Consequently

$$\dot{u}\dot{y}_1 + \dot{v}\dot{y}_2 = f(t). \tag{2.5.7}$$

The equations (2.5.6) and (2.5.7) are now solved for $\dot{u}$ and $\dot{v}$. Integration then supplies u, v and hence a particular integral.

It should be remarked that it is possible to solve (2.5.6) and (2.5.7) provided that $y_1\dot{y}_2 - \dot{y}_1 y_2 \neq 0$. Suppose that this is not true and that $y_1\dot{y}_2 - \dot{y}_1 y_2 = 0$. Then $\dot{y}_2/y_2 = \dot{y}_1/y_1$ which implies that $\ln y_2 = \ln y_1 + C$ or $y_2 = Ay_1$. Thus $C_1 y_1 + C_2 y_2 = (C_1 + AC_2)y_1$ and this cannot be the complementary function since it contains only one arbitrary constant. Therefore, so long as the genuine complementary function has been found, the solution of (2.5.6) and (2.5.7) can always be carried out.

Example 2.5.10
Solve

$$\ddot{y} - \dot{y} - 2y = \cosh t$$

by means of variation of parameters.

The complementary function is

$$y_c = C_1 e^{2t} + C_2 e^{-t}.$$

Therefore $y_1 = e^{2t}, y_2 = e^{-t}$ and (2.5.6) gives

$$\dot{u}e^{2t} + \dot{v}e^{-t} = 0$$

whereas (2.5.7) becomes

$$2\dot{u}e^{2t} - \dot{v}e^{-t} = \cosh t.$$

Hence

$$3\dot{u}e^{2t} = \cosh t, \quad \dot{v}e^{-t} = -\tfrac{1}{3}\cosh t$$

or

$$\dot{u} = \tfrac{1}{6}(e^{-t} + e^{-3t}), \quad \dot{v} = -\tfrac{1}{6}(1 + e^{2t}).$$

Consequently

$$u = -\tfrac{1}{6}e^{-t} - \tfrac{1}{18}e^{-3t} + C, \quad v = -\tfrac{1}{6}t - \tfrac{1}{12}e^{2t} + D$$

and

$$\begin{aligned}
y &= \left(-\tfrac{1}{6}e^{-t} - \tfrac{1}{18}e^{-3t} + C\right)e^{2t} + \left(-\tfrac{1}{6}t - \tfrac{1}{12}e^{2t} + D\right)e^{-t} \\
&= Ce^{2t} + \left(D - \tfrac{1}{18}\right)e^{-t} - \tfrac{1}{4}e^t - \tfrac{1}{6}te^{-t}.
\end{aligned}$$

This is the same general solution as derived in Examples 2.5.8 and 2.5.9. If we had placed $C = 0, D = 0$ earlier we would have obtained a particular integral but not the general solution. ∎

The method of variation of parameters is applicable for any right-hand side and works even if the characteristic equation has complex roots. It can also be generalised to other types of differential equations. Its disadvantage is that it often requires a lot of effort to carry through.

In summary, the strategy suggested is that the method of undetermined coefficients should be employed whenever the right-hand side has the right form. If it does not, try either factorisation of the operator or variation of parameters with preference for variation of parameters when the roots of the characteristic equation are complex.

2.6 Forced oscillations

A differential equation that arises frequently in practice is

$$\ddot{y} + 2b\Omega\dot{y} + \Omega^2 y = F\cos(\omega t + \beta), \tag{2.6.1}$$

where b, Ω, F, ω and β are real constants with $\Omega > 0$. It represents an oscillatory system subject to damping, when $b > 0$ as we shall assume, being vibrated by external means. In one application y represents the displacement of a particle subject to a restoring force $\Omega^2 y$ per unit mass and viscous damping $2b\Omega\dot{y}$ per unit mass acted on by a force of magnitude F per unit mass and circular frequency ω.

If $b \neq 1$, the complementary function is

$$y_c = C_1 e^{\delta_1 t} + C_2 e^{\delta_2 t}, \tag{2.6.2}$$

where $\delta_1 = -b\Omega + \Omega(b^2 - 1)^{1/2}$ and $\delta_2 = -b\Omega - \Omega(b^2 - 1)^{1/2}$. The quantities δ_1 and δ_2 are real if $b > 1$ and complex if $b < 1$. In either case, the real parts of δ_1 and δ_2 are negative.

If $b = 1$, the characteristic equation has the double root $-\Omega$ and

$$y_c = (A + Bt)e^{-\Omega t}. \tag{2.6.3}$$

To find a particular integral, we consider the right-hand side $Fe^{i(\omega t + \beta)}$. Since $i\omega$ is not a root of the characteristic equation, we try $Ce^{i(\omega t + \beta)}$ which leads to

$$(-\omega^2 + 2ib\Omega\omega + \Omega^2)C = F.$$

Therefore

$$\begin{aligned}
2y_p &= \frac{Fe^{i(\omega t + \beta)}}{\Omega^2 - \omega^2 + 2ib\Omega\omega} + \frac{Fe^{-i(\omega t + \beta)}}{\Omega^2 - \omega^2 - 2ib\Omega\omega} \\
&= \frac{2F\{(\Omega^2 - \omega^2)\cos(\omega t + \beta) + 2b\Omega\omega\sin(\omega t + \beta)\}}{(\Omega^2 - \omega^2)^2 + 4b^2\Omega^2\omega^2}.
\end{aligned}$$

Consequently, the general solution of (2.6.1) is

$$y = y_c + \frac{F\cos(\omega t + \beta - \phi)}{\{(\Omega^2 - \omega^2)^2 + 4b^2\Omega^2\omega^2\}^{1/2}}, \qquad (2.6.4)$$

where y_c is given by (2.6.2) or (2.6.3) depending on the value of b, and

$$\cos\phi = \frac{\Omega^2 - \omega^2}{\{(\Omega^2 - \omega^2)^2 + 4b^2\Omega^2\omega^2\}^{1/2}},$$

$$\sin\phi = \frac{2b\Omega\omega}{\{(\Omega^2 - \omega^2)^2 + 4b^2\Omega^2\omega^2\}^{1/2}}. \qquad (2.6.5)$$

The first term on the right-hand side of (2.6.4) is always present whether $F = 0$ or not. Its value depends upon what conditions are set at $t = 0$, but wherever it starts it will diminish to zero as t becomes large enough because of the form of (2.6.2) and (2.6.3). For this reason, the first term of (2.6.4) is often known as a **transient**.

The second term of (2.6.4) only occurs when $F \neq 0$ and varies at the same rate as the forcing device. Its amplitude is independent of the conditions at $t = 0$ and there is no decay as $t \to \infty$. It is known as the **forced oscillation**.

When one is concerned only with what happens for large t, the forced oscillation alone survives. Its amplitude Y can be expressed as

$$Y = \frac{F/\Omega^2}{[\{1 - (\omega^2/\Omega^2)\}^2 + 4b^2\omega^2/\Omega^2]^{1/2}}.$$

As ω varies, changes in Y occur only in the denominator and are dictated by the behaviour of

$$(1 - z^2)^2 + 4b^2z^2$$

with $z^2 = \omega^2/\Omega^2$. The derivative of this is

$$-4(1 - z^2)z + 8b^2z$$

which vanishes when $z = 0$ or $z^2 = 1 - 2b^2$. The second derivative has the value $4(2b^2 - 1)$ when $z = 0$ and the value $8(1 - 2b^2)$ when $z^2 = 1 - 2b^2$. Since z^2 cannot be negative, the denominator of Y has a minimum at $z = 0$ and no other stationary point if $2b^2 > 1$, but a maximum at $z = 0$ and a minimum at $z^2 = 1 - 2b^2$ if $2b^2 < 1$. Hence (i) if $2b^2 > 1, Y$ has a maximum at $\omega = 0$ and (ii) if $2b^2 < 1, Y$ has a minimum at $\omega = 0$ and a maximum at $\omega^2/\Omega^2 = 1 - 2b^2$. The value of Y at $\omega^2/\Omega^2 = 1 - 2b^2$ is $F/2\Omega^2 b(1 - b^2)^{1/2}$. So the maximum of the forced oscillation is larger the smaller b is, i.e., *with low damping it is possible to excite large vibrations provided that ω is chosen appropriately.*

It can be seen from (2.6.5) that $\phi = \frac{1}{2}\pi$ when $\omega = \Omega$, that $\phi \approx 0$ when $\omega/\Omega \ll 1$ and that $\phi \to \pi$ as $\omega/\Omega \to \infty$.

2.7 Differential equation of order n

Much of the discussion for the equation of second order carries over to the linear differential equation with constant coefficients of order n, namely

$$\frac{d^n y}{dt^n} + a_{n-1}\frac{d^{n-1} y}{dt^{n-1}} + \cdots + a_0 y = f(t). \tag{2.7.1}$$

Exactly as in Section 2.3, one may show that the general solution can be written

$$y = y_c + y_p$$

where y_p is a particular integral of (2.7.1) and y_c is the complementary function, i.e., the general solution of the associated homogeneous differential equation.

The complementary function is determined by solving the characteristic equation. Recalling from Section 2.4 that the characteristic equation could be reached by trying $y = e^{\lambda t}$, we do the same here and obtain

$$\lambda^n + a_{n-1}\lambda^{n-1} + \cdots + a_0 = 0. \tag{2.7.2}$$

Equation (2.7.2) has n roots $\lambda_1, \lambda_2, \ldots, \lambda_n$. For any λ_i which is different from all the rest, there is a contribution to the complementary function of $C_i e^{\lambda_i t}$. If, however, λ_j occurs m times, the complementary function acquires a term

$$(D_1 + D_2 t + \cdots + D_m t^{m-1})e^{\lambda_j t}.$$

Example 2.7.1
Find the general solution of

$$\frac{d^5 y}{dt^5} + 6\frac{d^4 y}{dt^4} + 15\frac{d^3 y}{dt^3} + 26\frac{d^2 y}{dt^2} + 36\frac{dy}{dt} + 24y = 0.$$

The characteristic equation is

$$\lambda^5 + 6\lambda^4 + 15\lambda^3 + 26\lambda^2 + 36\lambda + 24 = 0.$$

Now

$$\begin{aligned}
\lambda^5 &+ 6\lambda^4 + 15\lambda^3 + 26\lambda^2 + 36\lambda + 24 \\
&= (\lambda + 2)(\lambda^4 + 4\lambda^3 + 7\lambda^2 + 12\lambda + 12) \\
&= (\lambda + 2)^2(\lambda^3 + 2\lambda^2 + 3\lambda + 6) \\
&= (\lambda + 2)^3(\lambda^2 + 3).
\end{aligned}$$

Thus the roots of the characteristic equation are -2 (three times) and $\pm i\sqrt{3}$, once each. Therefore the general solution is

$$y = (C_1 + C_2 t + C_3 t^2)e^{-2t} + C_4 e^{it\sqrt{3}} + C_5 e^{-it\sqrt{3}}$$
$$= (C_1 + C_2 t + C_3 t^2)e^{-2t} + A\cos\sqrt{3}t + B\sin\sqrt{3}t$$

for a real solution. $\square$

Any of the three methods described for deriving a particular integral for the second-order differential equation may be employed for the general order in the right circumstances. For undetermined coefficients, the rules (I), (II) and (III) of Section 2.5(a) need to be supplemented because a root of the characteristic equation can be repeated more than twice. It is, however, clear from (I), (II) and (III) that the pertinent change is that *if a appears m times as a root of the characteristic equation, multiply by t^m the expansion that would have been tried for a solitary root*. For instance, if the right-hand side in Example 2.7.1 were te^{-2t}, the trial function would be $t^3(A + Bt)e^{-2t}$, leading to $A = 1/168, B = 2/147$.

There is nothing new to add to the method of factorisation of the operator, but now n equations of the first order have to be solved and the method becomes increasingly cumbersome as the size of n grows.

The principle of variation of parameters is unaltered but further detail is necessary. Let

$$y_c = C_1 y_1(t) + C_2 y_2(t) + \cdots + C_n y_n(t)$$

be the complementary function. Then a solution of (2.7.1) is sought in the form

$$y = u_1(t)y_1(t) + u_2(t)y_2(t) + \cdots + u_n(t)y_n(t) \tag{2.7.3}$$

with $u_1, \ldots, u_n$ subject to the $n - 1$ conditions

$$\frac{du_1}{dt}y_1 + \frac{du_2}{dt}y_2 + \cdots + \frac{du_n}{dt}y_n = 0,$$

$$\frac{du_1}{dt}\frac{dy_1}{dt} + \frac{du_2}{dt}\frac{dy_2}{dt} + \cdots + \frac{du_n}{dt}\frac{dy_n}{dt} = 0, \tag{2.7.4}$$

$$\cdots \quad \cdots \quad \cdots \quad \cdots$$
$$\cdots \quad \cdots \quad \cdots \quad \cdots$$

$$\frac{du_1}{dt}\frac{d^{n-2}y_1}{dt^{n-2}} + \frac{du_2}{dt}\frac{d^{n-2}y_2}{dt^{n-2}} + \cdots + \frac{du_n}{dt}\frac{d^{n-2}y_n}{dt^{n-2}} = 0.$$

If (2.7.3) is inserted in (2.7.1) and the conditions (2.7.4) imposed, the additional equation

$$\frac{du_1}{dt}\frac{d^{n-1}y_1}{dt^{n-1}} + \frac{du_2}{dt}\frac{d^{n-1}y_2}{dt^{n-1}} + \cdots + \frac{du_n}{dt}\frac{d^{n-1}y_n}{dt^{n-1}} = f(t) \tag{2.7.5}$$

is obtained. The equations (2.7.4) and (2.7.5) constitute n linear equations for the unknowns $\dot{u}_1, \ldots, \dot{u}_n$. Solving this system for $\dot{u}_1, \ldots, \dot{u}_n$, integrating and substituting in (2.7.3) then leads to the desired solution.

Example 2.7.2
Solve, by variation of parameters,

$$\frac{d^3y}{dt^3} + 5\frac{d^2y}{dt^2} - 6\frac{dy}{dt} = 9e^{3t}.$$

The characteristic equation is

$$\lambda^3 + 5\lambda^2 - 6\lambda = 0$$

with roots 0, 1, -6, so that

$$y_c = C_1 + C_2 e^t + C_3 e^{-6t}.$$

Now try

$$y = u_1(t) + u_2(t)e^t + u_3(t)e^{-6t}.$$

Equations (2.7.4) become

$$\dot{u}_1 + \dot{u}_2 e^t + \dot{u}_3 e^{-6t} = 0,$$
$$\dot{u}_2 e^t - 6\dot{u}_3 e^{-6t} = 0,$$

while (2.7.5) is

$$\dot{u}_2 e^t + 36\dot{u}_3 e^{-6t} = 9e^{3t}.$$

Therefore $\dot{u}_1 = -\frac{3}{2}e^{3t}, \dot{u}_2 = \frac{9}{7}e^{2t}, \dot{u}_3 = \frac{3}{14}e^{9t}$ and

$$u_1 = -\frac{1}{2}e^{3t} + D_1, \quad u_2 = \frac{9}{14}e^{2t} + D_2, \quad u_3 = \frac{1}{42}e^{9t} + D_3.$$

Consequently

$$y = \tfrac{1}{6}e^{3t} + D_1 + D_2 e^t + D_3 e^{-6t}$$

is the required general solution.

2.8 Uniqueness

The general solutions that have been constructed contain arbitrary constants and are therefore capable of assuming different values by assigning the constants in different ways. What we shall show now is that, if certain conditions are imposed, the values of the constants are fixed once and for all.

The general solution of

$$\dot{y} + a_0 y = 0 \tag{2.8.1}$$

is $y(t) = Ce^{-a_0 t}$. Suppose now that the extra condition $y(0) = 0$ is imposed. Then, there is no alternative to taking $C = 0$ and $y(t) \equiv 0$ throughout the interval where the differential equation holds.

The second order differential equation

$$\ddot{y} + a_1 \dot{y} + a_0 y = 0 \tag{2.8.2}$$

can be rewritten, according to Section 2.4, as

$$\dot{w} - \lambda_2 w = 0 \tag{2.8.3}$$

where

$$w = \dot{y} - \lambda_1 y. \tag{2.8.4}$$

Now require that $y = 0$ and $\dot{y} = 0$ at $t = 0$. It follows from (2.8.4) that $w = 0$ at $t = 0$. But (2.8.3) is of the same form as (2.8.1) and we conclude from the preceding paragraph that $w \equiv 0$. That makes (2.8.4) of the same type as (2.8.1) and since $y = 0$ at $t = 0, y \equiv 0$ is the only possibility. In other words, the solution of (2.8.2) such that $y = 0$ and $\dot{y} = 0$ at $t = 0$ vanishes throughout the interval.

Evidently, by factorising the operator in

$$\frac{d^n y}{dt^n} + a_{n-1} \frac{d^{n-1} y}{dt^{n-1}} + \cdots + a_0 y = 0 \tag{2.8.5}$$

and proceeding step by step as above we may deduce that *the solution of (2.8.5) such that*

$$y = 0, \quad \frac{dy}{dt} = 0, \quad \ldots, \quad \frac{d^{n-1} y}{dt^{n-1}} = 0 \tag{2.8.6}$$

at $t = 0$ vanishes identically.

There is an important consequence for the differential equation (2.7.1). Let it be desired to find a solution such that

$$y = D_0, \quad \frac{dy}{dt} = D_1, \quad \ldots, \quad \frac{d^{n-1} y}{dt^{n-1}} = D_{n-1} \tag{2.8.7}$$

at $t = 0$ with $D_0, \ldots, D_{n-1}$ some given constants. There could be two or more solutions which satisfy these conditions. Let y_1 and y_2 be two of them. Since y_1 and y_2 both comply with (2.8.7) their difference $y_1 - y_2$ satisfies (2.8.6) at $t = 0$. Also, since both are solutions of (2.7.1), their difference $y_1 - y_2$ is a solution of (2.8.5). By what has been established in the preceding paragraph $y_1 - y_2$ vanishes identically, i.e., there is no difference between y_1 and y_2.

This result may be stated as: *there is one, and only one, solution of*

$$\frac{d^n y}{dt^n} + a_{n-1}\frac{d^{n-1} y}{dt^{n-1}} + \cdots + a_0 y = f(t) \tag{2.8.8}$$

which meets the conditions (2.8.7) at $t = 0$.

To put it another way, the solution of (2.8.8) subject to (2.8.7) at $t = 0$ is *unique.*

Of course, it is not necessary to specify the conditions at the origin—that was chosen to fix ideas. If (2.8.7) held at $t = a$ the same argument would carry through and uniqueness would be valid still. However, it is essential that all the conditions are imposed at the same value of t. If some are enforced at one value of t and others at another, the situation is changed totally as will be seen in the next chapter.

Exercises

2.1 Find the general solutions of

   (a) $\ddot{y} - 8\dot{y} + 15y = 0$,

   (b) $\ddot{y} - 8\dot{y} + 16y = 0$,

   (c) $\ddot{y} + 2\dot{y} + 5y = 5t^2 + 5t$,

   (d) $\ddot{y} - \dot{y} - 2y = 30e^{2t}$,

   (e) $\ddot{y} + \omega^2 y = \cos \Omega t \ (\Omega^2 \neq \omega^2)$,

   (f) $\ddot{y} + \omega^2 y = \cos \omega t$,

   (g) $\ddot{y} + \dot{y} + y = 3\sin^2 t$.

2.2 Find the general solutions of

   (a) $\ddot{y} + 3\dot{y} + 2y = 4t^2 - 2t$,

   (b) $\ddot{y} + 2\dot{y} - 2y = \frac{3}{5}t^5$,

   (c) $\ddot{y} + 3\dot{y} = 6t^3 + 3t + 3$,

   (d) $\ddot{y} + \dot{y} + y = 26e^t \sin t$,

   (e) $\ddot{y} + 3\dot{y} = 15\sin t + 5$,

   (f) $\ddot{y} - 2\dot{y} + y = 6e^t + 2\sin t$,

   (g) $\ddot{y} + 9y = 3\sin 3t + 6\cos 3t$.

2.3 Give a suitable real form, but do not evaluate the coefficients, for a particular integral of

   (a) $\ddot{y} + y = 2t + t\sin t$,

   (b) $\ddot{y} - 4\dot{y} + 4y = t^2 + 3te^{2t} + 4t\sin 2t$,

(c) $\ddot{y} + 3\dot{y} + 2y = 2(1 + t^2)e^t \sin 2t - e^t \cos t + 3e^t,$

(d) $\ddot{y} + 4y = 2te^t + 3t \sin 2t.$

Do you think it would be simpler to employ exponentials with complex exponents in any of these cases?

2.4 Show that

$$t^2\ddot{y} + b_1 t\dot{y} + b_0 y = f(t),$$

where b_0 and b_1 are constants, can be converted to a linear ordinary differential equation with constant coefficients by the substitution $t = e^x$. Hence find the general solution of

$$t^2\ddot{y} - t\dot{y} + y = 2t.$$

2.5 Use factorisation of the operator to find the general solution of

$$\ddot{y} - \dot{y} - 2y = 15e^{2t}.$$

2.6 Find the general solution of $\ddot{y} + y = 3t$ by (a) undetermined coefficients and (b) factorisation of the operator.

2.7 By means of (a) undetermined coefficients and (b) variation of parameters find a particular integral of

$$\ddot{y} + 4y = e^t + \sin 2t.$$

explaining any difference between the answers.

2.8 (a) Find a particular integral of

$$\ddot{y} + 2\dot{y} + y = 5e^{-t}(1 + t)^{1/2}$$

by (i) factorisation of the operator and (ii) variation of parameters.

(b) Find the general solution of

$$\ddot{y} + y = 1/\sin t$$

by variation of parameters.

Would the method of undetermined coefficients be suitable for (a) or (b)? Would it be feasible to use factorisation of the operator for (b)?

2.9 The general solution of

$$\ddot{y} + p_1(t)\dot{y} + p_0(t)y = 0$$

is $y = C_1 y_1(t) + C_2 y_2(t)$. Show, by variation of parameters, that the general solution of

$$\ddot{y} + p_1(t)\dot{y} + p_0(t)y = f(t)$$

can be expressed as

$$y = -y_1(t) \int^t \frac{y_2(u)f(u)}{W(u)} du + y_2(t) \int^t \frac{y_1(u)f(u)}{W(u)} du$$

where $W(t) = y_1(t)\dot{y}_2(t) - \dot{y}_1(t)y_2(t)$.

2.10 Find the general solutions of

(a) $\dfrac{d^4 y}{dt^4} + 5\dfrac{d^2 y}{dt^2} + 4y = 0,$

(b) $\dfrac{d^3 y}{dt^3} + 7\dfrac{d^2 y}{dt^2} + 16\dfrac{dy}{dt} + 12y = 0,$

(c) $\dfrac{d^4 y}{dt^4} + 2\dfrac{d^3 y}{dt^3} + 10\dfrac{d^2 y}{dt^2} = 0,$

(d) $\dfrac{d^4 y}{dt^4} - 4\dfrac{d^3 y}{dt^3} + 6\dfrac{d^2 y}{dt^2} - 4\dfrac{dy}{dt} + y = 0.$

2.11 Prove that, if p and q are non-negative integers,

$$\left[\frac{d}{dt} - a\right]^p t^q e^{at} = \begin{cases} 0 & (p > q), \\ q(q-1)\cdots(q-p+1)t^{q-p}e^{at} & (p \le q). \end{cases}$$

2.12 Use the method of variation of parameters to solve

(a) $\dfrac{d^3 y}{dt^3} - \dfrac{dy}{dt} = \sin t,$

(b) $\dfrac{d^4 y}{dt^4} - y = 1 + t^2,$

(c) $\dfrac{d^4 y}{dt^4} - \dfrac{d^2 y}{dt^2} = 4te^t.$

2.13 (a) Show that, if $y_0(t)$ is a solution of the associated homogeneous differential equation of

$$\frac{d^n y}{dt^n} + p_{n-1}(t)\frac{d^{n-1}y}{dt^{n-1}} + \cdots + p_0(t)y = f(t),$$

the substitution $y = u(t)y_0(t)$ leads to a differential equation of order $n - 1$ for du/dt.

(b) Given that e^{-t^2} is a solution of

$$\ddot{y} + 4t\dot{y} + (4t^2 + 2)y = 0,$$

find the general solution.

(c) Given that $1/t$ satisfies the associated homogeneous differential equation of

$$t^2\ddot{y} + 4t\dot{y} + 2y = t\sin t,$$

find y so that $y = 1, \dot{y} = 0$ at $t = 1$.

Chapter 3

Simultaneous Equations with Constant Coefficients

3.1 Simultaneous equations of the first order

In studying natural phenomena, we are often interested in more than one quantity and the several quantities may well be connected by differential equations. We are therefore led to consider what happens when more than one differential equation has to be solved at a time. Suppose

$$a_1\dot{x} + b_1\dot{y} + c_1 x + d_1 y = f_1(t), \tag{3.1.1}$$
$$a_2\dot{x} + b_2\dot{y} + c_2 x + d_2 y = f_2(t) \tag{3.1.2}$$

where a_1, b_1, c_1, d_1, a_2, b_2, c_2, d_2 are constants and x, y are to be found. In other words, two simultaneous differential equations of the first order have to be solved.

Multiply (3.1.1) by b_2 and (3.1.2) by b_1. Then subtraction gives

$$(a_1 b_2 - a_2 b_1)\dot{x} + \alpha x + \beta y = F(t) \tag{3.1.3}$$

where $\alpha = c_1 b_2 - c_2 b_1$, $\beta = d_1 b_2 - d_2 b_1$ and $F(t) = b_2 f_1(t) - b_1 f_2(t)$.

There are two distinct cases to discuss according as $a_1 b_2 - a_2 b_1$ is or is not zero. We call $a_1 b_2 - a_2 b_1$ the **test determinant**.

(a)   Test determinant is non-zero

If $\beta \neq 0$, (3.1.3) can be solved to give y in terms of x and $\dot{x}$. If this expression is substituted in (3.1.1) or (3.1.2) a linear differential equation with constant coefficients of order 2 is obtained for x. This differential equation can be solved by techniques already described and its general solution will involve two arbitrary constants. Having found x we can determine y from (3.1.3). No further arbitrary constants are introduced and so the whole solution contains two arbitrary constants.

If $\beta = 0$, (3.1.3) is a differential equation of the first order for x which can be resolved by means of an integrating factor. Its general solution will possess one arbitrary constant. Once x is known, it can be substituted in (3.1.1) or

(3.1.2) resulting in a differential equation of the first order for y. Its general solution will bring in another arbitrary constant so again the whole solution contains two arbitrary constants.

(b)   Test determinant is zero

When $a_1b_2 = a_2b_1$, (3.1.3) reduces to

$$\alpha x + \beta y = F(t). \tag{3.1.4}$$

If $\beta \neq 0$, solve (3.1.4) for y and substitute in (3.1.1) or (3.1.2). The consequent differential equation for x is of the first order and so its general solution has one arbitrary constant. With x known, y is given by (3.1.4). Since no additional arbitrary constant is entailed, the whole solution possesses one arbitrary constant.

If $\beta = 0$, (3.1.4) immediately furnishes x provided that $\alpha \neq 0$. Then (3.1.1) or (3.1.2) supplies y with one arbitrary constant. Again the whole solution contains one arbitrary constant.

If $\beta = 0$ and $\alpha = 0$, (3.1.4) becomes $0 = F(t)$. There are now two possibilities. Either $F(t)$ is not zero over the interval of t under consideration when (3.1.4) cannot be satisfied and the original differential equations are inconsistent, or $F(t)$ is zero over the interval and then (3.1.2) is a constant multiple of (3.1.1).

To sum up, when the test determinant is non-zero, the whole solution can be found and contains two arbitrary constants. When the test determinant vanishes, either there is a solution and it includes one arbitrary constant, or there is no solution, or the two differential equations are not different.

Although $\dot{y}$ was eliminated to arrive at (3.1.3), a similar equation could be obtained by eliminating $\dot{x}$. The general conclusion concerning the role of the test determinant would remain unaltered. In practice, it is a matter of convenience whether $\dot{x}$ or $\dot{y}$ is eliminated.

Example 3.1.1
Solve

$$\dot{x} + 2\dot{y} + x - y = t, \tag{3.1.5}$$
$$\dot{x} - \dot{y} - x - 2y = 1. \tag{3.1.6}$$

The test determinant is $-1-2$, which is non-zero; so the whole solution should possess two arbitrary constants.

Eliminating $\dot{y}$ from (3.1.5) and (3.1.6) we obtain

$$3\dot{x} - x - 5y = t + 2$$

whence

$$y = \tfrac{1}{5}(3\dot{x} - x - t - 2). \tag{3.1.7}$$

Insertion of (3.1.7) in (3.1.6) gives

$$-\tfrac{3}{5}\ddot{x} - \tfrac{3}{5}x = -\tfrac{2}{5}t$$

or

$$\ddot{x} + x = \tfrac{2}{3}t.$$

It follows that

$$x = C_1 \cos t + C_2 \sin t + \tfrac{2}{3}t. \tag{3.1.8}$$

From (3.1.7)

$$y = \tfrac{1}{5}(3C_2 - C_1)\cos t - \tfrac{1}{5}(3C_1 + C_2)\sin t - \tfrac{1}{3}t.$$

Only two arbitrary constants C_1 and C_2 appear in the solution in accordance with earlier observations. □

Having found x in (3.1.8), one might have tried to find y from (3.1.6) instead of (3.1.7). Not only would this require more effort but also it would display another feature. Substitution of (3.1.8) in (3.1.6) gives

$$\dot{y} + 2y = (C_2 - C_1)\cos t - (C_1 + C_2)\sin t - \tfrac{2}{3}t - \tfrac{1}{3}$$

from which can be deduced

$$y = C_3 e^{-2t} + \tfrac{1}{5}(3C_2 - C_1)\cos t - \tfrac{1}{5}(3C_1 + C_2)\sin t - \tfrac{1}{3}t. \tag{3.1.9}$$

An extra arbitrary constant C_3 has made its presence known. However, because we did not find y from (3.1.7), there is no guarantee that (3.1.5) is satisfied. If (3.1.8) and (3.1.9) are put in (3.1.5) it will be discovered that x and y do not satisfy (3.1.5) unless $C_3 = 0$. It is therefore essential to minimise labour, to work through (3.1.7) or (3.1.3) in the general case.

There is another technique which suggests itself and which should be avoided, namely to try to treat (3.1.5) and (3.1.6) like algebraic equations and remove both $\dot{x}$ and x (or $\dot{y}$ and y) at the same time. Apply $(d/dt) - 2$ to (3.1.5) and $(d/dt) - 1$ to (3.1.6). There results

$$\ddot{x} + 3\dot{x} + 2x + 2\ddot{y} + 3\dot{y} - 2y = 1 + 2t,$$
$$2\ddot{x} - 3\dot{x} + x - 2\ddot{y} - 3\dot{y} + 2y = -1.$$

By addition

$$3(\ddot{x} + x) = 2t,$$

which supplies the same x as in (3.1.8). But, unless we re-derive (3.1.7), we have to determine y by substituting for x in one of the original differential equations. This will run into the same trouble as in the preceding paragraph and recover (3.1.9).

Example 3.1.2

Find the solution of

$$\dot{x} - 2\dot{y} - x = e^t, \tag{3.1.10}$$

$$-2\dot{x} + 4\dot{y} + y = 1. \tag{3.1.11}$$

Here the test determinant vanishes and one of three possibilities may occur. Elimination of $\dot{y}$ provides

$$2x + y = 2e^t + 1. \tag{3.1.12}$$

Thus the equations are consistent and different; consequently, there will be a general solution with one arbitrary constant only.

Substitute for y from (3.1.12) in (3.1.10). Then

$$5\dot{x} + x = 5e^t$$

whence

$$x = Ce^{-t/5} + \tfrac{5}{6}e^t.$$

From (3.1.12)

$$y = \tfrac{1}{3}e^t + 1 - 2Ce^{-t/5}$$

and the whole solution contains the single arbitrary constant C.   ⬜

Finally, note that, when the test determinant does not vanish, (3.1.3) can be divided by $a_1 b_2 - a_2 b_1$ with the result

$$\dot{x} = \alpha_1 x + \beta_1 y + F_1(t). \tag{3.1.13}$$

Similarly, by getting rid of $\dot{x}$ from (3.1.1) and (3.1.2), we obtain

$$\dot{y} = \alpha_2 x + \beta_2 y + F_2(t). \tag{3.1.14}$$

Thus (3.1.1) and (3.1.2) could be replaced by (3.1.13) and (3.1.14), if desired, so long as the test determinant is non-zero.

3.2   Replacement of one differential equation by a system

The second-order differential equation

$$\ddot{y} + a_1 \dot{y} + a_0 y = f(t) \tag{3.2.1}$$

can be represented as a system of first-order equations. Put

$$y(t) = x_1(t), \quad \dot{y}(t) = x_2(t). \tag{3.2.2}$$

From the first of (3.2.2), $\dot{y} = \dot{x}_1$ and so, from the second of (3.2.2),

$$\dot{x}_1 = x_2. \tag{3.2.3}$$

Furthermore, (3.2.1) can be written as

$$\dot{x}_2 = f(t) - a_1 x_2 - a_0 x_1. \tag{3.2.4}$$

Thus (3.2.1) gives rise to the system of first-order equations (3.2.3) and (3.2.4). Any solution y of (3.2.1) provides a solution of the system via the identification (3.2.2). Conversely, given a solution of (3.2.3) and (3.2.4), we can substitute from (3.2.3) in (3.2.4) to obtain

$$\ddot{x}_1 + a_1 \dot{x}_1 + a_0 x_1 = f(t)$$

so that a solution of (3.2.1) is obtained by putting $y = x_1$. Therefore the differential equation (3.2.1) and the system (3.2.3)–(3.2.4) are equivalent. It is thereby possible to deduce properties of a second-order differential equation from those of a first-order system or vice versa.

These notions can be extended to the differential equation of order n

$$\frac{d^n y}{dt^n} + a_{n-1} \frac{d^{n-1} y}{dt^{n-1}} + \cdots + a_0 y = f(t) \tag{3.2.5}$$

by placing

$$y(t) = x_1(t), \quad \dot{y} = x_2(t), \quad \ldots, \quad \frac{d^{n-1}}{dt^{n-1}} y(t) = x_n(t).$$

It is evident from these last relations that

$$\dot{x}_1 = x_2,$$
$$\dot{x}_2 = x_3,$$
$$\vdots$$
$$\dot{x}_{n-1} = x_n.$$

Moreover, (3.2.5) can be expressed as

$$\dot{x}_n = f(t) - a_0 x_1 - \cdots - a_{n-1} x_n.$$

Again a system (of n first-order equations) has been produced. The equivalence of the system and (3.2.5) can be demonstrated in the same manner as for the second-order differential equation.

The systems derived here have the same structure as (3.1.13) and (3.1.14) but the latter are more general than the former. From now on it will be assumed that the test determinant of our system is non-zero so that (3.1.13) and (3.1.14) are valid. Their generalisation will be investigated in the next section.

3.3   The general system

A system of differential equations,

$$\dot{x}_1 = \sum_{j=1}^{n} a_{ij} x_j + f_i(t) \quad (i = 1, \ldots, n), \tag{3.3.1}$$

where every a_{ij} is a constant, is known as **a system of linear differential equations of the first order with constant coefficients**. The system

$$\dot{x}_i = \sum_{j=1}^{n} a_{ij} x_j \quad (i = 1, \ldots, n) \tag{3.3.2}$$

is called the **associated homogeneous system**. A general solution of (3.3.1) or of (3.3.2) must determine the n quantities $x_1, x_2, \ldots, x_n$. It is convenient to use the abbreviated notation $\mathbf{x}$ for the n quantities $x_1, x_2, \ldots, x_n$ and say that $\mathbf{x}$ is a solution of (3.3.2) when the $x_1, x_2, \ldots, x_n$ satisfy (3.3.2).

Suppose now that $\mathbf{x}^{(k)}$ is a solution of (3.3.2) for $k = 1, \ldots, n$. Consider

$$x_i = \sum_{k=1}^{n} C_k x_i^{(k)} \quad (i = 1, \ldots, n) \tag{3.3.3}$$

where the C_k are constants. Then

$$\dot{x}_i = \sum_{k=1}^{n} C_k \dot{x}_i^{(k)} = \sum_{k=1}^{n} C_k \sum_{j=1}^{n} a_{ij} x_j^{(k)}$$

$$= \sum_{j=1}^{n} a_{ij} \sum_{k=1}^{n} C_k x_j^{(k)}$$

$$= \sum_{j=1}^{n} a_{ij} x_j$$

so that (3.3.3) also furnishes a solution of the associated homogeneous system. This explains why the system is called linear.

The formula (3.3.3) is a candidate for the general solution. It will be satisfactory if we can choose $C_1, \ldots, C_n$ so that $x_i(t_0) = x_{i0}$ $(i = 1, \ldots, n)$ for any t_0 in the interval under consideration and for any selection of the constants x_{i0} that we care to make. The choice is possible if

$$\sum_{k=1}^{n} C_k x_i^{(k)}(t_0) = x_{i0} \quad (i = 1, \ldots, n).$$

These constitute n linear algebraic equations for the unknowns $C_1, \ldots, C_n$. They can be solved for arbitrary right-hand sides if, and only if, the determinant of the coefficients is non-zero, i.e.,

$$
\begin{vmatrix}
x_1^{(1)} & x_1^{(2)} & \cdots & x_1^{(n)} \\
x_2^{(1)} & x_2^{(2)} & \cdots & x_2^{(n)} \\
\vdots & \vdots & & \vdots \\
x_n^{(1)} & x_n^{(2)} & \cdots & x_n^{(n)}
\end{vmatrix} \neq 0.
$$

The determinant is known as the **Wronskian** and written $W(\mathbf{x}^{(1)}, \ldots, \mathbf{x}^{(n)})$ for brevity. If the Wronskian is non-zero throughout the interval, the solutions $\mathbf{x}^{(1)}, \ldots, \mathbf{x}^{(n)}$ are said to form a **fundamental system**. We conclude that if $\mathbf{x}^{(1)}, \ldots, \mathbf{x}^{(n)}$ form a fundamental system, the general solution of (3.3.2) can be expressed as

$$
x_i = \sum_{k=1}^{n} C_k x_i^{(k)} \quad (i = 1, \ldots, n)
$$

where $C_1, \ldots, C_n$ are arbitrary constants.

Given the general solution of the associated homogeneous system, the general solution of (3.3.1) is

$$
x_i = \sum_{k=1}^{n} C_k x_i^{(k)} + X_i \quad (i = 1, \ldots, n) \tag{3.3.4}
$$

where $X_1, \ldots, X_n$ is some particular solution of (3.3.1). The proof is the same as for the second-order differential equation, the first term on the right of (3.3.4) corresponding to the complementary function and the second to the particular integral.

The most reliable way of finding $\mathbf{X}$ for general values of n is the method of variation of parameters. With the solutions $\mathbf{x}^{(1)}, \ldots, \mathbf{x}^{(n)}$ forming a fundamental system, we look for a solution of (3.3.1) of the type

$$
x_i(t) = \sum_{k=1}^{n} u_k(t) x_i^{(k)}(t) \quad (i = 1, \ldots, n). \tag{3.3.5}
$$

Then

$$
\dot{x}_i = \sum_{k=1}^{n} \left(\dot{u}_k x_i^{(k)} + u_k \dot{x}_i^{(k)} \right)
$$

$$
= \sum_{k=1}^{n} \dot{u}_k x_i^{(k)} + \sum_{k=1}^{n} u_k \sum_{j=1}^{n} a_{ij} x_j^{(k)}
$$

because $\mathbf{x}^{(k)}$ satisfies (3.3.2). Thus (3.3.5) implies that

$$\dot{x}_i - \sum_{j=1}^{n} a_{ij}x_j = \sum_{k=1}^{n} \dot{u}_k x_i^{(k)},$$

and (3.3.1) is satisfied if

$$\sum_{k=1}^{n} \dot{u}_k x_i^{(k)} = f_i \quad (i = 1, \ldots, n). \tag{3.3.6}$$

There are n equations in (3.3.6) for the n unknowns $\dot{u}_1, \ldots, \dot{u}_n$. They can always be solved because the determinant of the coefficients is the same as the Wronskian, which is non-zero because $\mathbf{x}^{(1)}, \ldots, \mathbf{x}^{(n)}$ make a fundamental system. Integration supplies $u_1, \ldots, u_n$ and a particular integral has been found.

The reader should verify that, for a single differential equation of order n, the equations (3.3.6) do go over to (2.7.4) and (2.7.5) when the substitutions of Section 3.2 are made.

3.4   The fundamental system

It is evident from the preceding section that the general solution of a system can be elicited if a fundamental system can be unearthed. For the single differential equation, searching for solutions proportional to $e^{\lambda t}$ was profitable and so the same device may be effective for a system. Therefore try $x_i = c_i e^{\lambda t}$ for $i = 1, \ldots, n$. The system (3.3.2) will be satisfied if

$$\lambda c_i e^{\lambda t} = \sum_{j=1}^{n} a_{ij} c_j e^{\lambda t}$$

or

$$\sum_{j=1}^{n} a_{ij} c_j = \lambda c_i \quad (i = 1, \ldots, n). \tag{3.4.1}$$

The linear equations (3.4.1) will force every c_i to be zero unless the determinant of the coefficients vanishes. To obtain a non-zero solution, at least one c_i must be different from zero. Hence the determinant of the coefficients must be made to vanish, i.e.,

$$\begin{vmatrix} a_{11} - \lambda & a_{12} & \cdots & a_{1n} \\ a_{21} & a_{22} - \lambda & \cdots & a_{2n} \\ \vdots & \vdots & & \vdots \\ a_{n1} & a_{n2} & \cdots & a_{nn} - \lambda \end{vmatrix} = 0. \tag{3.4.2}$$

When expanded, the determinant becomes a polynomial of degree n in λ, which may be expressed as

$$\lambda^n + p_{n-1}\lambda^{n-1} + \cdots + p_0 = 0. \tag{3.4.3}$$

This polynomial has a special property (needed later) which is usually designated as the Cayley-Hamilton Theorem. Denote the $n \times n$-matrix (a_{ij}), with entries as in (3.4.2) when $\lambda = 0$, by A. Then, if I is the unit $n \times n$-matrix,

$$A^n + p_{n-1}A^{n-1} + \cdots + p_0 I = 0. \tag{3.4.4}$$

There are n values of λ which satisfy (3.4.2). Let these roots be denoted by $\lambda_1, \ldots, \lambda_n$; they are called the **eigenvalues** of the matrix (a_{ij}). Put $\lambda = \lambda_k$ in (3.4.1) and solve for c_i to obtain $c_i^{(k)}$, say. Then $x_i = c_i^{(k)} e^{\lambda_k t}$ is a solution of the associated homogeneous system.

We must now investigate whether the identification $x_i^{(k)} = c_i^{(k)} e^{\lambda_k t}$ for $k = 1, \ldots, n$ will construct a fundamental system. Suppose that $\lambda_1, \ldots, \lambda_n$ are all different. Let there be n constants $B_1, \ldots, B_n$ such that

$$\sum_{k=1}^{n} B_k c_i^{(k)} = 0 \quad (i = 1, \ldots, n). \tag{3.4.5}$$

Then

$$\sum_{j=1}^{n} a_{ij} \sum_{k=1}^{n} B_k c_j^{(k)} = 0 \quad (i = 1, \ldots, n)$$

or, from (3.4.1),

$$\sum_{k=1}^{n} \lambda_k B_k c_i^{(k)} = 0 \quad (i = 1, \ldots, n). \tag{3.4.6}$$

Multiply (3.4.5) by λ_n and subtract from (3.4.6). Then

$$\sum_{k=1}^{n-1} (\lambda_k - \lambda_n) B_k c_i^{(k)} = 0. \tag{3.4.7}$$

Since $\lambda_k \neq \lambda_n$ for $k \neq n$, we have equations of the same form as (3.4.6) except that $c_i^{(n)}$ has been removed. Starting from (3.4.7) we can repeat the process and strike off $c_i^{(n-1)}$. Continuing in this way we arrive at

$$B_1' c_i^{(1)} = 0 \quad (i = 1, \ldots, n) \tag{3.4.8}$$

where B_1' is a non-zero multiple of B_1. At least one of $c_1^{(1)}, \ldots, c_n^{(1)}$ is not zero so that $B_1' = 0$, which implies that $B_1 = 0$. But now the stage before (3.4.8) will enforce $B'_2 c_i^{(2)} = 0$, which entails $B_2 = 0$. Repetition of the procedure leads to the conclusion that, if (3.4.5) holds, $B_k = 0$ $(k = 1, \ldots, n)$. However, that is possible only if the determinant of the coefficients is non-zero, i.e.,

$$
\begin{vmatrix}
c_1^{(1)} & c_1^{(2)} & \cdots & c_1^{(n)} \\
c_2^{(1)} & c_2^{(2)} & \cdots & c_2^{(n)} \\
\vdots & \vdots & & \vdots \\
c_n^{(1)} & c_n^{(2)} & \cdots & c_n^{(n)}
\end{vmatrix} \neq 0.
$$

Remembering that the Wronskian is $e^{(\lambda_1 + \cdots + \lambda_n)t}$ times this determinant, we see that the Wronskian does not vanish. Thus, in this case, a fundamental system has been educed.

What has been shown is that, *if all the eigenvalues are distinct, the general solution of the associated homogeneous system is*

$$
x_i = \sum_{k=1}^{n} c_i^{(k)} e^{\lambda_k t} \qquad (i = 1, \ldots, n). \tag{3.4.9}
$$

Example 3.4.1
Find the general solution of

$$
\dot{x} = 4x + y,
$$
$$
\dot{y} = 3x + 2y.
$$

The determinantal equation (3.4.2) is

$$
\begin{vmatrix}
4 - \lambda & 1 \\
3 & 2 - \lambda
\end{vmatrix} = 0
$$

or

$$
\lambda^2 - 6\lambda + 5 = 0.
$$

Thus we can take $\lambda_1 = 1$ and $\lambda_2 = 5$.
For $\lambda = 1$, (3.4.1) becomes

$$
3c_1^{(1)} + c_2^{(1)} = 0
$$

twice. Consequently $c_1^{(1)} = C_1, c_2^{(1)} = -3C_1$ where C_1 is an arbitrary constant.
For $\lambda = 5$, (3.4.1) goes over to

$$
-c_1^{(2)} + c_2^{(2)} = 0
$$

so that $c_1^{(2)} = C_2, c_2^{(2)} = C_2$ where C_2 is an arbitrary constant.

The desired general solution is

$$x = C_1 e^t + C_2 e^{5t},$$
$$y = -3C_1 e^t + C_2 e^{5t}.$$

When some of the eigenvalues are repeated, the situation is much more complicated. Assume that λ_1 occurs p times. When λ is placed equal to λ_1 in (3.4.1) there may be p solutions $c_i^{(1)}, \ldots, c_i^{(p)}$ such that

$$\sum_{k=1}^{p} B_k c_i^{(k)} = 0 \quad (i = 1, \ldots, n)$$

enforces $B_k = 0$ $(k = 1, \ldots, p)$. In that case our earlier analysis ensures that the contribution of this eigenvalue has the same form as in (3.4.9).

It may happen that p solutions cannot be found. Additional solutions must now be generated. Drawing on our experience with the single differential equation we try

$$x_i = (d_i + c_i t)e^{\lambda t}.$$

This will satisfy (3.3.2) if

$$(\lambda d_i + c_i + \lambda c_i t)e^{\lambda t} = \sum_{j=1}^{n} a_{ij}(d_j + c_j t)e^{\lambda t} \quad (i = 1, \ldots, n).$$

These can be true for an interval of t only if

$$\sum_{j=1}^{n} a_{ij} c_j = \lambda c_i, \tag{3.4.10}$$

$$\sum_{j=1}^{n} a_{ij} d_j = \lambda d_i + c_i \tag{3.4.11}$$

for $i = 1, \ldots, n$. Equation (3.4.10) is the same as (3.4.1) and may be solved in the same way as before. Once c_i has been determined, we solve (3.4.11) for d_i. In this manner, extra solutions of the differential system may be created.

There may still not be enough to fill the p slots available. If so, quadratic and possibly higher powers of t can be added into the exprerssion for x_i. In fact, it can be asserted that the trial solution

$$x_i = (r_i + \cdots + d_i t^{p-2} + c_i t^{p-1})e^{\lambda t}$$

is bound to produce enough solutions corresponding to the eigenvalue λ_1.

An alternative method for manufacturing solutions is discussed in the next section.

Example 3.4.2

Find the general solution of

$$\dot{x} = 5x + 3y,$$
$$\dot{y} = -3x - y.$$

In this case (3.4.2) is

$$\begin{vmatrix} 5 - \lambda & 3 \\ -3 & -1 - \lambda \end{vmatrix} = 0$$

or

$$\lambda^2 - 4\lambda + 4 = 0.$$

The eigenvalue 2 occurs twice. With $\lambda = 2$ in (3.4.1)

$$3c_1^{(1)} + 3c_2^{(1)} = 0,$$
$$-3c_1^{(1)} - 3c_2^{(1)} = 0.$$

Now $c_1^{(1)} = C_1, c_2^{(1)} = -C_1$ is the only possible solution and so is insufficient for our purposes. Invoking (3.4.11) we have

$$3d_1 + 3d_2 = c_1,$$
$$-3d_1 - 3d_2 = -c_1$$

since $c_2 = -c_1$. A solution is $d_1 = C_2, d_2 = 0, c_1 = 3C_2$. No other solution is necessary since it will differ from this only by a solution of (3.4.10). Two solutions are known now and

$$x = C_1 e^{2t} + C_2(1 + 3t)e^{2t},$$
$$y = -C_1 e^{2t} - 3C_2 t e^{2t}.$$

The reader should confirm that the Wronskian is non-zero, consistent with the derivation of a fundamental system. $\quad\square$

3.5 Matrix notation

The system (3.3.1) can be expressed in terms of matrices by introducing the column vector $\mathbf{x}$ with components $x_1, \ldots, x_n$, the column vector $\mathbf{f}$ with components $f_1, \ldots, f_n$ and the matrix $A = (a_{ij})$. Then

$$\dot{\mathbf{x}} = A\mathbf{x} + \mathbf{f} \tag{3.5.1}$$

and the associated homogeneous system is

$$\dot{\mathbf{x}} = A\mathbf{x}. \tag{3.5.2}$$

The similarity of (3.5.2) to the single first-order differential equation suggests that it ought to be possible to write the general solution as

$$\mathbf{x} = e^{tA}\mathbf{C} \tag{3.5.3}$$

where $\mathbf{C}$ is an arbitrary column vector. However, (3.5.3) has no significance until a meaning is attributed to the matrix e^{tA}. A suitable definition is

$$e^{tA} = I + tA + \frac{1}{2!}t^2 A^2 + \frac{1}{3!}t^3 A^3 + \cdots \tag{3.5.4}$$

where I is the unit $n \times n$-matrix. The presence of an infinite series in the definition means that (3.5.3) has a deceptive air of simplicity. It conceals the fact that it may be quite difficult to calculate e^{tA}.

By putting $t = 0$ in (3.5.4) we see at once that

$$e^0 = I. \tag{3.5.5}$$

Furthermore, by taking derivatives of (3.5.4) with respect to t term-by-term, without worrying about the legitimacy, we obtain

$$\frac{d}{dt}e^{tA} = Ae^{tA} = e^{tA}A. \tag{3.5.6}$$

Next, notice that the polynomial (3.4.3) would occur when considering solutions of the differential equation

$$\frac{d^n w}{dt^n} + p_{n-1}\frac{d^{n-1}w}{dt^{n-1}} + \cdots + p_0 w = 0. \tag{3.5.7}$$

This suggests that the solutions of (3.5.2) are related in some way to those of (3.5.7). In fact, a derivative of (3.5.2) gives

$$\ddot{\mathbf{x}} = A\dot{\mathbf{x}} = A^2\mathbf{x}$$

by virtue of (3.5.2). Clearly

$$\frac{d^m \mathbf{x}}{dt^m} = A^m\mathbf{x} \tag{3.5.8}$$

in general. Observe that this is consistent with (3.5.3) as can be seen by invoking (3.5.6). From (3.5.8)

$$\frac{d^n \mathbf{x}}{dt^n} + p_{n-1}\frac{d^{n-1}\mathbf{x}}{dt^{n-1}} + \cdots + p_0\mathbf{x} = (A^n + p_{n-1}A^{n-1} + \cdots + p_0 I)\mathbf{x} = 0 \tag{3.5.9}$$

on account of (3.4.4). Comparison of (3.5.9) with (3.5.7) reveals that *each element of* $\mathbf{x}$ *is a solution of the differential equation* (3.5.7).

Suppose now that $\mathbf{x} = \mathbf{0}$ at $t = 0$. Then (3.5.8) implies that the derivatives of $\mathbf{x}$ are also zero at $t = 0$. Hence each element of $\mathbf{x}$ is a solution of (3.5.7) which, together with its derivatives, vanishes at $t = 0$. The theory of Section 2.8 tells us that such a solution must be identically zero. Hence, *if* $\mathbf{x} = \mathbf{0}$ *at* $t = 0$, *the solution of* (3.5.2) *is zero throughout the interval under consideration*. It follows, as in Section 2.8, that the solution of (3.5.1) *such that* $\mathbf{x} = \mathbf{C}$ *at* $t = 0$ *is unique*. Naturally, the point $t = 0$ can be replaced by some other if desired.

Now consider some special solutions $w_1, w_2, \ldots, w_n$ of (3.5.7). We choose w_1 so that

$$w_1 = 1, \quad dw_1/dt = 0, \quad d^2 w_1/dt^2 = 0, \quad \ldots, \quad d^{n-1} w_1/dt^{n-1} = 0$$

at $t = 0$. For w_2 we take

$$w_2 = 0, \quad dw_2/dt = 1, \quad d^2 w_2/dt^2 = 0, \quad \ldots, \quad d^{n-1} w_2/dt^{n-1} = 0$$

at $t = 0$ and generally we select $d^m w_i/dt^m$ to vanish at $t = 0$ except for $m = i - 1$ when the value is to be unity. The functions $w_1, w_2, \ldots, w_n$ are uniquely defined (Section 2.8) and any solution of (3.5.7) can be expressed in terms of them by adding appropriate multiples to reproduce values specified for the solution and its derivatives at $t = 0$. Accordingly, there are constant vectors $\mathbf{c}_1, \ldots, \mathbf{c}_n$ such that

$$\mathbf{x} = w_1 \mathbf{c}_1 + w_2 \mathbf{c}_2 + \cdots + w_n \mathbf{c}_n. \tag{3.5.10}$$

Let $\mathbf{x} = \mathbf{C}$ at $t = 0$. Putting $t = 0$ in (3.5.10) we have $\mathbf{c}_1 = \mathbf{C}$. Take a derivative of (3.5.10) and put $t = 0$. Then $\dot{\mathbf{x}} = \mathbf{c}_2 = A\mathbf{C}$ by virtue of (3.5.8). Repeating the process we obtain $\mathbf{c}_3 = A^2\mathbf{C}$ and generally $\mathbf{c}_m = A^{m-1}\mathbf{C}$. Consequently, the solution of (3.5.2) such that $\mathbf{x} = \mathbf{C}$ at $t = 0$ is

$$\mathbf{x} = (w_1 I + w_2 A + \cdots + w_n A^{n-1})\mathbf{C}.$$

On the other hand, (3.5.5) indicates that (3.5.3) is the solution that is $\mathbf{C}$ at $t = 0$. By the uniqueness property, which has been demonstrated already, the two solutions must be the same, i.e.,

$$e^{tA}\mathbf{C} = (w_1 I + w_2 A + \cdots + w_n A^{n-1})\mathbf{C}.$$

But $\mathbf{C}$ is arbitrary since no particular values have been assigned to it and so

$$e^{tA} = w_1 I + w_2 A + \cdots + w_n A^{n-1}. \tag{3.5.11}$$

The formula (3.5.11) has the advantage over (3.5.4) of being a finite series rather than an infinite one. However, it does entail the determination of n

solutions of (3.5.7). Whether it is more effective in practice than the method described in earlier sections is more difficult to assess. Probably, the earlier method is best when all the eigenvalues of A are distinct. With repeated eigenvalues the scales will tend to tilt towards (3.5.11). To aid the reader in forming an assessment some of the preceding examples will be tackled by the method of this section.

Example 3.5.1

Find the general solution of

$$\dot{x} = 4x + y,$$
$$\dot{y} = 3x + 2y.$$

The analogue of (3.5.7) is, from Example 3.4.1,

$$\ddot{w} - 6\dot{w} + 5 = 0.$$

Consequently, $w_1 = (5e^t - e^{5t})/4$ and $w_2 = (e^{5t} - e^t)/4$. Substitute in (3.5.11) with $n = 2$. Then

$$4e^{tA} = (5e^t - e^{5t}) \begin{pmatrix} 1 & 0 \\ 0 & 1 \end{pmatrix} + (e^{5t} - e^t) \begin{pmatrix} 4 & 1 \\ 3 & 2 \end{pmatrix}$$
$$= \begin{pmatrix} 3e^{5t} + e^t & e^{5t} - e^t \\ 3e^{5t} - 3e^t & e^{5t} + 3e^t \end{pmatrix}.$$

Consequently, if $\mathbf{C}$ has elements C_1, C_2

$$x = (3C_1 + C_2)e^{5t}/4 + (C_1 - C_2)e^t/4,$$
$$y = (3C_1 + C_2)e^{5t}/4 - 3(C_1 - C_2)e^t/4.$$

This has the same structure as in Example 3.4.1 although here C_1 and C_2 are the values of x and y at $t = 0$. □

Example 3.5.2

Find the general solution of

$$\dot{x} = 5x + 3y,$$
$$\dot{y} = -3x - y.$$

The differential equation to be solved is, by Example 3.4.2,

$$\ddot{w} - 4\dot{w} + 4w = 0.$$

Therefore $w_1 = (1 - 2t)e^{2t}$ and $w_2 = te^{2t}$. Hence

$$e^{tA} = e^{2t} \begin{pmatrix} 1 + 3t & 3t \\ -3t & 1 - 3t \end{pmatrix}$$

resulting in

$$x = \{C_1 + 3(C_1 + C_2)t\}e^{2t},$$
$$y = \{C_2 - 3(C_1 + C_2)t\}e^{2t}.$$

Again there is consistency with Example 3.4.2 on adjusting the constants. ▯

Example 3.5.3
Find the general solution of

$$\dot{x} = 3x + y,$$
$$\dot{y} = 3y + z,$$
$$\dot{z} = 3z.$$

In this case $A = \begin{pmatrix} 3 & 1 & 0 \\ 0 & 3 & 1 \\ 0 & 0 & 3 \end{pmatrix}$ and $A^2 = \begin{pmatrix} 9 & 6 & 1 \\ 0 & 9 & 6 \\ 0 & 0 & 0 \end{pmatrix}$. The relevant differential equation is

$$\frac{d^3 w}{dt^3} - 9\frac{d^2 w}{dt^2} + 27\frac{dw}{dt} - 27w = 0.$$

Hence $w_1 = (1 - 3t + 9t^2/2)e^{3t}$, $w_2 = (t - 3t^2)e^{3t}$, $w_3 = t^2 e^{3t}/2$ leading to

$$e^{tA} = e^{3t} \begin{pmatrix} 1 & t & t^2/2 \\ 0 & 1 & t \\ 0 & 0 & 1 \end{pmatrix}$$

and

$$x = (C_1 + C_2 t + C_3 t^2/2)e^{3t},$$
$$y = (C_2 + C_3 t)e^{3t},$$
$$z = C_3 e^{3t}.$$
▯

Matrix notation offers a neat way of representing a particular integral of the system

$$\dot{\mathbf{x}} = A\mathbf{x} + \mathbf{f}(t). \tag{3.5.12}$$

When $\mathbf{f}$ is absent we know that a solution of (3.5.12) can be expressed as $\mathbf{x} = e^{tA}\mathbf{x}_0$. Therefore, as in the method of variation of parameters, try

$$\mathbf{x} = e^{tA}\mathbf{y}(t) \tag{3.5.13}$$

as a solution of (3.5.12). Since, from (3.5.6),

$$\dot{\mathbf{x}} = Ae^{tA}\mathbf{y}(t) + e^{tA}\dot{\mathbf{y}}(t) = A\mathbf{x} + e^{tA}\dot{\mathbf{y}}(t)$$

(3.5.12) is satisfied provided that

$$e^{tA}\dot{\mathbf{y}}(t) = \mathbf{f}(t).$$

Hence $\dot{\mathbf{y}}(t) = e^{-tA}\mathbf{f}(t)$ and

$$\mathbf{y}(t) = \int^{t} e^{-uA}\mathbf{f}(u)du.$$

It follows from (3.5.13) that a particular integral of (3.5.12) is

$$\mathbf{x}(t) = e^{tA} \int^{t} e^{-uA}\mathbf{f}(u)du. \tag{3.5.14}$$

The meaning of $\int \mathbf{g}(u)du$ is specified by

$$\int \mathbf{g}(u)du = \begin{pmatrix} \int g_1(u)du \\ \int g_2(u)du \\ \vdots \\ \int g_n(u)du \end{pmatrix}$$

when $\mathbf{g}^T = (g_1, g_2, \ldots, g_n)$ where $\mathbf{g}^T$ is the transpose of $\mathbf{g}$.

Observe that e^{-tA} differs from e^{tA} only in the sign of t. Consequently, e^{-tA} can be written down as soon as e^{tA} has been calculated.

Example 3.5.4

Find a particular integral for

$$\dot{x} = 5x + 3y + 2te^{2t},$$
$$\dot{y} = -3x - y + 4.$$

We know from Example 3.5.2 that

$$e^{tA} = e^{2t} \begin{pmatrix} 1 + 3t & 3t \\ -3t & 1 - 3t \end{pmatrix}$$

and so

$$e^{-tA} = e^{-2t} \begin{pmatrix} 1 - 3t & -3t \\ 3t & 1 + 3t \end{pmatrix}.$$

Hence, with $\mathbf{f}^T = (2te^{2t}, 4)$,

$$\int^{t} e^{-uA}\mathbf{f}(u)du = \begin{pmatrix} t^2 - 2t^3 + 3(2t + 1)e^{-2t} \\ 2t^3 - (6t + 5)e^{-2t} \end{pmatrix}.$$

We do not need to include arbitrary constants since they add only multiples of the complementary function.

After insertion in (3.5.14) we obtain the particular integral

$$x = (t^2 + t^3)e^{2t} + 3,$$
$$y = -t^3 e^{2t} - 5.$$

⬚

3.6 Initial and boundary value problems

It has been mentioned from time to time that solutions of differential equations are often subject to extra conditions. This section will be devoted to a discussion of two types of conditions that are of frequent occurrence in practice.

Suppose a solution of

$$\frac{d^n y}{dt^n} + a_{n-1}\frac{d^{n-1} y}{dt^{n-1}} + \cdots + a_0 y = f(t) \tag{3.6.1}$$

is required such that

$$y = y^{(0)}, \quad \frac{dy}{dt} = y^{(1)}, \quad \ldots, \quad \frac{d^{n-1} y}{dt^{n-1}} = y^{(n-1)}$$

at $t = t_0$, with $y^{(0)}, \ldots, y^{(n-1)}$ prescribed constants. This is termed an **initial value problem**.

The general solution of (3.6.1) is

$$y = C_1 y_1 + \cdots C_n y_n + y_p$$

where y_p is a particular integral and the remaining terms represent the complementary function. Then the imposed conditions can be complied with if

$$C_1 \frac{d^m y_1}{dt^m} + \cdots C_n \frac{d^m y_n}{dt^m} = y^{(m)} - \frac{d^m y_p}{dt^m} \quad (m = 0, 1, \ldots, n-1)$$

when $t = t_0$. These n equations for $C_1, \ldots, C_n$ can always be solved if

$$\begin{vmatrix} y_1 & y_2 & \cdots & y_n \\ \dfrac{dy_1}{dt} & \dfrac{dy_2}{dt} & \cdots & \dfrac{dy_n}{dt} \\ \vdots & \vdots & & \vdots \\ \dfrac{d^{n-1} y_1}{dt^{n-1}} & \dfrac{d^{n-1} y_2}{dt^{n-1}} & \cdots & \dfrac{d^{n-1} y_n}{dt^{n-1}} \end{vmatrix} \neq 0.$$

This determinant is the same as that derived from the Wronskian of Section 3.3 when the differential equation is converted to a first-order system and, accordingly, is also known as a Wronskian. The non-vanishing of the Wronskian warrants the statement that

$$B_1 y_1(t) + \cdots + B_n y_n(t) = 0 \qquad (3.6.2)$$

for an interval of t, which necessitates the constants $B_1, \ldots, B_n$ all being zero. For $n - 1$ derivatives of (3.6.2) give a set of equations for $B_1, \ldots, B_n$ with non-zero determinant and zero right-hand side. Expressed in other words, the non-vanishing of the Wronskian makes $y_1, \ldots, y_n$ **linearly independent** over the interval. To put it another way, it makes certain that the complementary function has been determined correctly.

Since there is only one set of $C_1, \ldots, C_n$ that satisfies the equations for non-zero Wronskian, it has been demonstrated that *the initial value problem always possesses a solution and there is only one which satisfies the imposed conditions*. This constitutes another verification of the uniqueness property.

The initial value problem is characterised by all the restrictions being applied at a single value of t. In some instances the conditions refer to more than one value of t—we then have a **boundary value problem**. In contrast to the initial value problem, it is by no means certain that a boundary value problem has a solution. Consider

$$\ddot{y} + y = 0$$

of which the general solution is

$$y = C_1 \cos t + C_2 \sin t.$$

Let the conditions be $y(0) = 0, y(1) = 0$. The first requires $C_1 = 0$ and the second $C_2 \sin 1 = 0$. Since $\sin 1 \neq 0$ we must have $C_2 = 0$ and the only solution is the trivial one which vanishes everywhere. Now change the conditions to $y(0) = 0, y(\pi) = 0$. In this event, $y = C_2 \sin t$ is a solution with C_2 arbitrary. Thus, boundary value problems may have many solutions or none (if the trivial one is discounted). It is also obvious that the interval of t has a critical role to play.

Instead of varying the interval, it is usual to fix it and incorporate a parameter in the differential equation. A typical problem might be to solve

$$\frac{d}{dt}\left(p(t)\frac{dy}{dt}\right) + \{q(t) + \lambda\}y = 0 \qquad (3.6.3)$$

subject to $y(a) = 0, y(b) = 0$. The values of λ, which is independent of t, are crucial. For some there will be only the trivial solution and for others there will be many solutions. Those λ for which non-trivial solutions exist are called **eigenvalues** and the corresponding solutions **eigenfunctions**.

Example 3.6.1
Consider

$$\ddot{y} + \lambda y = 0$$

under the conditions $y(0) = 0, y(\pi) = 0$.

If $\lambda = 0$, the general solution is $y = A + Bt$, which satisfies the boundary conditions only if $A = 0$ and $B = 0$. Therefore $\lambda = 0$ is not an eigenvalue.

If $\lambda \neq 0$, the general solution is

$$y = C_1 \cos \sqrt{\lambda} t + C_2 \sin \sqrt{\lambda} t.$$

To comply with the boundary conditions we must have $C_1 = 0$ and $C_2 \sin \sqrt{\lambda} \pi = 0$. For a non-trivial solution $C_2 \neq 0$ and $\lambda = m^2$ where m is a positive integer. The eigenvalues are real and infinite in number. They may be designated $\lambda_1, \lambda_2, \ldots$ where $\lambda_m = m^2$. The eigenfunction corresponding to λ_m is $C_m \sin mt$ where C_m is arbitrary. ⬜

The discussion of (3.6.3) will assume that $p(t)$ is a continuously differentiable real function that does not change sign for any t in (a, b). No loss of generality is incurred in taking it to be positive. The function q will be assumed to be real and continuous in (a, b). We shall also suppose that there is an infinite set of eigenvalues $\lambda_1, \lambda_2, \ldots$ with associated eigenfunctions $Y_1, Y_2, \ldots$.

With these assumptions the first thing to be shown is that *the eigenvalues are real*. Y_m satisfies

$$\frac{d}{dt}\left(p(t)\frac{dY_m}{dt}\right) + \{q(t) + \lambda_m\}Y_m = 0 \tag{3.6.4}$$

and $Y_m(a) = 0, Y_m(b) = 0$. By taking a complex conjugate

$$\frac{d}{dt}\left(p(t)\frac{dY_m^*}{dt}\right) + \{q(t) + \lambda_m^*\}Y_m^* = 0 \tag{3.6.5}$$

and $Y_m^*(a) = 0, Y_m^*(b) = 0$. Multiply (3.6.4) by Y_m^*, (3.6.5) by Y_m and subtract. There results

$$Y_m^*\frac{d}{dt}\left(p(t)\frac{dY_m}{dt}\right) - Y_m\frac{d}{dt}\left(p(t)\frac{dY_m^*}{dt}\right) + (\lambda_m - \lambda_m^*)|Y_m|^2 = 0.$$

Hence

$$(\lambda_m^* - \lambda_m)\int_a^b |Y_m|^2 dt = \int_a^b \left\{Y_m^*\frac{d}{dt}\left(p(t)\frac{dY_m}{dt}\right) - Y_m\frac{d}{dt}\left(p(t)\frac{dY_m^*}{dt}\right)\right\}dt$$

$$= \left[Y_m^* p(t)\frac{dY_m}{dt} - Y_m p(t)\frac{dY_m^*}{dt}\right]_a^b \tag{3.6.6}$$

by integration by parts. The right-hand side of (3.6.6) is zero because of the conditions on Y, Y^* at $t = a, t = b$. The integral on the left is positive because Y_m is a non-trivial solution. Consequently, $\lambda_m = \lambda_m^*$ and λ_m is real.

The reality of the eigenvalues means that there is no loss of generality in taking the eigenfunctions to be real. With this understood note that

$$\frac{d}{dt}\left(p(t)\frac{dY_n}{dt}\right) + \{q(t) + \lambda_n\}Y_n = 0 \tag{3.6.7}$$

and $Y_n(a) = 0, Y_n(b) = 0$. Multiply (3.6.4) by Y_n, (3.6.7) by Y_m, subtract and proceed as above. Then

$$(\lambda_n - \lambda_m)\int_a^b Y_m Y_n dt = \left[p(t)\left(Y_n\frac{dY_m}{dt} - Y_m\frac{dY_n}{dt}\right)\right]_a^b. \tag{3.6.8}$$

The right-hand side is zero on account of the values of Y_m, Y_n at the endpoints. Therefore, if $\lambda_m \neq \lambda_n$,

$$\int_a^b Y_m Y_n dt = 0. \tag{3.6.9}$$

Functions that satisfy (3.6.9) are said to be *orthogonal*, i.e., the eigenfunctions of distinct eigenvalues are orthogonal. If, in addition, the eigenfunctions are **normalised** so that $\int_a^b Y_m^2 dt = 1$, the eigenfunctions are called **orthonormal**.

It will not have escaped the reader's notice that the right-hand sides of (3.6.6) and (3.6.8) can vanish for conditions other than those delineated. For example, if $y(a) = 0$ is replaced by

$$\alpha_1 y(a) + \alpha_2 \dot{y}(a) = 0, \tag{3.6.10}$$

where at least one of the real α_1, α_2 is non-zero, the right-hand sides are still zero. A similar remark is true if $y(b) = 0$ is changed to

$$\beta_1 y(b) + \beta_2 \dot{y}(b) = 0 \tag{3.6.11}$$

where β_1, β_2 are real with at least one non-zero. Equations (3.6.10) and (3.6.11) can be deemed **standard boundary conditions** (they include the previous ones by putting $\alpha_2 = 0, \beta_2 = 0$). What has been shown is that the *eigenvalues are real and the eigenfunctions orthogonal for standard boundary conditions.*

Example 3.6.2
Find the eigenfunctions of $\ddot{y} + \lambda y = 0$ subject to $\dot{y}(0) = 0, \dot{y}(\pi) = 0$.

For $\lambda = 0$, the solution $y = A + Bt$ meets the boundary conditions if $B = 0$. The eigenfunction is $y = A$.

For $\lambda \neq 0$, proceed as in Example 3.6.1 to show that there is an eigenvalue n^2 with eigenfunction $C_n \cos nt$.

The first eigenfunction can be subsumed in the second group by allowing $n = 0$. Thus the eigenfunctions are $C_n \cos nt$ for $n = 0, 1, 2, \ldots$. ▯

Eigenvalues can also occur for **periodic boundary conditions** where $p(a) = p(b)$, $y(a) = y(b)$, $\dot{y}(a) = \dot{y}(b)$. Again the eigenvalues are real and the eigenfunctions orthogonal.

3.7 Solving the inhomogeneous differential equation

This section is concerned with the boundary value problem in which

$$\frac{d}{dt}\left(p(t)\frac{dy}{dt}\right) + q(t)y = f(t). \tag{3.7.1}$$

For simplicity, the conditions $y(a) = 0, y(b) = 0$ will be imposed, although it will be clear that the technique is equally valid for the standard boundary conditions.

Let $\lambda_1, \lambda_2, \ldots$ and $Y_1, Y_2, \ldots$ be the eigenvalues and eigenfunctions determined in the section before. Assume that we can write

$$f(t) = \sum_{m=1}^{\infty} b_m Y_m(t).$$

Multiply by Y_n and integrate from a to b. Then, by virtue of the orthogonality of the eigenfunctions,

$$b_n \int_a^b Y_n^2 dt = \int_a^b f(t)Y_n(t)dt$$

which specifies the coefficient b_n. Putting

$$y = \sum_{m=1}^{\infty} a_m Y_m(t) \tag{3.7.2}$$

in (3.7.1) we obtain, provided that derivatives can be taken term-by-term,

$$-\sum_{m=1}^{\infty} a_m \lambda_m Y_m(t) = \sum_{m=1}^{\infty} b_m Y_m(t),$$

which suggests that (3.7.2) is the desired solution of the boundary value problem when $a_m = -b_m/\lambda_m$.

Example 3.7.1
Find the solution of

$$\ddot{y} = t$$

such that $y(0) = 0, y(\pi) = 0$.

It should first be pointed out that this problem can be solved easily without eigenfunctions but for many problems there is no option to finding the solution as a series of eigenfunctions.

The eigenfunctions are $\sin nt$ and so

$$b_n = \int_0^\pi t \sin nt \, dt \Big/ \int_0^\pi \sin^2 nt \, dt = (-1)^{n+1} 2/n.$$

Since $\lambda_n = n^2$ our solution is

$$y = 2 \sum_{m=1}^\infty \frac{(-1)^m}{m^3} \sin mt. \qquad \Box$$

The derivation of the series solution rested on a number of assumptions, namely

(a) f can be expanded in a series of eigenfunctions,

(b) $a_m = -b_m/\lambda_m$ for $m = 1, 2, \ldots$,

(c) $\sum_{m=1}^\infty b_m Y_m/\lambda_m$ is a continuous function that possesses two derivatives which can be calculated by taking derivatives of the series term-by-term.

There is no difficulty about (b) when $\lambda_m \neq 0$ for $m = 1, 2, \ldots$ and a unique solution is obtained. If, however, one of the eigenvalues, say λ_1, is zero, the boundary value problem has no solution when $b_1 \neq 0$ and an infinite number of solutions when $b_1 = 0$.

Both (a) and (c) raise delicate matters because they require knowledge of the properties of expansions of functions in terms of eigenfunctions. It would take us too far afield to derive these properties; so we content ourselves with a few observations without proof. Generally speaking, the smoother f is, i.e., the more derivatives it has, the more likely is the process to be legitimate. In fact, for the standard boundary conditions, the series for a piecewise smooth function converges uniformly and absolutely to the function on any closed interval in which the function is continuous. Nevertheless, even in the absence of a specific theorem, there is nothing to prevent one from carrying out the process formally and then attempting to confirm that the resulting series has the desired properties.

3.8 Appendix: symbolic computation

This appendix contains some illustrations of the solutions of differential equations by means of the program MATHEMATICA. You will see that, although MATHEMATICA is very powerful, it still needs guidance from someone with a knowledge of differential equations on occasion. There are many variants of MATHEMATICA; so the details on your machine may be somewhat different from those given below but the principles should be the same.

Often in the following the same symbol will be used for different things but any necessary clearing operations will not be indicated.

(a)   Separable equations

The first example is of the separable differential equation

$$\frac{dy}{dt} = \frac{2t}{(1+t^2)y}.$$

The input to and output from MATHEMATICA could then be

```
In[1]:= equ=y'[t]==2 t/ ((1+t^2)y[t])
```

```
                      2 t
Out[1] = y'[t]==   -------
                   (1 + t²) y[t]
```

```
In[2] := sol=DSolve[equ,y[t],t]
```

```
Out[2] = {{y[t]->-Sqrt[C[1]+2 Log[1+t²]]},
```

```
>{y[t]->Sqrt[C[1]+2 Log[1+t²]]}}
```

```
In[3]:= y[t]/.sol
```

```
Out[3] = {-Sqrt[C[1]+2 Log[1+t²]], Sqrt[C[1]+
```

```
> 2 Log[1+t²]]}
```

Explicit forms involving an arbitrary constant C[1] are obtained for y. They can be separated off into a list if desired for future purposes.

In the next example MATHEMATICA cannot find y explicitly and leaves the answer as a Solve. Nevertheless the appropriate form can be derived by using the instruction to take part of an expression.

```
In[8] := equ=y'[t]==2 t/ ((1+t^2)y[t] Sin[y[t]])
```

```
                   2 t Csc[y[t]]
Out[8] = y'[t]==   -------
                   (1+t²) y[t]
```

```
In[9] := sol=DSolve[equ,y[t],t]
```

```
Solve::tdep: The equations appear to involve transcendental
functions of the variables in an essentially non-algebraic way.

Out[9] = Solve[-Log[1+t^2]+Sin[y[t]]-Cos[y[t]] y[t]==

> C[1],y[t]]

In[10] := sol[[1]]

Out[10] = -Log[1+t^2]+Sin[y[t]]-Cos[y[t]] y[t]==C[1]
```

(b)   Homogeneous equations

To see what happens with homogeneous equations Example 1.6.1 is considered.

```
In[12] := equ=y'[t]==(3 y[t]-t)/(3 t-y[t])

                      -t + 3 y[t]
Out[12] = y'[t]==     -----
                      3 t - y[t]

In[13] := sol=DSolve[equ,y[t],t]

                         -t + 3 y[t]
Out[13] = DSolve[ y'[t]==  -----   ,y[t],t]
                         3 t - y[t]
```

In this case the program does not know how to solve the differential equation and merely repeats the last instruction. However, we can ask it to follow the same route as in Example 1.6.1.

```
In[14]:= equ1=equ/.{y[t]->t z[t],y'[t]->D[t z[t],t]}

                         -t + 3 t z[t]
Out[14] = z[t]+t z'[t]==  ------
                         3 t - t z[t]

In[15] := sol1=DSolve[equ1,z[t],t]

Solve::tdep: The equations appear to involve transcendental
functions of the variables in an essentially non-algebraic way.

Out[15] = Solve[Log[t]-Log[1-z[t]]+2 Log[1+z[t]]==

> C[1],z[t]]
```

```
In[16] := sol1[[1]]
```

```
Out[16] = Log[t]-Log[1-z[t]]+2 Log[1+z[t]]==C[1]
```

```
In[17] := Expand[Exp[Sol1[[1,1]]]]
```

$$\text{Out}[17]= \frac{t\,(1 + z[t])^2}{1 - z[t]}$$

Notice that the substitution for y[t] does not automatically carry over to y'[t]; the information for y' has to be supplied. Notice also that the logarithms are not combined automatically. To put the solution in the form of Example 1.6.1 an additional instruction has to be supplied.

(c)   Linear equation of the first order

The program can cope with a differential equation like (1.7.1) because the general solution is built in. At first sight the output does not look anything like the correct answer. You need to realise that **Integrate** is used instead of the usual symbol for integration because the integrals cannot be evaluated explicitly. Also, when an integral occurs within an integral the symbol **DSolve't** is used as an integrating variable.

```
In[19] := equ=y'[t]+f[t] y[t]==g[t]
```

```
Out[19] = f[t] y[t]+y'[t]==g[t]
```

```
In[20] := sol=DSolve[equ,y[t],t]
```

$$\text{Out}[20] = \{\{y[t] \rightarrow \frac{C[1]}{E^{\text{Integrate}[f[t],t]}} +$$

$$> \text{Integrate}[\frac{\text{Integrate}[f[DSolve't],DSolve't]}{E}$$

$$> g[DSolve't], \{Dsolve't,0,t\}]/$$

$$> \frac{\text{Integrate}[f[DSolve't],DSolve't]}{E}\}\}$$

With the explanation above this is the same result as is obtained by means of an integrating factor.

In fact, it is not even necessary to put the differential equation in the form (1.7.1). Example 1.7.1 will serve to illustrate the point.

```
In[21] := equ1=(t^2+1)y'[t]+t y[t]==1/2
```

$$Out[21] = t\ y[t]+(1+t^2)y'[t]== \frac{1}{2}$$

```
In[22] := sol1=DSolve[equ1,y[t],t]
```

$$Out[22] = \{\{y[t]-> \frac{ArcSinh[t]}{2\ Sqrt[1 + t^2]} + \frac{C[1]}{Sqrt[1 + t^2]} \}\}$$

This is not quite how it appears in Example 1.7.1 because `ArcSinh` is used rather than a logarithm. It is easy to check that the answers do coincide.

```
In[24] := Simplify[Sinh[Log[t+Sqrt[1+t^2]]]]

Out[24] = t
```

When the integrals cannot be calculated in terms of known functions numerical methods can be called on. A first approach to tackling the differential equation might be Euler's method of Section 1.5 but more sophisticated techniques are built into `NDSolve`. It provides an answer as an interpolating function which, in essence, gives the values of the solution at various points in the interval of interest. In general, these values are not displayed but the range covered is. Sometimes the program cannot find an interpolant to cover the whole of the range you desire. In that event, you go part of the way and then restart `NDSolve` with the information you have acquired.

Here is an example of `NDSolve` in which the solution is started from zero:

```
In[27] := equ=y'[t]+t y[t]/(2+cos[t])==Exp[-t] Sin[t]
```

$$Out[27] = \frac{t\ y[t]}{2 + Cos[t]} +y'[t]== \frac{Sin[t]}{E^t}$$

```
In[28] := sol=NDSolve[{equ,y[0]==0},y[t],{t,0,2 Pi}]
```

$$Power::infy: Infinite\ expression.\ \frac{1}{0.}\ --encountered.$$

```
Out[28] = {{y[t]->InterpolatingFunction[{0.,6.28319},

> <>][t]}}
```

The solution does cover the range required and now can be shown graphically.

```
In[29] := Plot[y[t]/.sol,{t,0,2 Pi}]

Out[29] = -Graphics-
```

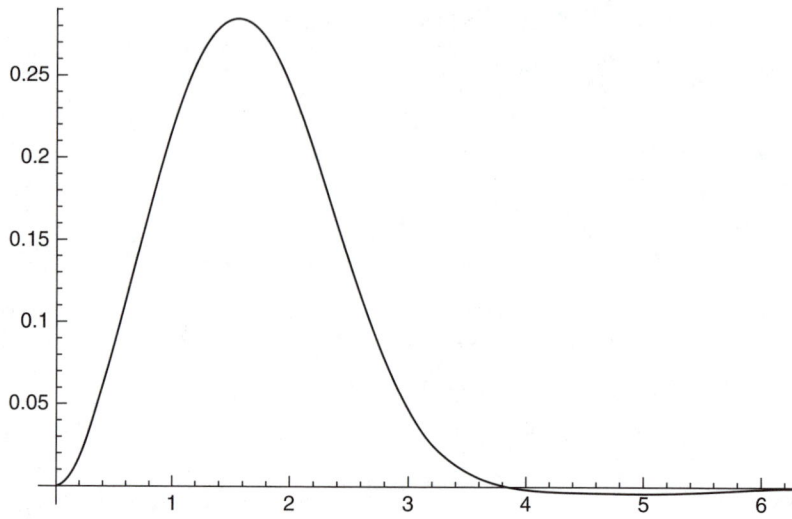

FIGURE 3.8.1: Curve of interpolant.

Note that the curve (Figure 3.8.1) does have zero slope at the origin as it should according to the differential equation.

You can, if you wish, employ numerical integration in the known form of the solution instead of applying NDSolve to the differential equation. Usually, this takes more effort and is much slower in execution than NDSolve. Nevertheless, it does offer a method of checking on the performance of NDSolve. For the differential equation just considered you could proceed as follows:

```
In[30]:= intf[x_]:= intf[x]=NIntegrate[t/(2+Cos[t]),
{t,0,x}]

In[31]:= calc[x_]:= NIntegrate[Exp[intf[t]] Exp[-t]
Sin[t],{t,0,x}]

In[32]:= sol1[x_]:= calc[x]/Exp[intf[x]]

In[33]:= Plot[sol1[x],{x,0,2 Pi}]
```

The resulting graph should coincide with the one already drawn.

(d) Other equations

An important part is played in higher order differential equations (and other contexts) by functions that are linearly independent. Checking linear dependence or otherwise, when it is not immediately clear, entails evaluating the Wronskian determinant. As soon as four or more functions are involved this can be extremely tedious especially if the functions are at all complicated. The computer can give a helping hand.

Since $\cos 2t = 2\cos^2 t - 1$ we know that $\cos 2t$ and $2\cos^2 t - 1$ are linearly dependent. Let us see if the computer agrees.

```
In[36]:= mat={{Cos[2 t],2 Cos[t]^2-1},{-2 Sin[2 t],
-4 Cos[t] Sin[t]}}

Out[36]={{Cos[2 t],-1+2 Cos[t]^2},{-2 Sin[2 t],

> -4 Cos[t] Sin[t]}}

In[37]:= wronsk=Det[mat]

Out[37]= -4 Cos[t] Cos[2 t] Sin[t]-2 Sin[2 t]+

> 4 Cos[t]^2 Sin[2 t]
```

It is not by any means obvious that this is zero and is another illustration that MATHEMATICA may not simplify expressions as much as we would wish unless told to do so. Accordingly, the next step attempts to combine the terms.

```
In[38]:= Expand[wronsk,Trig->True]

Out[38]= 0
```

The linear dependence of $\cos 2t$ and $2\cos^2 t - 1$ has been confirmed.

With this experience in mind a way of finding the value of the Wronskian of a list of functions is:

```
In[41]:= wronsk[f_List]:=Module[{row,mat},

row[1] = f;

row[p_].=row[p]=D[row[p-1],t];

mat=Table[row[i],{i,1,Length[f]}];

Expand[Det[mat],Trig->True]

]
```

A quick confirmation that this works even if the functions are not trigono-metric is provided by

```
In[42]:= wronsk[{Cos[2 t],2 Cos[t]^2-1}]

Out[42]= 0

In[43]:= wronsk[{Exp[t],Exp[-t]}]

Out[43]= -2
```

When MATHEMATICA solves a system of differential equations with con-stant coefficients for the complementary function it does so by means of the formula (3.5.3). It can do so because it has a routine for calculating the expo-nential of a matrix. An illustration is provided by the system

$$\dot{x} = 5x + 6y - 6z,$$
$$\dot{y} = x + 4y - 5z,$$
$$\dot{z} = -2x + 2y - 3z.$$

In order that you may see how you could obtain your own solution via (3.5.3) the matrix of the coefficients on the right-hand side is constructed first. Then the appropriate exponential is determined by taking advantage of the instruction `MatrixExp`.

```
In[54]:= mat={{5,6,-6},{1,4,-5},{-2,2,-3}}

In[55]:= mat1=MatrixExp[t mat]
```

$$
\begin{aligned}
Out[55]= \{\{ &\frac{1}{3\,E^t} + \frac{2\,E^{8t}}{3}, \quad \frac{2\,E^{8t}}{3}, \quad \frac{2}{3\,E^t} - \frac{2\,E^{8t}}{3} \}, \\
\{ &\frac{-7}{27\,E^t} + \frac{7\,E^{8t}}{27} - \frac{4\,t}{3\,E^t}, \quad \frac{20}{27\,E^t} + \frac{7\,E^{8t}}{27} + \frac{8\,t}{3\,E^t}, \\
&\frac{7}{27\,E^t} - \frac{7\,E^{8t}}{27} - \frac{8\,t}{3\,E^t} \}, \\
\{ &\frac{2}{27\,R^t} - \frac{2\,E^{8t}}{27} - \frac{4\,t}{3\,E^t}, \quad \frac{2}{27\,E^t} - \frac{2\,E^{8t}}{27} + \frac{8\,t}{3\,E^t}, \\
&\frac{25}{27\,E^t} + \frac{2\,E^{8t}}{27} - \frac{8\,t}{3\,E^t} \}\}
\end{aligned}
$$

The formula (3.5.3) can be implemented now via

```
In[56]:= ans=mat1.{c1,c2,c3}
```

The resulting output, which runs to several lines, will be omitted. The form of the general solution furnished by **ans** is convenient when you know that $(x, y, z) = (c1, c2, c3)$ at $t = 0$. It is not, however, the simplest form of the general solution. A simpler version is obtained by transforming the constants $c1, c2, c3$ to others.

```
In[57]:= Simplify[ans/.{c1->-2 d2-18,c2->d1+2 d2-7 d3,
c3->d1+2 d3}]
```

$$\text{Out[57]} = \left\{ \frac{-2\ d2}{E^t} - 18 - d3\ E^{8t}, \quad \frac{d1 + 2\ d2 - 7\ d3\ E^{9t} + 8\ d2\ t}{E^t}, \right.$$

$$\left. > \quad \frac{d1 + 2\ d3\ E^{9t} + 8\ d2\ t}{E^t} \right\}$$

While the structure of this form is much simpler than that of **ans** it has the disadvantage of requiring the solution of three simultaneous equations for $d1, d2$ and $d3$ when initial conditions have to be satisfied.

Exercises

3.1 Find the general solution of

(a) $\dot{x} + 2x + 3y = 0,$
$\dot{y} + 3x + 2y = 2e^{2t};$

(b) $\dot{x} + \dot{y} - 5x + 3y = 15t^{1/2}e^t,$
$\dot{x} - 2\dot{y} + x = -30t^{1/2}e^t.$

3.2 Find the general solution of

$$x - \dot{x} + \dot{y} + y = 1,$$
$$2\dot{x} + x - 2\dot{y} - y = 2$$

by eliminating (a) y, (b) $\dot{y}$.

3.3 Find the general solution of

(a) $\dot{x} = 9x - 8y,$
$\dot{y} = 24x - 19y;$

(b) $\dot{x} = 2y - x,$
$\dot{y} = -2x - y;$

(c) $\dot{x} = x + y,$
$\dot{y} = y;$

(d) $\dot{x} = x - 2y - z,$
$\dot{y} = -x + y + z,$
$\dot{z} = x - z;$

(e) $\dot{x} = x - y + z,$
$\dot{y} = x + y - z,$
$\dot{z} = 2z - y.$

3.4 Show that $x = e^t$, $y = 2e^t$ and $x = te^t$, $y = (2t - 1)e^t$ form a fundamental system for

$$\dot{x} = 3x - y + 1, \qquad \dot{y} = 4x - y + t$$

and hence find the solution of the system such that $x = 1$, $y = 0$ at $t = 0$.

3.5 Find the solutions of the initial value problems

(a) $\dot{x} + 2x + 3y = 0$, $\dot{y} + 3x + 2y = 2e^{2t}$ with $x = 0$, $y = 0$ at $t = 0$;

(b) $\dot{x} + 3x + 2y = 3t - 1$, $\dot{y} + 3x - 2y = 3t - 10$ with $x = 1$, $y = 0$ at $t = 0$.

3.6 Find a particular integral by matrix methods of

$$\dot{x} = 3x + y + 3t^2 e^{3t},$$
$$\dot{y} = 3y + z + 9t,$$
$$\dot{z} = 3z + 18.$$

3.7 The function $f(t)$ is defined by

$$f(t) = \begin{cases} t & (0 \le t \le \pi), \\ \pi e^{\pi - t} & (t \ge \pi). \end{cases}$$

Find the solution of the initial value problem

$$\ddot{y} + y = f(t)$$

which is continuous, with a continuous derivative, for all $t \ge 0$ and such that $y = 0$, $\dot{y} = 1$ at $t = 0$.

3.8 If $f(t)$ is defined as in Exercise 3.7, show that the boundary value problem with $y(0) = 0$, $y(2\pi) = a$ has no continuous solution with continuous derivative on $0 \le t \le 2\pi$ if $a \ne \frac{1}{2}\pi(e^{-\pi} - 1)$, but has infinitely many such solutions if $a = \frac{1}{2}\pi(e^{-\pi} - 1)$.

3.9 Find the eigenvalues and eigenfunctions of $\ddot{y} + \lambda y = 0$ subject to the boundary conditions

(a) $\dot{y}(-\pi) = 0$, $\dot{y}(\pi) = 0$;

(b) $y(0) = 0$, $ay(b) + \dot{y}(b) = 0$ $(a > 0, b > 0)$.

3.10 Show that

$$\frac{d^4 y}{dt^4} - \mu^4 y = 0$$

has non-trivial solutions satisfying the boundary conditions $y(0) = 0$, $\dot{y}(0) = 0$, $y(1) = 0$, $\dot{y}(1) = 0$ if, and only if, $\cos \mu \cosh \mu = 1$ $(\mu \neq 0)$.

3.11 Find series solutions in terms of eigenfunctions of

(a) $\ddot{y} = t(t - 2\pi)$ subject to $y(0) = 0$, $\dot{y}(\pi) = 0$;

(b) $\ddot{y} = \sin(\pi t/b)$ subject to $y(0) = 0$, $\dot{y}(b) = 0$;

(c) $t\dfrac{d}{dt}\left(t\dfrac{dy}{dt}\right) + 5y = 3\sin(5\ln t)$ subject to $y(1) = 0$, $y(e^\pi) = 0$.

3.12 Show that, for

$$\frac{d}{dt}\left(p(t)\frac{dy}{dt}\right) + \{q(t) + \lambda r(t)\}y = 0$$

subject to the standard boundary conditions and r positive, the eigenvalues are real and the eigenfunctions are orthogonal with respect to the weight function r, i.e.,

$$\int_a^b r Y_m Y_n \, dt = 0$$

when Y_m and Y_n correspond to distinct eigenvalues.

3.13 The eigenfunctions $Y_1(t), Y_2(t), \ldots$ satisfy

$$t^2 \frac{d^2 Y_m}{dt^2} - t\frac{dY_m}{dt} + \left(\tfrac{3}{4} + \lambda_m t^2\right) Y_m = 0$$

on $\pi < t < 2\pi$ under the boundary conditions $Y_m(\pi) = 0$, $Y_m(2\pi) = 0$. Show that the eigenfunctions are orthogonal with respect to a suitable weight function.

By means of the substitution $y(t) = t^{1/2} u(t)$, obtain as a series of eigenfunctions the solution of

$$t^2 \ddot{y} - t\dot{y} + \left(\tfrac{3}{4} - t^2\right) y = t^{5/2}$$

such that $y(\pi) = 0$, $y(2\pi) = 0$.

Chapter 4

Modelling Biological Phenomena

4.1 Introduction

Mathematical modelling of physical phenomena, such as the dynamics of a rigid body, the deformation of an elastic material or the propagation of electromagnetic waves in the atmosphere, is based, to the best of our present scientific knowledge, on sound physical laws. Thus to describe the dynamical behaviour of a rigid body undergoing the influence of external forces we have the fundamental Newtonian laws of motion at our disposal. The behaviour of deforming elastic materials is governed by the constitutive equations of continuum mechanics; Maxwell's equations are the fundamental postulates that govern electromagnetic waves.

These basic laws have been the result of centuries of experiment, observation and inspiration of mathematicians and scientists including Sir Isaac Newton, Leonard Euler and James Clerk Maxwell.

In biology and the life sciences in general this interplay between the observed phenomenon and its mathematical description is still in the early stages of development and apart from the **Hardy-Weinberg law** associated with the Mendelian theory of genetics there are few sound postulates to guide us. Instead the philosophy is to develop mathematical models that in the first instance describe in only a qualitative way the observed biological process. As in all scientific endeavours, the real test of the model is that it not only agrees qualitatively with the biological process but has the ability to suggest new experiments and bring deeper insight to the biological situation. If enough experience is gained by this philosophy then hopefully, together with a better understanding of the life sciences, sound postulates will emerge upon which a mathematical theory can be developed.

This philosophy of qualitative description will be exploited in the topics treated in this chapter and throughout a major portion of this book.

4.2 Heart beat

The heart is a complex but robust pump (see the simplified illustration in Figure 4.2.1). It consists of four chambers and four valves. There are essentially

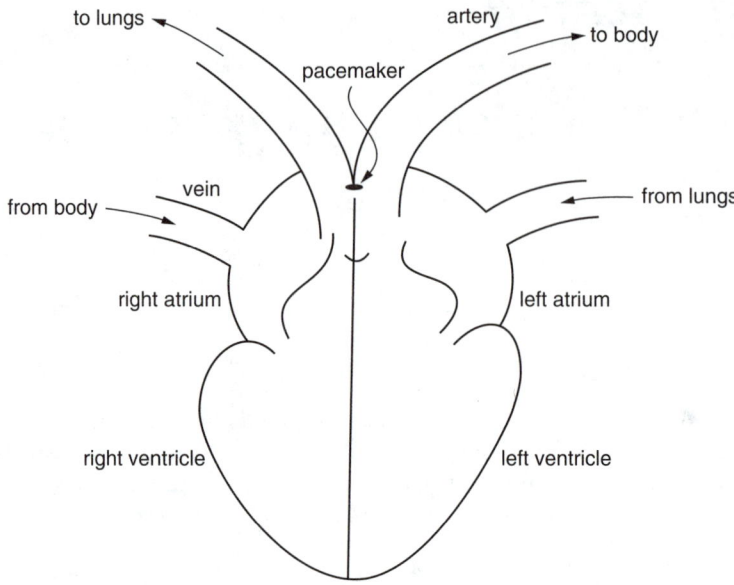

FIGURE 4.2.1: Schematic description of the heart as viewed from the front of the body.

two circuits for the blood, one which spreads through the lungs to pick up oxygen and the other which spreads through the body to deliver the oxygenated blood. The first circuit is a low-pressure circuit so as not to damage the delicate membrane in the lungs, whereas the second is a high-pressure circuit in order for the blood to get down to the feet and up again. From Figure 4.2.1 it is apparent that the right side of the heart is the low-pressure pump to the lungs while the left side is the high-pressure pump to the body.

Each pump has a main pumping chamber called the ventricle with an inlet and an outlet valve. The purpose of the inlet valve is to prevent flow back up the veins while pumping, and the outlet valve is to prevent flow back from the arteries while filling. Since the heart is made of non-rigid tissue it only has the power to push out and no power to suck in. Thus to get a good pump of blood it is necessary to fill the ventricle completely, and to aid this there is a small chamber called the atrium whose job is to pump gently beforehand, just enough to fill the ventricle but not enough to cause any flow back.

During the heart beat cycle there are two extreme equilibrium states, namely **diastole** which is the relaxed state and **systole** which is the contracted state. What makes the heart beat is the presence of a pacemaker which is located on the top of the atrium. The pacemaker causes the heart to contract into systole. That is, it triggers off an electrochemical wave which spreads slowly over the atria causing the muscle fibres to contract and push blood into the ventricles and then spreads rapidly over the ventricles causing the whole ventricle to contract into systole and deliver a big pump of blood down the arteries. The muscle fibres then rapidly relax and return the heart to diastole; the process is then repeated.

In order to develop a mathematical model that reflects the behaviour of the heart beat action described above, we choose to single out the following features. First of all, the model should exhibit an equilibrium state corresponding to diastole. Secondly, there must be a threshold for triggering the electrochemical wave emanating from the pacemaker causing the heart to contract into systole. Thirdly, the model must reflect the rapid return to the equilibrium state.

We begin by doing a little mathematical experimentation. Suppose we let x denote muscle fibre length referred to some convenient origin, say $x = 0$, which corresponds to the equilibrium state. Let b be an electrical control variable which governs the electrochemical wave. As far as the muscle fibres are concerned, we look for a differential equation which has $x = 0$ as an equilibrium state and at least for small times has a rapidly decreasing solution. An appropriate equation exhibiting these features is

$$\epsilon \frac{dx}{dt} = -x, \tag{4.2.1}$$

where ϵ is a small positive parameter. When $\frac{dx}{dt}$, the velocity of the fibre, is zero we have the equilibrium state $x = 0$. Furthermore we know (Chapter 1) that (4.2.1) has the general solution

$$x = A \exp(-t/\epsilon) \tag{4.2.2}$$

which is rapidly decreasing in time. Thus (4.2.1) seems to be a good candidate to represent the behaviour initially of the muscle fibres causing contraction into systole.

Turning now to the electrochemical wave, we need the control b to represent initially the relatively slow spread of this wave over the atria. A simple model which does this is

$$\frac{db}{dt} = -b. \tag{4.2.3}$$

Here $b = 0$ is an equilibrium state and (4.2.3) has the solution

$$b = B \exp(-t), \tag{4.2.4}$$

which, in comparison with (4.2.2), represents a relatively slow decay time.

The features that are not covered by this simple model obtained from (4.2.1) and (4.2.3) are (i) the threshold or trigger and (ii) the rapid return to equilibrium. At this stage, our knowledge of differential equations is insufficient to include these features and the discussion must be deferred until Chapter 6.

The model that incorporates the desired features is the coupled non-linear first-order system

$$\epsilon\frac{dx}{dt} = -(x^3 + ax + b),$$

$$\frac{db}{dt} = x - x_a. \tag{4.2.5}$$

Here x represents the length of the muscle fibre, $-a$ represents tension, b represents the chemical control and x_a represents a typical fibre length when the heart is in diastole. The model (4.2.5) is due to E.C. Zeeman.

That Zeeman chose to single out the above three qualities of the heart beat cycle and to attempt to model them through the system of equations (4.2.5) should not lead the reader to assume that such a description is the only one. Indeed, the model can only be considered a reasonable one if it reflects the basic features of the heart beat cycle well.

It is appropriate to remark, however, that the model (4.2.5) has been quite successful in distinguishing between some extreme forms of heart beat behaviour, for example, the effects of high blood pressure or an excess of adrenalin in the bloodstream due to rage or vigorous exercise. Likewise there is the situation when the heart beats in a feeble manner and does not contract into systole.

4.3   Blood flow

If we consider for a moment a simplified concept of the circulatory blood system in man, we can imagine that we have a pump delivering blood to a complicated network of pipes, which has innumerable connections. To develop an adequate mathematical model of this system and its behaviour is an almost impossible task. Thus, in order to make any progress, we attempt to model parts of the system separately. Here we concentrate on a small section of this circuit, say in the region of the aorta as shown in Figure 4.3.1. Indeed, we shall consider the relatively straight section between A and B. One can imagine that blood flow in this section behaves in much the same way as water in a cylindrical tube. This, however, is a gross oversimplification of the situation. To see this, let us consider some salient facts regarding blood flow. First of all, unlike water, blood does not have constant viscosity and this varies with velocity. Thus blood may be claimed to be non-Newtonian; indeed the properties of blood change rapidly if removed from the system and so it is extremely difficult to perform experiments on it under laboratory conditions.

If we now consider the type of flow in an artery, it is apparent that because the heart delivers blood in short bursts during contraction into systole, the flow is pulsatile and not uniform. Furthermore, we do not know the velocity profile of the flow entering A in Figure 4.3.1 and consequently the velocity

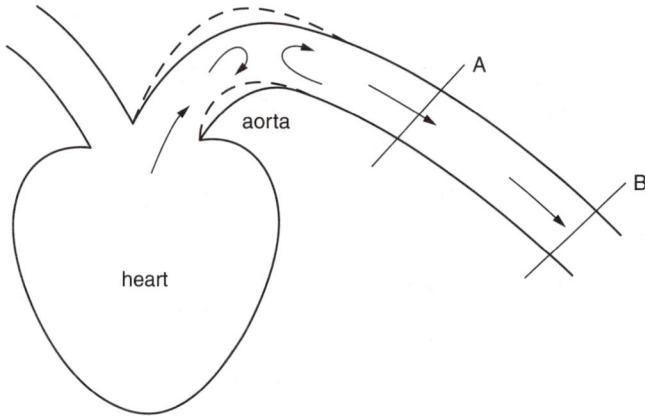

FIGURE 4.3.1: Schematic description of an aorta.

profile at B is also unknown. This observation is of fundamental importance in the mathematical description of blood flow. On the other hand, the hydrodynamic problem of considering the change of an initial velocity profile of a Newtonian fluid in a rigid pipe is fairly well understood and is based on the fundamental theory of Poiseuille (1846). One should remark here that Poiseuille, whose contributions to hydrodynamics are well known to engineers and mathematicians, was in fact a physician and his interest was precisely the problem we are considering here, namely, the study of blood flow.

Let us now focus on the arteries themselves. We know them to be elastic and a typical cross section may change significantly with time due to the pulsating nature of the flow of blood. Thus once again it may be unreasonable to treat the arteries as rigid tubes. Nevertheless we find it necessary to assume this as a first appoximation.

Referring to Figure 4.3.1 consider the flow of blood delivered into an aorta. The blood is pumped in an asymmetrical fashion and there are large cross-channel components of velocity in the arch region and consequently large cross-channel components in the pressure gradient. This is well known from thoracic surgery on animals. However, away from the arch itself, say in section $A - B$, the cross-channel components of velocity are considerably reduced and the flow is almost entirely longitudinal but, of course, still pulsatile. In the arch region it is found in thoracic surgery that the arch is very pliant and yields easily to the cross-channel pressure gradients. Thus it is reasonable to assume that the recurrent upward yielding of the arch region in response to pressure changes and the "general give" radially of all cross sections of the aorta cause changes in pressure to be dampened, especially the radial components. We shall assume then that, as the blood proceeds down the trunk of the aorta, radial velocity components may be neglected. This assumption is known to physiologists as the **Windkessel effect assumption**, an idea introduced by the German physiologist Otto Frank.

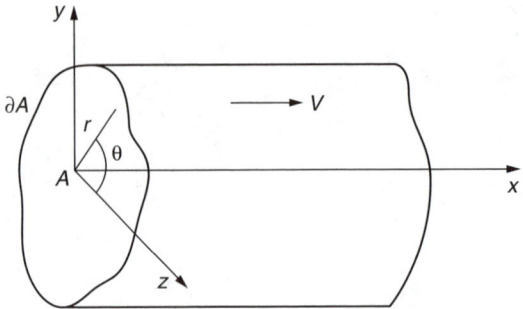

FIGURE 4.3.2:   Section of a rigid tube.

To begin our development of a mathematical model, consider pulsatile flow in a rigid tube of constant cross section A and boundary ∂A. A typical section is shown in Figure 4.3.2.

At time t assume that the velocity of flow along the tube is $V(x, y, z, t)$. In accordance with the Windkessel assumption above we assume that there are no velocity components across the tube and that the pressure P depends only on x and t. That is, P does not vary radially with position, but only longitudinally. To derive a mathematical model we proceed, as Poiseuille did, to balance inertial forces

$$F_I = \rho \frac{dV}{dt}$$

where ρ is the density of the blood, with the drag F_D due to viscous shear and the pressure force F_P, which can be written in terms of $P(x, t)$ as

$$F_P = -\frac{\partial P}{\partial x}.$$

Note that F_P is a surface force and body forces such as gravity are neglected. Equating these forces and dividing the resulting expression by ρ gives

$$\frac{F_D}{\rho} - \frac{dV}{dt} = -\frac{1}{\rho}\frac{\partial P}{\partial x}. \tag{4.3.1}$$

Now

$$\begin{aligned} \frac{dV}{dt} &= \frac{d}{dt}V(x(t), y(t), z(t)) \\ &= \frac{\partial V}{\partial t} + \frac{\partial V}{\partial x}\frac{dx}{dt} + \frac{\partial V}{\partial y}\frac{dy}{dt} + \frac{\partial V}{\partial z}\frac{dz}{dt}, \end{aligned}$$

using the chain rule for differentiation of a function of several variables. Since we assume that the velocity V is always in the x direction, the velocity components $\frac{dy}{dt}$ and $\frac{dz}{dt}$ are zero and $\frac{dx}{dt} = V$. Thus (4.3.1) now takes the form

$$\frac{1}{\rho}F_D - \frac{\partial V}{\partial t} - V\frac{\partial V}{\partial x} = -\frac{1}{\rho}\frac{\partial P}{\partial x}. \qquad (4.3.2)$$

We now assume that the drag F_D is proportional to the sum of the second partial derivatives of the velocity, that is

$$F_D = \nu\left(\frac{\partial^2 V}{\partial x^2} + \frac{\partial^2 V}{\partial y^2} + \frac{\partial^2 V}{\partial z^2}\right), \qquad (4.3.3)$$

where ν is the coefficient of viscosity. Using (4.3.3) in (4.3.2) gives the final form

$$\frac{\nu}{\rho}\left(\frac{\partial^2 V}{\partial x^2} + \frac{\partial^2 V}{\partial y^2} + \frac{\partial^2 V}{\partial z^2}\right) - \frac{\partial V}{\partial t} - V\frac{\partial V}{\partial x} = -\frac{1}{\rho}\frac{\partial P}{\partial x}. \qquad (4.3.4)$$

For a precise development of this model from the fluid mechanical point of view we refer the reader to Chorin and Marsden (1979).

If the pressure gradient $-\frac{\partial P}{\partial x}$ is known and ν and ρ are also known then (4.3.4) defines a partial differential equation satisfied by the velocity $V(x, y, z, t)$. In fact (4.3.4) is a second order partial differential equation because it involves second order partial derivatives. It is also *non-linear* since it contains the non-linear term

$$V\frac{\partial V}{\partial x} = \frac{1}{2}\frac{\partial V^2}{\partial x}.$$

The term in parentheses on the left-hand side of (4.3.4) is called the "Laplacian" of V after the French mathematician Laplace, and we often use the symbol ∇^2 (written Δ by some authors) to denote the Laplacian operator, i.e.,

$$\nabla^2 = \frac{\partial^2 V}{\partial x^2} + \frac{\partial^2 V}{\partial y^2} + \frac{\partial^2 V}{\partial z^2}.$$

Under the assumption that the walls of the tube are rigid and the pressure is the only driving force directed along the tube, the velocity does not change with position x along the tube, only with position across the tube, i.e., V depends only on y, z and t. In this case (4.3.4) simplifies and reduces to the *linear equation*

$$\frac{\nu}{\rho}\left(\frac{\partial^2 V}{\partial x^2} + \frac{\partial^2 V}{\partial y^2}\right) - \frac{\partial V}{\partial t} = -\frac{1}{\rho}\frac{\partial P}{\partial x}, \qquad (4.3.5)$$

where $\frac{\partial P}{\partial x}$ depends only on t.

The boundary condition to be applied here is that

$$V(y, z, t) = 0 \qquad (4.3.6)$$

on ∂A. Notice that we cannot, as discussed above, provide an "initial" velocity profile, i.e.,

$$V(y, z, 0) \qquad (4.3.7)$$

is not given. If (4.3.7) was known then it can be proved that (4.3.5) together with the boundary condition (4.3.6) leads to a unique solution $V(y, z, t)$. That is, the problem (4.3.5)–(4.3.7) is said to be well posed. In the case of blood flow we seek a different type of uniqueness which says essentially that if $u(y, z, t)$ and $w(y, z, t)$ are two solutions of (4.3.5) with each satisfying (4.3.6), then, as t becomes large, the two flows become indistinguishable from one another. This condition may be conveniently called the **Windkessel condition**.

To give some idea of the types of solution to be expected from (4.3.5) and (4.3.6) let us suppose the coefficient of viscosity ν and the density ρ are constant. Suppose also that the rigid tube is a circular cylinder of radius a. Introducing polar co-ordinates we have, from Figure 4.3.2,

$$z = r\cos\theta, \quad y = r\sin\theta. \qquad (4.3.8)$$

The lateral surface of the cylinder is then described by $r = a$. Using the transformation (4.3.8) and writing $\frac{\partial P}{\partial x} = f(t)$ we can write (4.3.5) as

$$\frac{1}{r^2}\frac{\partial^2 V}{\partial\theta^2} + \frac{\partial^2 V}{\partial r^2} + \frac{1}{r}\frac{\partial V}{\partial r} - \frac{\rho}{\nu}\frac{\partial V}{\partial t} = -\frac{1}{\nu}f(t) \qquad (4.3.9)$$

and furthermore

$$V(r, \theta, t) = 0, \quad r = a. \qquad (4.3.10)$$

If we assume the flow is axially symmetric then the velocity V is independent of θ and (4.3.9) reduces to

$$\frac{\partial^2 V}{\partial r^2} + \frac{1}{r}\frac{\partial V}{\partial r} - \frac{\rho}{\nu}\frac{\partial V}{\partial t} = -\frac{1}{\nu}f(t).$$

That is

$$\frac{1}{r}\frac{\partial}{\partial r}\left(r\frac{\partial V}{\partial r}\right) = \frac{\rho}{\nu}\frac{\partial V}{\partial t} - \frac{f(t)}{\nu}. \qquad (4.3.11)$$

Integrating (4.3.11) with respect to r gives

$$\left[r\frac{\partial V}{\partial r}\right]_{r=0}^{r} = \frac{\rho}{\nu}\int_0^r \tau\frac{\partial V(\tau, t)}{\partial t}d\tau - \frac{r^2}{2\nu}f(t)$$

i.e., under the assumption that $\frac{\partial V}{\partial r}$ exists and is continuous at $r = 0$ we have

$$r\frac{\partial V}{\partial r} = \frac{\rho}{\nu}\int_0^r \tau\frac{\partial V(\tau, t)}{\partial t}d\tau - \frac{r^2}{2\nu}f(t)$$

or

$$\frac{\partial V}{\partial r} = \frac{\rho}{\nu r} \int_0^r \tau \frac{\partial V(\tau, t)}{\partial t} d\tau - \frac{r}{2\nu} f(t).$$

Integrating once more we arrive at

$$V(a, t) - V(r, t) = \frac{\rho}{\nu} \int_r^a \frac{1}{\xi} \int_0^\xi \tau \frac{\partial V(\tau, t)}{\partial t} d\tau \, d\xi - \frac{(a^2 - r^2)}{4\nu} f(t),$$

which on using the boundary condition (4.3.10) gives

$$V(r, t) = \frac{(a^2 - r^2)}{4\nu} f(t) - \frac{\rho}{\nu} \int_r^a \frac{1}{\xi} \int_0^\xi \tau \frac{\partial V(\tau, t)}{\partial t} d\tau \, d\xi. \tag{4.3.12}$$

In order to make further progress towards a solution of (4.3.12) let us assume as a first approximation that

$$V(r, t) = V^0(r, t) = \frac{(a^2 - r^2)}{4\nu} f(t). \tag{4.3.13}$$

This approximation is precisely the parabolic velocity profile well known in Poiseuille flow. As a better approximation, we substitute V^0 under the integral sign in (4.3.12) to arrive at the next approximation

$$V^{(1)}(r, t) = \frac{(a^2 - r^2)}{4\nu} f(t) - \frac{\rho}{\nu} \int_r^a \frac{1}{\xi} \int_0^\xi \tau \frac{\partial V^0(\tau, t)}{\partial t} d\tau \, d\xi.$$

Repeating this idea we arrive at the iterative method:

$$V^{(n+1)}(r, t) = \frac{(a^2 - r^2)}{4\nu} f(t) - \frac{\rho}{\nu} \int_r^a \frac{1}{\xi} \int_0^\xi \tau \frac{\partial V^{(n)}(\tau, t)}{\partial t} d\tau \, d\xi. \tag{4.3.14}$$

If this scheme converges as $n \to \infty$, then we obtain a solution to the given problem. Of course this is not the only possible solution; indeed the method defined by (4.3.14) will in general only converge if $\frac{\rho}{\nu}$ is sufficiently small. If this is not the case then we must seek solutions by other means. Thus, for example, it may be possible to use the method of separation of variables as discussed in Chapter 11.

Let us now return to the problem of pulsating blood flow in a section of the trunk of the aorta. The model governing the velocity V is again (4.3.4) but now we cannot neglect the non-linear term $V \frac{\partial V}{\partial x}$. In addition, the situation is further complicated by the fact that both ρ and ν depend both on position and time t; also due to the elasticity of the aorta and the pulsating nature of the flow, the boundary ∂A of the tube is time dependent and may also vary with position. Thus in summary, we are asked to solve a non-linear partial differential equation subject to a moving boundary constraint, a problem well beyond the scope of this book. Finally we should not forget the Windkessel uniqueness requirement.

4.4   Nerve impulse transmission

The axon portion of a nerve cell (see Figure 4.4.1) is made up of a conducting material called axoplasm which is contained in a roughly cylindrical membrane between 50 and 70 Å thick. The membrane is permeable to potassium ions, K^+, concentrated in the interior, and to sodium ions, Na^+, concentrated in the exterior. Also present, but to a much lesser extent, are other ions such as chlorine Cl^-. In nature, the high concentration of sodium ions is maintained by the organism in the fluid medium exterior to the nerve cell. In its resting state there is a potential difference across the axon membrane of between -50 and -70 millivolts (mV).

Suppose we place a segment of axon in a bath containing a sodium concentration similar to the one usually present in the exterior fluid medium and apply a potential difference across the membrane. In the laboratory, this is usually done by inserting a fine micropipette into the axon and injecting sodium ions. The induced sodium gives rise to an applied current. It is observed that if a small potential difference, which is positive relative to the resting potential, is applied across the membrane, the sodium and potassium ionic currents are briefly disturbed but quickly return to their zero resting state and the membrane settles back to the resting potential. If a much larger positive membrane potential is applied (between 7 and 10 mV) the equilibrium

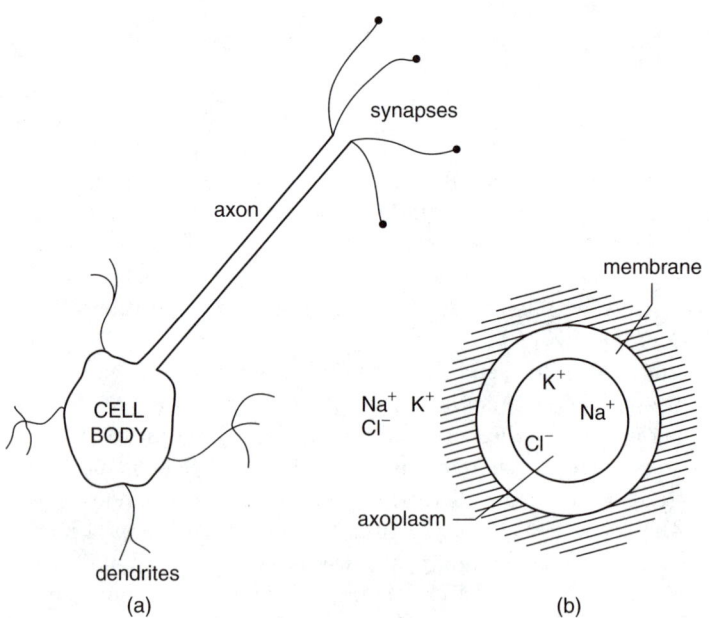

FIGURE 4.4.1:   Schematic description of the nerve axon.

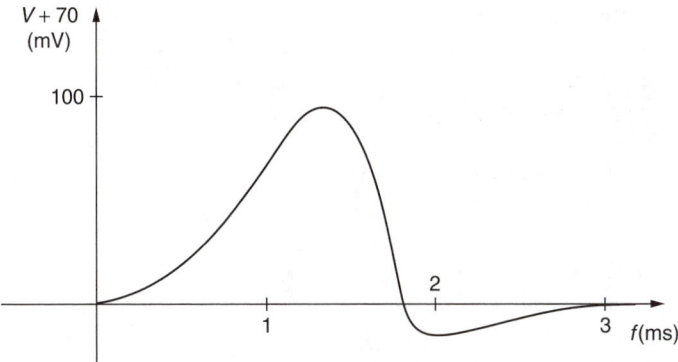

FIGURE 4.4.2: Membrane potential.

state is exceeded and the sodium currents become active. What happens now is that the axon membrane becomes permeable to positive sodium ions which flow inwards, making the membrane potential increase and this causes the membrane to become even more permeable to sodium ions. If circumstances are just right in that the inflow of sodium ions keeps the membrane potential increasing, there is a critical level in membrane permeability which we call the **threshold**, which results in a rapid impulsive rise in potential difference across the membrane to about 100 mV relative to the negative resting potential.

Following this "firing" of the axon two things happen. First the sodium ion permeability begins to decrease slowly and secondly the potassium ion permeability rapidly rises. Potassium ions thus flow outwards and eventually restore the membrane potential once again to its resting state after an overshoot of about 5 mV. The impulse lasts about 3 to 5 ms (see Figure 4.4.2).

The question to ask now is how an impulse is transmitted along the axon during this process. Near the point of stimulation, an impulse is created, which is shot off down the axon along the membrane, being renewed at each point, to approximately 100 mV as the membrane potential at each point achieves a value that initiates the active phase of sodium. As we mentioned above, a threshold is involved, that is, either this voltage is not achieved and no impulse is propagated or it is achieved and at least one impulse is propagated.

Except at a discrete set of points called **nodes of Ranvier**, which are about 1 mm apart, every vertebrate nerve axon is covered with a sheath that electrically insulates the axoplasm and containing membrane from the exterior medium. A vertebrate nerve fibre is said to be **myelinated** and, although current in a vertebrate nerve fibre can easily pass freely along the axoplasm or the exterior fluid as happens in an **invertebrate unmyelinated nerve**, it can pass through the membrane only at certain points. Currents circulate on paths around the boundary of a section of arbitrary location for the membrane of an unmyelinated axon, but only on paths that pass through the nodes of Ranvier for a myelinated axon.

Our understanding of the mechanism governing the action potential is due chiefly to the inspired and carefully executed experiments of the physiologists A.L. Hodgkin and A.F. Huxley in 1952. This work, which led to the award of a Nobel prize, was largely performed on the large axon to be found in the squid *Loligo*, and culminated in the development of a mathematical model. This model not only agrees in a qualitative way with the experimental results but gives remarkably accurate quantitative results. Since the development of this model, others have been subsequently formulated which reflect current experimental findings with good agreement. However the so-called **Hodgkin-Huxley** model is still regarded as the fundamental model governing nerve impulse transmissions. The experiments performed by Hodgkin and Huxley were set up in the same way as described above, except that the induced current was achieved not by injection of sodium ions but by inserting a fine current-carrying wire into the axon. This induced current, which we denote by I, is found to give rise to a membrane potential E which is the same at each point of the segment of axon and depends only on the time t. That is, E is independent of x, the position of a point along the axon relative to some convenient origin. This configuration is known as **space clamp**.

The basic assumption of the Hodgkin-Huxley theory is that there are separate channels for the sodium, potassium and other ions like chlorine. We can envisage these channels in terms of the electrical circuit shown in Figure 4.4.3. Thus, each channel is described in terms of a voltaic cell E in series with a conductance g together with a capacitance C_m across the whole ensemble.

The transmembrane current is then given by

$$I = C_m \frac{dE}{dt} + g_{Na}(E - E_{Na}) + g_K(E - E_K) + g_l(E - E_l), \qquad (4.4.1)$$

where I is the current density, E is the membrane potential, C_m is the membrane capacity, g_{Na} is the sodium conductance, g_K is the potassium conductance, g_l is the leakage conductance, E_{Na} is the sodium equilibrium potential,

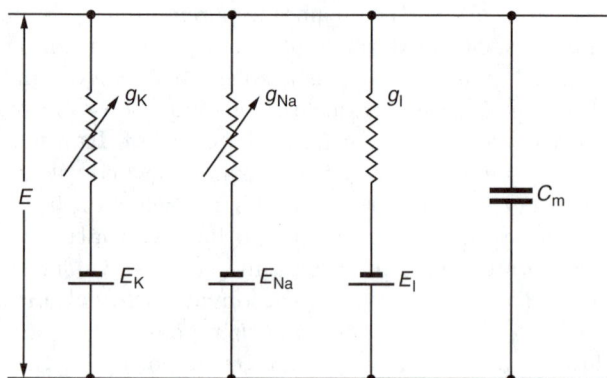

FIGURE 4.4.3: Conductance model of the nerve membrane.

E_K is the potassium equilibrium potential and E_l is the leakage equilibrium potential. The conductances g_{Na} and g_K are assumed to vary with time and the potential E, while g_l is assumed constant. To describe the variation in g_{Na} and g_K Hodgkin and Huxley assumed that g_K is described by

$$g_K = \bar{g_K} n^4, \qquad (4.4.2)$$

and that g_{Na} is determined by

$$g_{Na} = \bar{g_{Na}} m^3 h. \qquad (4.4.3)$$

In these expressions, $\bar{g_K}$ is the maximum potassium conductance and $\bar{g_{Na}}$ is the maximum sodium conductance. The quantities n, m and h are dimensionless quantities that vary between 0 and 1 and are functions of E and t. Their precise forms are determined as solutions of the system of ordinary differential equations:

$$\frac{dm}{dt} = \alpha_m(E)(1 - m) + \beta_m(E)m,$$

$$\frac{dh}{dt} = \alpha_h(E)(1 - h) + \beta_h(E)h,$$

$$\frac{dn}{dt} = \alpha_n(E)(1 - h) + \beta_n(E)n. \qquad (4.4.4)$$

In these equations the coefficient functions $\alpha_j, \beta_j, j = m, h, n$, are functions of the membrane potential E, and were found by careful empirical fitting with the experimental results. The exact forms of these coefficient functions are complicated formulae involving $\exp E$. The mathematical model developed from the Hodgkin-Huxley theory is the formidable system of linked differential equations given by (4.4.1)–(4.4.4). That Hodgkin and Huxley were able to solve these equations at all is quite remarkable, especially if one appreciates the limited computational facilities available in (1952). It should be remarked, however, that although this system admits to numerical solution, the underlying analytical structure is by no means fully understood.

If the space clamp is removed in the sense that the membrane potential is allowed to vary with position x along the axon, then Hodgkin and Huxley assumed Kelvin's cable theory to assert that the current I is given by

$$I = \frac{a}{2R} \frac{\partial^2 E}{\partial x^2} \qquad (4.4.5)$$

where a is the radius of the axon and R is the specific resistivity of the axoplasm. Since E now depends on both x and t all derivatives must be replaced by partial derivatives, and so by incorporating (4.4.5) in (4.4.1) we are led to consider the partial differential equation

$$\frac{a}{2R} \frac{\partial^2 E}{\partial x^2} = C_m \frac{\partial E}{\partial t} + g_{Na}(E - E_{Na}) + g_K(E - E_K) + g_l(E - E_l), \quad (4.4.6)$$

together with the system (4.4.2)–(4.4.4).

As we have mentioned before, several alternative mathematical models of nerve impulse transmission have been developed since 1952. Some of these are of the same complexity as the Hodgkin-Huxley model while others, by making certain additional assumptions about the behaviour of the various ionic conductances, are considerably simpler. These simplified models nevertheless retain the main features characteristic of the Hodgkin-Huxley theory. One such model that has attracted much interest is the **FitzHugh-Nagumo** model, originally proposed by R. FitzHugh in 1961 and subsequently developed by J. Nagumo and his co-workers in 1962. This model is developed in analogy with the **Van der Pol oscillator**, well known to electrical engineers and physicists, and takes the form

$$\frac{\partial^2 u}{\partial x^2} = \frac{\partial u}{\partial t} - u(1-u)(u-a) + w,$$
$$\frac{\partial w}{\partial t} = bu - \gamma w, \tag{4.4.7}$$

where a, b and γ are positive constants and $0 < a < 1$. In this simplified model, u represents the membrane potential E and, as before, x measures distance along the axon and t is time. The cubic term $u(1-u)(u-a)$ in the first equation in (4.4.7) is analogous to an instantaneous turning on of sodium permeability and can be thought of as playing the role of the variable m in the Hodgkin-Huxley equations. w is a recovery variable and is analogous to the turning on of potassium permeablity and so behaves like the variable n in the Hodgkin-Huxley model. There is no counterpart to inactivation of sodium permeability.

There is a further simplification which can be adopted and which comes about from the following observation. The simplified model cannot be expected to give quantitative comparisons with experiment, as the Hodgkin-Huxley model does. Thus we can only expect a qualitative comparison and, consequently, if we could simplify the non-linearity in (4.4.7) without destroying the desired behaviour of solutions, then the methods of solution could be greatly simplified. To this end H.P. McKean in 1970 proposed that the term $u(1-u)(u-a)$ in (4.4.7) could be replaced, for example, by the piecewise linear term illustrated in Figure 4.4.4, where the angle θ can take any value in the semi-open interval $0 < \theta \leq \pi/2$. The benefit of incorporating this simplification is that, along each line segment, the set of equations in (4.4.7) are linear and in general linear equations are much easier to solve than non-linear ones.

The mathematical models of nerve impulse transmission are incomplete without some notion of the solutions to expect or the appropriate initial and boundary conditions. If we consider the Hodgkin-Huxley model as applied to the giant axon of the squid, we know that the length of the axon is large compared with its radius a. Thus if we put

$$X = x/\sqrt{a} \tag{4.4.8}$$

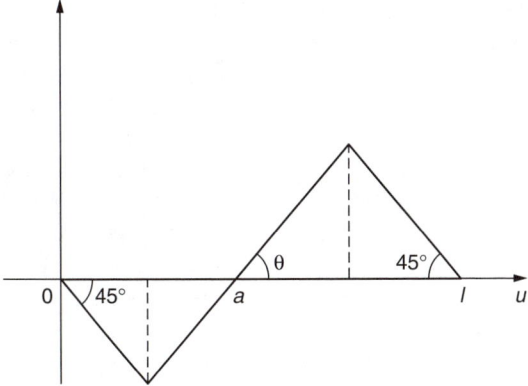

FIGURE 4.4.4: The piecewise linear term.

in (4.4.5) and (4.4.6) we can approximate the range of X to be infinitely large. In this way we take the range of x in the system (4.4.7) also to be $0 \le x < \infty$.

Both of the models we have discussed have solutions that depend only on $x + ct$ and are called **travelling waves**. That is, there are membrane potentials $E(x, t) = E(x+ct)$ that move along the axon with no loss of strength and in a direction determined by the sign of c. The rate of propagation is $|c|$ and the problem of finding this value is very important.

Suppose the axon is stimulated from one end defined, for example, by $x = 0$. Then the appropriate conditions are:

$$E(0, t) = P(t), \quad t > 0,$$
$$E(x, 0) = 0, \quad x \ge 0. \tag{4.4.9}$$

Here $P(t)$ is a function of t which defines the stimulus emanating at $x = 0$. An important and largely unresolved problem is to classify those stimuli $P(t)$ so that $E(x, t)$ approaches a travelling wave $E(x + ct)$ as $t \to \infty$. The study of travelling waves will be taken up in Chapter 7, while some idea of the behaviour of stimulated membrane potentials will be considered in Chapter 12.

4.5 Chemical reactions

The chemical reaction described here is not part of any living system and is certainly not biological. However, its interest and importance lies in its ability to "oscillate," forming patterns such as parallel bands, concentric rings and cell-like structures. Such phenomena are well known in biology: for example, in morphogenesis (discussed later in Chapter 12), the regulating processes involved in living cells and organisms, and in information transmission. A further reason for studying pattern formation in non-living systems is that the constituent forces are more limited than those of even the simplest biological systems where electrical forces, surface tensions, colloid properties and crystallising forces, and complex chemical reactions may play a significant role.

The reaction we shall consider is the usually dramatic oscillatory reaction discovered by Belousov in 1958 and Zhabotinskii in 1964. The **Belousov-Zhabotinskii** reaction, as it is now called, is essentially the oxidation of malonic acid by bromate in a sulphuric acid medium, in the presence of a cerium catalyst. In the reaction, two overall processes I and II can be identified, and the chemistry involved can be described as follows. In the reaction when the bromide ion (Br^-) is above some critical concentration, process I occurs. Here the bromate ion (BrO_3^-) is reduced to bromine (Br_2), with bromous acid $(HBrO_2)$ as an intermediary, and the malonic acid, $CH_2(COOH)_2$, is brominated. During this process there is little oxidation of the cerium ion $Ce(III)$. Process I thus uses up the bromide. When the concentration of bromide becomes sufficiently low, process II takes over. In this the bromous acid and the bromate ion produce a radical bromate species (BrO_2), which oxidises the cerium ion $Ce(III)$ to the $Ce(IV)4$ form with bromous acid generated autocatalytically. When all of the $Ce(III)$ has been oxidised to $Ce(IV)$ and the bromide ion concentration is low, the $Ce(IV)$ then reacts with the bromomalonic acid to produce the cerium ion $Ce(III)$ and bromide again. When the bromide passes a critical concentration, process I takes over again and the cycle is repeated. During the process, one sees the reagent oscillate in colour, turning alternatively bright blue and reddish purple if ferroin is used as an indicator.

On the bases of the mechanism described above, Field and Noyes have suggested the following mathematical model which involves three intermediaries which oscillate and which are associated with the bromide ion (Br^-), bromous acid $(HBrO_2)$ and the cerium ion $Ce(IV)$.

As is standard in Chemistry we write

$$A + B \underset{k'}{\overset{k}{\rightleftharpoons}} C$$

to mean that a molecule each of A and B combine reversibly to form the molecule C. The quantities k and k' are rate constants.

If $X = HBrO_2, Y = Br^-$ and $Z = Ce(IV)$, then the Field-Noyes system is

$$A + Y \overset{k_1}{\rightarrow} X, \tag{4.5.1}$$

$$X + Y \overset{k_2}{\rightarrow} P, \tag{4.5.2}$$

$$A + X \overset{k_3}{\rightarrow} 2X + Z, \tag{4.5.3}$$

$$2X \overset{k_4}{\rightarrow} Q, \tag{4.5.4}$$

$$Z \overset{k_5}{\rightarrow} fY. \tag{4.5.5}$$

In this system, A is a maintained constant reactant (namely BrO_3^-), P and Q are products and f is a stoichiometric factor. To arrive at the mathematical model, we invoke the law of mass action, which states that the rate of a reaction is proportional to the active concentrations of the reactants. Thus for the reaction

$$A + B \xrightarrow{k} C$$

the law of mass action (using A to mean concentration of reactant A, etc.) gives

$$\frac{dC}{dt} = kAB = -\frac{dA}{dt} = -\frac{dB}{dt}.$$

From the reaction (4.5.1) and the law of mass action, we see that X is produced at the rate $k_1 AY$. However at the same time X is being used up in reaction (4.5.2) to produce P at the rate $-k_2 XY$, and by reaction (4.5.4) at the rate $-2k_4 X^2$ to produce Q. Finally, reaction (4.5.3) gives rise to a further concentration of X at the rate $k_3 AX$. Thus the total rate of change of the concentration of X is

$$\frac{dX}{dt} = k_1 AY - k_2 XY + k_3 AX - 2k_4 X^2. \tag{4.5.6}$$

If we analyse the rates of change of Y and Z in the same way, the law of mass action gives

$$\frac{dY}{dt} = -k_1 AY - k_2 XY + fk_5 Z, \tag{4.5.7}$$

and

$$\frac{dZ}{dt} = -k_3 AX - k_5 Z. \tag{4.5.8}$$

The system of coupled first-order differential equations (4.5.6)–(4.5.8) is the mathematical model of Field, Körös and Noyes representing the Belousov-Zhabotinskii (B-Z) reaction.

Before we proceed further with the mathematical models of the B-Z reaction, let us describe one or two of the experiments performed on this reaction and the patterns that arise.

Under certain conditions, the reagent is capable of organising itself into spatially inhomogeneous structures that are seen as coloured patterns. The physical reasons for the pattern formations as well as the patterns themselves are different according to the experimental set-up. If the reagent is placed in a thin layer and allowed to convect, say, by heating the fluid from below or cooking it from above by evaporation, then a reaction-diffusion process resembling the Bénard phenomenon of cellular convection occurs. Here the fluid, when viewed from above, organises itself into hexagonal or rectangular cells, which are outlined in reddish purple. It seems that the boundaries of the convection cells appear to contain most of their cerium in the reduced state, while the rest of each cell contains more cerium in an oxidised state. However, the precise behaviour of this hydrodynamic and chemical phenomenon is not fully understood.

An interesting pattern formation is observed when the reagent is placed in a vertical container and a gradient of temperature or of one of the concentrations is imposed on the fluid. In the latter case, sulphuric acid is carefully added after the other ingredients have been mixed. Under these circumstances, horizontal bands form and propagate vertically through the container. In this case there is no fluid motion at all; it is only the lines of constant phase of the oscillation that are moving through the fluid. The explanation of this phenomenon is as follows: the concentration gradient or temperature gradient produces a vertical gradient in the frequency of the oscillation. This frequency gradient can account in detail for the space-time behaviour of the patterns that emerge. Diffusion plays a negligible role unless the pattern has a very small spatial scale.

Perhaps the most striking pattern formations that can be observed are those found in the experiments of Zaikin and Zhabotinskii using ferroin for the catalyst and malonic acid in the Belousov reaction. Here the reagent is spread thinly, about 2 mm thick on a petri dish. Circular chemical waves are observed propagating outwards. These waves, essentially oxidation bands, are blue and they propagate through the reddish background fluid. When two waves collide both disappear. When a faster one catches up with a slower one the latter is entrained. These waves have been designated "trigger" waves since diffusion combines with the chemical reaction to trigger the waves. To include diffusion effects in our model (4.5.6)–(4.5.8) we suppose the intermediaries X, Y and Z can diffuse with diffusion coefficients D_X, D_Y, D_Z and hence are functions of the space variables x, y, z and time t. The system (4.5.6)–(4.5.8) is modified to read

$$\frac{\partial X}{\partial t} = k_1 AY - k_2 XY + k_3 AX - 2k_4 X^2 + D_X \Delta X, \qquad (4.5.9)$$

$$\frac{\partial Y}{\partial t} = -k_1 AY - k_2 XY + f k_5 Z + D_Y \Delta Y, \qquad (4.5.10)$$

$$\frac{\partial Z}{\partial t} = -k_3 AX - k_5 Z + D_Z \Delta Z. \qquad (4.5.11)$$

This model is a system of coupled partial differential equations and, under certain circumstances, admits travelling wave solutions. For instance, if we take $Z = 0$ and assume $X = X(x + ct), Y = Y(x + ct)$, then the reduced system (4.5.9), (4.5.10) can be studied in the same way as the travelling wave solutions of nerve axon equations are considered. Of course boundary and initial conditions are somewhat different. These considerations will be taken up in Chapters 8 and 12.

4.6 Predator-prey models

The problems we shall consider here are of fundamental importance in ecology, i.e., the study of the interactions between living organisms and their environment. Let us consider two organisms or species characterised, for example, by their respective population densities, say X and Y. Thus X may be the population density of a carnivore occupying a certain habitat and Y may be the population density of a herbivore occupying the same habitat as X and considered as a food source for X. Alternatively, X and Y may be used to represent parasite and host, or herbivore and plant.

It is convenient to classify the direct interaction between a pair of species into the following categories:

(a) **Competition**: each species has an inhibiting effect on the growth of the other.

(b) **Commensalism**: each species has an accelerating effect on the growth of the other.

(c) **Predation**: one species, the "predator," has an inhibiting effect on the growth of the other; the "prey" has an accelerating effect on the growth of the predator.

Throughout this section, we consider the interaction between X and Y to be that of predation. Furthermore, we shall make the following simplifying assumptions.

(a) The density of a species—that is, the number of individuals per unit area—can be represented as a function of a single variable. Thus we ignore possible age differences and differences of sex or genotype.

(b) Changes in density are deterministic; that is, we assume there are no random effects in the environment influencing the interaction between X and Y. Obviously this is a severe limiting assumption in many realisable situations.

(c) The effects of interactions within and between species are instantaneous. Thus, in the predator-prey interaction, this means that the delay between the moment a predator eats a prey, and the moment when the ingested material is converted into part of a new predator is ignored.

We begin by considering, in a rather informal way, some simple models of predator-prey interaction. Let X be the predator density and Y the prey density. In the absence of predators we expect no inhibition in the growth of the prey. A simple growth relation Y could be

$$\frac{dY}{dt} = kY, \tag{4.6.1}$$

where t is time and k is a positive rate constant. We know that the ordinary differential equation (4.6.1) has the solution

$$Y = Y_0 \exp kt, \qquad (4.6.2)$$

where Y_0 is the initial population density. Thus if (4.6.1) is used to describe the population growth of Y, then Y will increase exponentially in time. Such a growth behaviour is reasonable for a limited time, but ultimately an increasing population will exhaust its resources. Consequently we expect in practice that Y will either settle down to some steady state value, fluctuate between various levels or decline. If the first possibility arises then we could replace (4.6.1) by the logistic equation

$$\frac{dY}{dt} = aY - bY^2, \qquad (4.6.3)$$

which has the general solution

$$Y = \frac{aY_0}{bY_0 + (a - bY_0)exp(-at)}. \qquad (4.6.4)$$

The justification for this type of growth is that

(a) when Y is small, (4.6.3) formally reduces to (4.6.1) and the growth is exponential;

(b) as t increases Y approaches the value a/b steadily and without oscillation.

In (4.6.3) it is standard to call a the intrinsic rate of increase and $k = a/b$ the carrying capacity.

Before we introduce into either (4.6.1) or (4.6.3) the effects of predation by X, let us consider the behaviour of the predator species X in the absence of prey. Without prey, the predators X are expected to decrease and so their decline could be represented in terms of the exponentially decaying solution $X_0 \exp(-et)$ of the differential equation

$$\frac{dX}{dt} = -eX, \qquad (4.6.5)$$

where e is a positive constant. Again, like (4.6.2), the exponentially declining population of predators X is only reasonable for a limited period of time, since one would expect X to become extinct in a finite rather than an infinite amount of time when starved of prey Y.

If we assume, as Volterra and Lotka did, that in the absence of predation, Y follows a logistic growth curve and that the rate at which prey are eaten is proportional to the product of the densities of predator and prey, then we are to the model

$$\frac{dX}{dt} = -eX + fXY, \qquad (4.6.6)$$

$$\frac{dY}{dt} = aY - bY^2 - cXY. \qquad (4.6.7)$$

The assumptions of Volterra and Lotka are valid under the following conditions:

(a) one or both species move at random;

(b) when they meet, there is a constant probability that the predator will kill the prey;

(c) the time taken by the predator in consuming the prey is negligible.

There are a number of variants of the Volterra-Lotka model that have been proposed and that attempt to take into account other effects that may influence the predator-prey interaction. For example, when a constant number $\bar{Y}$ of the prey can find some cover or refuge, which makes them inaccessible to the predator, then the system (4.6.6), (4.6.7) is modified to the form

$$\frac{dX}{dt} = -eX + fX(Y - \bar{Y}), \qquad (4.6.8)$$

$$\frac{dY}{dt} = aY - bY^2 - cX(Y - \bar{Y}). \qquad (4.6.9)$$

In a much more general way, we can follow the model of Rosenzweig and MacArthur (1963) and assume a general growth rate $f(Y)$ for the prey in the absence of predators and assume that prey are eaten at a rate proportional to some function $\phi(X, Y)$ by the predators. In this case we obtain the general model:

$$\frac{dX}{dt} = -eX + k\phi(X, Y), \qquad (4.6.10)$$

$$\frac{dY}{dt} = f(Y) - \phi(X, Y). \qquad (4.6.11)$$

This model will be examined in some depth in Chapter 9, under certain simplifying but nevertheless realistic assumptions regarding the functions $f(Y)$ and $\phi(X, Y)$.

To conclude this section, we introduce a somewhat different but also interesting ecological interacting two-species model. Suppose we have two species X and Y competing to exist in the same habitat. We assume that, in the absence of Y, X grows according to a logistic law and, similarly in the absence of X, that Y does the same. However, when X and Y are both present, it is natural to assume that each has an inhibiting effect on the growth of the other. Suppose the inhibition is the same as that in the Volterra-Lotka model. Then we have the system

$$\frac{dX}{dt} = X(e - fY - gX), \qquad\qquad (4.6.12)$$

$$\frac{dY}{dt} = Y(a - bY - cX). \qquad\qquad (4.6.13)$$

4.7 Notes

Heart beat

Much of the material here is based on the work of E.C. Zeeman and can be found in his article "Differential equations for the heart beat and nerve impulse," which appeared in *Towards a Theoretical Biology*, vol. 4, C.H. Waddington, Ed., Edinburgh University Press, Edinburgh, 1972.

Blood flow

For a general background to the study of blood, particularly blood flow in arteries, we recommend the book by D.A. McDonald, *Blood Flow in Arteries*, 2nd ed., Edward Arnold, 1974 and that by T.J. Pedley, *The Fluid Mechanics of Large Blood Vessels*, Cambridge University Press, London, 1980. The mathematical model developed here follows along the ideas suggested in the book by H. Melvin Lieberstein, *Mathematical Physiology*, Elsevier, Amsterdam, 1973.

For a precise development of the model (4.3.4) from the fluid mechanical point of view, see A.J. Chorin and J.E. Marsden, *A Mathematical Introduction to Fluid Mechanics*, Springer-Verlag, Heidelberg, 1979.

Nerve impulse transmission

It is highly recommended that the reader consult the fundamental papers of A.L. Hodgkin and A.F. Huxley, A quantitative description of membrane current and its application to conduction and excitation in nerve, *J. Physiol.*, **117**, 500–544. There are several sources of good background material relating to nerve modelling. See, for example, B. Katz, *Nerve, Muscle and Synapse,* McGraw-Hill, New York, 1966; D. Junge, *Nerve and Muscle Excitation*, Sinauer Associates, Sunderlund, MA, 1976; H.C. Tuckwell, *Introduction to Theoretical Neurobiology*, vols. 1, 2, Cambridge University Press, London, 1988; and J. Cronin, *Mathematical Aspects of Hodgkin-Huxley Neural Theory*, Cambridge University Press, London, 1987.

Chemical reactions

A comprehensive treatment of the Belousov-Zhabotinskii reaction is to be found in J.D. Murray, *Mathematical Biology*, Springer-Verlag, Heidelberg, 1993.

Predator-prey models

The book by J. Maynard Smith, *Models in Ecology*, Cambridge University Press, London, 1974 is a good source of information relating to predator-prey models as well as some general models of species interaction.

Further Reading

Belousov, B. P., An oscillating reaction and its mechanism, *Sborn. Referat. Radiat. Med. Medgiz.*, Moscow, 145, 1959.

Field, R. J., Körös, E., and Noyes, R. M., Oscillations in chemical systems. II. Thorough analysis of temporal oscillation in bromate-cerium malonic acid systems, *J. Am. Chem. Soc.*, **94**, 8649–8664, 1972.

Field, R. J. and Noyes, R. M., Oscillations in chemical systems. IV. Limit cycle behaviour in a model of a real chemical reaction, *J. Chem. Phys.*, **60**, 1877–1884, 1974.

FitzHugh, R., Impulses and physiological states in theoretical models of nerve membrane, *Biophys. J.*, **1**, 445–466, 1961.

Kolmogorov, A., Petrovskii, I., and Piskounov, N., Etude de l'equation de la diffusion avec croissance de la quantité de matiére et son application à une problème biologique, *Bull. Univ. Moscow. Ser. Int.*, A, I, **6**, 1–25, 1937.

Lotka, A. J., *Elements of Physical Biology*, Williams & Wilkins, Baltimore, 1925.

McKean, H. P., Nagumo's equation, *Adv. Math.*, **4**, 209–223, 1970.

Nagumo, J., Arimoto, S., and Yoshizawa, S., An active pulse transmission line simulating nerve axon, *Proc. IRE*, **50**, 2061–2071, 1962.

Poiseuille, J. L. M., Recherches expérimentales sur le mouvement de liquides dans les tubes de trés petits diametres, *Mémoirs présentés par divers savants à l'académie royale des sciences de l'Institut de France*, **9**, 433–545, 1846.

Rosenzweig, M. L. and MacArthur, R. H., Graphical representation and stability conditions of predator-prey interactions, *Am. Nat.*, **97**, 209–223, 1963.

Rybak, B. and Béchet, J. J., Recherches sur l'électromécanique cardiaque, *Path. Biol.*, **9**, 2035–2054, 1961.

Volterra, V., Variazione e fluttuazini del numero d'individui in specie animali conviventi, *Mem. Accad., Nazionale Lincei*(ser. 6) **2**, 31–113, 1926.

Zaikin, A. N. and Zhabotinskii, A. M., Concentration wave propagation in two-dimensional liquid-phase self-oscillating systems, *Nature*, **225**, 535–537, 1970.

Zhabotinskii, A. M., Periodic process of the oxidation of malonic acid in solution (Study of the kinetics of Belousov's), *Biofizika*, **9**, 306–311, 1964.

Exercises

4.1 Show that Zeeman's heart beat equations have a unique resting state $x = x_a, b = -(x_a^3 + ax_a)$ and derive a single differential equation satisfied by the muscle fibre length x.

4.2 In the differential equation satisfied by the muscle fibre length x of Exercise 4.1, let $x = x_a + y$ and assume that y is small and that $y\frac{dy}{dt}$ and $y^2\frac{dy}{dt}$ can be neglected. Show that, if $x_a < \sqrt{-a/3}, y$ grows exponentially with time but decays exponentially with time if $x_a > \sqrt{-a/3}$. What happens when $x_a = \sqrt{-a/3}$? Give possible interpretations of these three cases.

4.3 Using the polar co-ordinates defined by (4.3.8) show that the blood flow equation (4.3.5) takes the form (4.3.9).

4.4 Show that $V(r, t) = \phi(r)f(t)$ satisfies (4.3.11) provided $\phi(r)$ satisfies the ordinary differential equation

$$\frac{d}{dr}\left(r\frac{d\phi}{dr}\right) - \lambda r\phi = -\frac{r}{\nu}$$

(λ is a constant), and the condition $\phi(a) = 0$ and $f(t) = \exp(\nu\lambda t/\rho)$.

4.5 Use the iterative scheme (4.3.14) to calculate the first three approximations to $V(r, t)$ of Exercise 4.4.

4.6 In the Hodgkin-Huxley model of nerve impulse transmission, assume the potential $E(x, t)$ has the form of a travelling wave as well as the conductances m, h and n and show that such solutions satisfy the system

$$\frac{a}{2R}\frac{d^2E}{d\xi^2} = cC_m\frac{dE}{d\xi} + g_{Na}(E - E_{Na}) + g_K(E - E_K) + g_l(E - E_l),$$

$$c\frac{dm}{d\xi} = \alpha_m(E)(1 - m) + \beta_m(E)m,$$

$$c\frac{dh}{d\xi} = \alpha_h(E)(1 - h) + \beta_h(E)h,$$

$$c\frac{dn}{d\xi} = \alpha_n(E)(1 - h) + \beta_n(E)n,$$

where $\xi = x + ct$ and c is the wave number.

4.7 Assume the FitzHugh-Nagumo equations (4.4.7) admit travelling wave solutions $u(x,t) = \phi(x+ct)$, $v(x,t) = \psi(x+ct)$. Deduce the equations to be satisfied by ϕ and ψ.

4.8 Verify that if $b = 0 = \gamma$ and (4.4.7) have travelling wave solutions then

$$\phi(x+ct) \equiv \phi(\xi) = \frac{1}{1 + e^{-\xi/\sqrt{2}}}$$

is a solution provided $c = \sqrt{2(\frac{1}{2} - a)}, 0 < a < \frac{1}{2}$.

4.9 For what values of the wave speed c is

$$\phi(x+ct) = 3\left(\sqrt{\left[(2-a)\left(\frac{1}{2}-a\right)\right]}\cosh\sqrt{ax} + (1+1/a)\right)^{-1},$$

$$0 < a < \frac{1}{2}$$

a solution of (4.4.7) when $b = \gamma = 0$?

4.10 If the FitzHugh-Nagumo equations have travelling wave solutions, show that the system governing these solutions has a unique rest state if and only if

$$(1-a)^2 < 4b/\gamma.$$

4.11 In the simplified McKean model of nerve impulse transmission (see Figure 4.4.4), let $b = 0 = \gamma$ and determine the forms of travelling waves as functions of $\xi = x+ct$, $-\infty < \xi < \infty$ and which satisfy the conditions $\phi(\xi) \to 0$ as $\xi \to -\infty$, $\phi(\xi) \to 1$ as $\xi \to \infty$.
Determine the wave speed c in each of the cases $\theta = \frac{\pi}{2}$ and $\tan\theta = \frac{1}{8}$.

4.12 Determine the rest states in the Belousov-Zhabotinskii reaction governed by the model (4.5.6)–(4.5.8).

4.13 Using the law of mass action, derive a mathematical model governed by the intermediaries X and Y in the trimolecular reaction:

$$A \to X,$$
$$B + X \to Y + D,$$
$$2X + Y \to 3X,$$
$$X \to E,$$

where A, B, D and E are initial and final products, and all rate constants are taken to be unity.

4.14 Determine the rest states in the Volterra-Lotka model of predator-prey interaction (4.6.6), (4.6.7). If predator and prey are present in the steady state, show that the coefficients a, b, c, e and f must satisfy the constraints

$$\frac{a}{b} > \frac{e}{f}, \quad c \neq 0.$$

4.15 Discuss the same problem as in Exercise 4.14 in relation to the models (4.6.8), (4.6.9) and (4.6.12), (4.6.13).

Chapter 5

First-order Systems of Ordinary Differential Equations

5.1 Existence and uniqueness

In Chapters 2 and 3 certain types of differential equations have been discussed and methods for deriving their solutions have been described. When more general differential equations are considered it is not by any means obvious that they possess solutions. Spending a lot of time trying to solve a differential equation, which does not have a solution, can be very frustrating to say the least. Therefore, we shall give one theorem, which guarantees that a differential equation that satisfies its conditions possesses a solution, and say something about its ramifications. The proof of the theorem is given in the Appendix to this chapter as well as a method for finding the solution.

EXISTENCE THEOREM I
Let $f(t, y)$ be a single-valued continuous function of t and y in $t_0 \leq t \leq t_0 + h$, $|y - y_0| \leq k$ that satisfies:

$$(a) \quad |f(t, y)| < M,$$
$$(b) \quad |f(t, y) - f(t, y')| < K|y - y'|$$

for any (t, y) and (t, y') that comply with the above inequalities. Then, for $h < k/M$, the differential equation

$$\dot{y} = f(t, y) \tag{5.1.1}$$

possesses one, and only one, continuous solution $y(t)$ in $t_0 \leq t \leq t_0 + h$ such that $y(t_0) = y_0$.

The constant h determines the range of t for which the solution is valid while the constant k sets a limit on how far $y(t)$ deviates from its initial value. It would be ideal if h could be made as large as desired. However, the restriction $h < k/M$ means that h cannot be increased beyond a certain point without a corresponding increase in k. But, larger h and k may entail an increase in the bound M to meet condition (a) and this increase may be sufficient to prevent any improvement in k/M.

Nevertheless, it may be possible to extend the solution to larger t by taking $y(t_0 + h)$ as the initial value at $t = t_0 + h$ provided that suitable new h, k, M can be found with this starting point.

Suppose that f satisfies the conditions of the theorem and, by some means, two continuous solutions $y_1(t)$, $y_2(t)$ of (5.1.1) have been found such that $y_1(t_0) = y_0$ and $y_2(t_0) = y_0$. Suppose, further, it is known that $y_1(t)$ is valid for $t_0 \leq t \leq t_0 + h_1$ whereas $y_2(t)$ holds for $t_0 \leq t \leq t_0 + h_2$ with $h_2 > h_1$. The uniqueness part of the theorem then says that $y_1(t) = y_2(t)$ for $t_0 \leq t \leq t_0 + h_1$. The same assertion cannot be made for larger values of t unless it can be demonstrated that $y_1(t)$ can be continued beyond $t = t_0 + h_1$. For example, $y_1(t) = 1 - t + t^2 - \cdots$ and $y_2(t) = 1/(1 + t)$ are solutions of $(1 + t)\dot{y} = -y$, which are unity at $t = 0$ so long as $h_1 < 1$, but the series in y_1 is not valid in $t > 1$.

Generally, it is not difficult to recognise when f is continuous and to assess M. Checking (b) can require more effort but there is one case when (b) holds for sure and that is when

$$\left| \frac{\partial}{\partial y} f(t, y) \right| \leq N \tag{5.1.2}$$

for $t_0 \leq t \leq y_0 + h, |y - y_0| \leq k$ and N finite.

For then

$$|f(t, y) - f(t, y')| = \left| \int_{y'}^{y} \frac{\partial}{\partial u} f(t, u) du \right|$$

$$= \left| \int_{0}^{y - y'} \frac{\partial}{\partial u} f(t, u - y') du \right|$$

$$\leq \int_{0}^{|y - y'|} \left| \frac{\partial}{\partial u} f(t, u - y') \right| du$$

$$\leq N|y - y'|.$$

Example 5.1.1

The differential equation

$$\dot{y} = g(t)y^2$$

in which g is continuous clearly has $f(t, y)$ continuous and, because

$$\frac{\partial}{\partial y} g(t)y^2 = 2g(t)y.$$

satisfies the condition (5.1.2) so long as t and y are bounded. Therefore the differential equation has one and only one continuous solution such that $y(t_0) = y_0$. It remains valid as t increases as long as t and y remain finite.

Consider, in particular, $\dot{y} = y^2$. The solution of this such that $y(t_0) = 0$ is $y(t) = 0$ for all t. On the other hand, if $y(t_0) = y_0$ with $y_0 \neq 0$, $y(t) = y_0/\{1 + (t_0 - t)y_0\}$. If $y_0 < 0$ this solution holds for all $t \geq t_0$. In contrast, if y_0 is positive, $y(t)$ becomes unbounded as t approaches $t_0 + 1/y_0$; in this case the region of validity is confined to $t_0 \leq t < t_0 + 1/y_0$. $\qquad$ ▯

The conditions of Existence Theorem I are sufficient but not necessary. There are differential equations that do not satisfy the conditions but which possess a unique continuous solution. For example,

$$\dot{y} = \begin{cases} (1 - 2t)y & (t > 0) \\ (2t - 1)y & (t < 0) \end{cases} \tag{5.1.3}$$

subject to $y = y_0 \, (\neq 0)$ at $t = 0$. In this case, f is discontinuous at $t = 0$ when $y_0 \neq 0$ so the conditions of Existence Theorem I are not met. Nevertheless there is a unique continuous solution, namely

$$y(t) = \begin{cases} y_0 e^{t - t^2} & (t \geq 0) \\ y_0 e^{t^2 - t} & (t \leq 0). \end{cases}$$

There may, however, be values of t_0 or y_0 for which the initial value problem

(i) has no solution;

(ii) has a discontinuous solution;

(iii) has more than one continuous solution.

For instance, the differential equation

$$y\dot{y} = -t \tag{5.1.4}$$

has solution $y^2 + t^2 = C$ where C is a constant. An example when there is no solution is to take $y = 0$ at $t = 0$. Then $C = 0$ and $y^2 + t^2 = 0$. This forces $t = 0$ and there is no solution for $t > 0$.

An illustration of more than one solution is provided by $y = 0$ at $t = t_0 \neq 0$. Then $C = t_0^2$ and $y = \pm(t_0^2 - t^2)^{1/2}$ giving two solutions while $t^2 \leq t_0^2$.

Thus the initial value $y = 0$ originates difficulties for (5.1.4). For it, $f(t, y)$ in (5.1.1) is $-t/y$, which is infinite at $y = 0$ for any non-zero t, and so the conditions of Existence Theorem I cannot be met. Thus there is no warranty of a unique continuous solution. Yet, there is no problem if $y(t_0) = y_0$ with $y_0 \neq 0$. Now Existence Theorem I applies and there is the unique continuous solution $y = (t_0^2 + y_0^2 - t^2)^{1/2}$ so long as $t^2 \leq t_0^2 + y_0^2$.

Any point (t_0, y_0) at which (i), (ii) or (iii) is true is known as a **singular point**. For example, any point $(t_0, 0)$ is a singular point of (5.1.4). At a singular point Existence Theorem I must fail, but the converse is false as (5.1.3) shows. Thus, while places where f does not abide by the conditions of Existence

Theorem I are candidates for singular points, a special investigation has to be undertaken to check whether or not they are actually singular points.

The same nomenclature of singular points is used in connection with systems and with the differential equation of order n when (i), (ii) or (iii) occurs. Existence Theorems II and III (given in the Appendix) are invalid at singular points but the points where their conditions are unsatisfied are not necessarily singular points.

The linear system

$$\dot{y}_i = \sum_{j=1}^{n} a_{ij}(t)y_j + f_i(t) \quad (i = 1, \ldots, n) \tag{5.1.5}$$

conforms to Existence Theorem II except at those values of t where a_{ij} or f_i are discontinuous. Apart from these values, the initial value problem has a unique continuous solution. In particular, the linear system with constant coefficients possesses a unique continuous solution except, perhaps, for those t where f_i is not continuous.

5.2 Epidemics

The simplest model of the spread of an epidemic in a population stipulates that at time t there are x susceptible individuals and y infected who may transmit the disease. It is assumed that the mixing of these two groups passes on the illness and that, in the short time δt, $\mu x y \delta t$ new infections occur. Also some of those infected will die, or stop mixing, or recover and become immune; suppose that $\nu y \delta t$ disappear in this way. Then, if $\rho \delta t$ new susceptibles arrive in the interval δt,

$$\dot{x} = -\mu xy + \rho, \qquad \dot{y} = \mu xy - \nu y. \tag{5.2.1}$$

Normally μ, ν and ρ are taken as non-negative constants and it is convenient to assume that they are positive. The right-hand sides of (5.2.1) vanish for $x = x_0$, $y = y_0$ where

$$-\mu x_0 y_0 + \rho = 0, \qquad \mu x_0 y_0 - \nu y_0 = 0 \tag{5.2.2}$$

or $x_0 = \nu/\mu$, $y_0 = \rho/\nu$. If $x(t) = x_0$, $y(t) = y_0$ then $\dot{x} = 0, \dot{y} = 0$ and the differential equations (5.2.1) are satisfied. In other words, (x_0, y_0) is an *equilibrium state* in which the numbers of susceptibles and infected do not vary.

Let us now address the question of whether equilibrium is approached from a nearby state and, if so, in what manner. Put

$$x = x_0(1 + \xi), \qquad y = y_0(1 + \eta)$$

where ξ and η are so small that their products may be neglected. Then (5.2.1) become

$$\dot{\xi} = -\sigma(\xi + \eta), \qquad \dot{\eta} = \nu\xi$$

where $\sigma = \mu\rho/\nu$. Substituting for ξ from the second equation we obtain

$$\ddot{\eta} + \sigma\dot{\eta} + \sigma\nu\eta = 0$$

of which the solution such that $\eta = \eta_0$, $\dot{\eta} = \nu\xi_0$ at $t = 0$ is

$$\eta = e^{-\sigma t/2} \left\{ \eta_0 \cos \omega t + \frac{1}{\omega}(\nu\xi_0 + \tfrac{1}{2}\sigma\eta_0) \sin \omega t \right\}$$

where $\omega^2 = \sigma\nu - \sigma^2/4$. Hence

$$\xi = e^{-\sigma t/2} \left\{ \xi_0 \cos \omega t - \frac{\sigma}{2\omega}(\xi_0 + 2\eta_0) \sin \omega t \right\}.$$

When $\omega^2 > 0$, i.e., $4\nu > \sigma$ or $4\nu^2 > \mu\rho$, the population, after a small departure from equilibrium, returns to equilibrium in an oscillatory fashion with exponential decay. If $\omega^2 < 0$ or $4\nu^2 < \mu\rho$ the fact that $|\omega| < \sigma/2$ ensures exponential decay again but there is no accompanying oscillation. In either case the population returns to equilibrium, the approach being more rapid when oscillations are present.

5.3 The phase plane

If less specific assumptions are made about the mechanism of propagation of epidemics, the most that can be said is that a system of the type

$$\dot{x} = f(x, y), \qquad \dot{y} = g(x, y) \tag{5.3.1}$$

will need to be solved. Any point (x_0, y_0) such that

$$f(x_0, y_0) = 0, \qquad g(x_0, y_0) = 0 \tag{5.3.2}$$

is called a *critical point* or *fixed point* or *equilibrium point*. A solution that starts at an equilibrium point never leaves it because $\dot{x}$ and $\dot{y}$ both vanish there provided that f and g satisfy the conditions of Existence Theorem II.

When the solution of (5.3.1) has been found, say $x = h_1(t), y = h_2(t)$, the point (x, y) can be plotted in the (x, y)-plane at time t. As t varies, (x, y) will trace a curve in the (x, y)-plane. This curve is known as a *trajectory* and the (x, y)-plane is called the **phase plane**. By attaching an arrow to each trajectory the direction in which (x, y) moves as t increases can be indicated.

The phase plane then contains all the information in (5.3.1) except the rate at which the trajectory is traversed. The slope of a trajectory is given from (5.3.1) by

$$\frac{dy}{dx} = \frac{\dot{y}}{\dot{x}} = \frac{g(x, y)}{f(x, y)}. \tag{5.3.3}$$

A trajectory is vertical at any (x, y) where $g(x, y) \neq 0$, $f(x, y) = 0$ and horizontal where $g(x, y) = 0$, $f(x, y) \neq 0$. The trajectory corresponding to an equilibrium point reduces to a single point.

When f and g satisfy the conditions of Existence Theorem II, the initial value problem has a single continuous solution in the neighbourhood of $t = t_0$. Therefore, in this case, only one trajectory passes through a given point of the phase plane, i.e., under the conditions of Existence Theorem II, *two trajectories do not intersect in general.*

All these notions can be generalised to a system of n equations. A solution still describes a trajectory, which is now a curve in a space of n dimensions, and we talk of a **phase space** rather than a phase plane. Diagrams are, however, much more difficult to draw.

The same question about behaviour near equilibrium that was asked for epidemics can be raised here. Put

$$x = x_0 + \xi, \quad y = y_0 + \eta$$

where (x_0, y_0) is in conformity with (5.3.2). Because of the smallness of ξ and η, it will be assumed that $f(x, y)$ can be approximated by the first terms in its Taylor expansion, namely

$$\xi \left(\frac{\partial f}{\partial x} \right)_0 + \eta \left(\frac{\partial f}{\partial y} \right)_0$$

where $()_0$ means calculate the value at $x = x_0$, $y = y_0$. The approximation to (5.3.1) is then

$$\dot{\xi} = \xi \left(\frac{\partial f}{\partial x} \right)_0 + \eta \left(\frac{\partial f}{\partial y} \right)_0, \tag{5.3.4}$$

$$\dot{\eta} = \xi \left(\frac{\partial g}{\partial x} \right)_0 + \eta \left(\frac{\partial g}{\partial y} \right)_0, \tag{5.3.5}$$

a linear system with constant coefficients. The behaviour of such systems in the phase plane will be examined in succeeding sections.

The behaviour near equilibrium is a matter of **local stability**. The larger question of what happens when the initial state is not near equilibrium is one of **global stability**. This will be tackled in Section 5.6 and, in the meantime, we merely remark that for systems of three or more equations global behaviour is very varied and imperfectly understood.

5.4 Local stability

It has been discovered in the preceding section that local stability reduces to a discussion of

$$\dot{x} = ax + by, \qquad \dot{y} = cx + dy \qquad (5.4.1)$$

where a, b, c and d are real constants. The goal of this section is to determine the trajectories of this system.

Let us first remark that, if α is a real constant, $x(t + \alpha)$, $y(t + \alpha)$ occupies the same points in the phase plane as t varies as $x(t)$, $y(t)$ though at a time α earlier. So both points describe the same trajectory despite being different solutions. More than one solution can lie on one trajectory.

In finding the trajectories we shall ignore the degenerate case in which $ad = bc$; should it arise, the equations can be integrated directly without trouble. It would, in any case, be necessary to reconsider the validity of (5.3.4) and (5.3.5) as an adequate prescription for local stability in these circumstances. Therefore, from now on, it will be assumed that

$$ad \neq bc \qquad (5.4.2)$$

so that there is a single critical point at the origin.

According to Section 3.4, the first attempt at a solution is $x = \alpha e^{\lambda t}$, $y = \beta e^{\lambda t}$. For the satisfaction of (5.4.1) we require

$$(a - \lambda)\alpha + b\beta = 0, \qquad (5.4.3)$$
$$c\alpha + (d - \lambda)\beta = 0. \qquad (5.4.4)$$

These give non-zero α, β only if

$$(a - \lambda)(d - \lambda) - bc = 0$$

or

$$\lambda^2 - (a + d)\lambda + ad - bc = 0.$$

The roots are λ_1 λ_2 where

$$2\lambda_1 = a + d + \{(a - d)^2 + 4bc\}^{1/2},$$
$$2\lambda_2 = a + d - \{(a - d)^2 + 4bc\}^{1/2}.$$

Several cases have to be studied.

(a) $(a - d)^2 + 4bc > 0$

In this case the values of λ_1 and λ_2 are real and distinct; the special procedure for multiple roots does not have to be called on.

Assume first that $b \neq 0$. Then, from (5.4.3), we can choose $\alpha = b$, $\beta = \lambda_1 - a$ corresponding to λ_1 and $\alpha = b$, $\beta = \lambda_2 - a$ corresponding to λ_2. Consequently

$$x = b(C_1 e^{\lambda_1 t} + C_2 e^{\lambda_2 t}), \tag{5.4.5}$$

$$y = (\lambda_1 - a)C_1 e^{\lambda_1 t} + (\lambda_2 - a)C_2 e^{\lambda_2 t} \tag{5.4.6}$$

where C_1 and C_2 are arbitrary constants. These equations may be rearranged to give

$$(\lambda_2 - a)x - by = b(\lambda_2 - \lambda_1)C_1 e^{\lambda_1 t}, \tag{5.4.7}$$

$$(\lambda_1 - a)x - by = b(\lambda_1 - \lambda_2)C_2 e^{\lambda_2 t}. \tag{5.4.8}$$

From (5.4.7), $(\lambda_2 - a)x - by$ cannot change sign as t varies. Therefore the trajectory cannot go over the line $(\lambda_2 - a)x = by$. Similarly, from (5.4.8), the trajectory cannot trespass across $(\lambda_1 - a)x = by$. These lines are displayed in Figure 5.4.1 as well as the regions to which the trajectory is confined for various choices of C_1, C_2 when $b > 0$ and both λ_1, λ_2 are negative.

Suppose now that both λ_1 and λ_2 are negative. It is evident that $x \to 0$, $y \to 0$ as $t \to \infty$. Also $\lambda_1 > \lambda_2$ so that $(\lambda_1 - a)x \sim by$ as $t \to \infty$, so long as $C_1 \neq 0$. Moreover, as $t \to -\infty$, $|x|$ and $|y|$ become large and $(\lambda_2 - a)x \sim by$ provided $C_2 \neq 0$, x approaching $-\infty$ when $b > 0$, $C_2 < 0$. The trajectories, therefore, have the shape depicted in Figure 5.4.2 for $b > 0$. The exclusion so far of $C_1 = 0$ or $C_2 = 0$ can be remedied immediately because their trajectories are the dividing straight lines on account of (5.4.7) and (5.4.8). The arrows

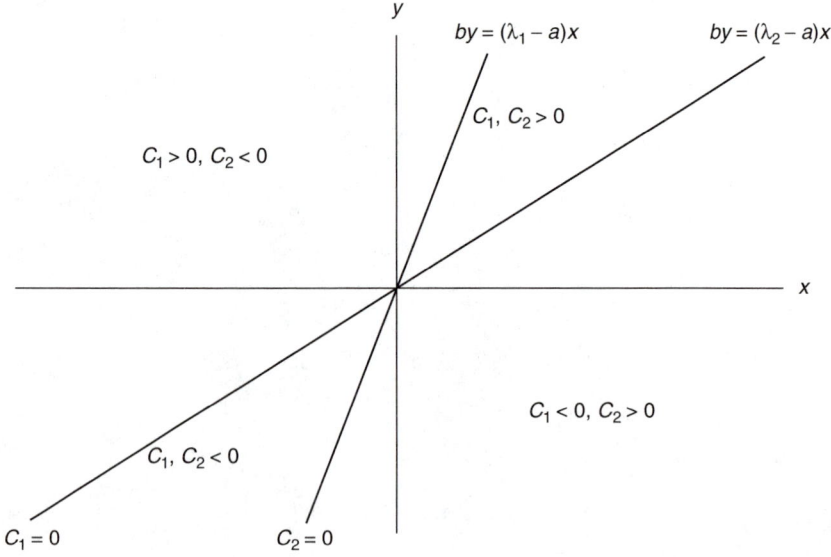

FIGURE 5.4.1: Lines that cannot be crossed by trajectories.

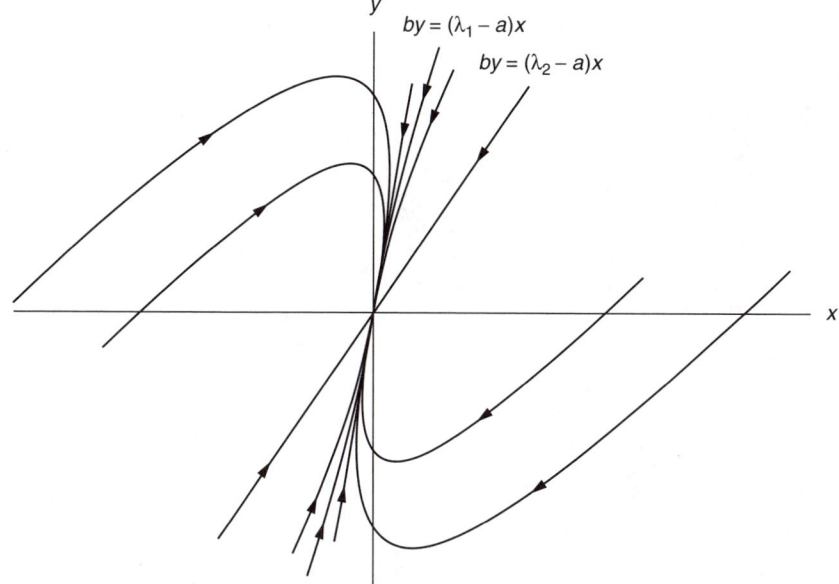

FIGURE 5.4.2:   The stable node.

on the curves indicate the direction in which (x, y) moves as t increases. A critical point of this type is known as a **stable node**.

When λ_1 and λ_2 are both positive, the curves are similar in character to those in Figure 5.4.2 but the directions of the arrows are reversed because (x, y) moves away from the origin as t increases. We have an **unstable node**.

The remaining possibility is that λ_1 and λ_2 have opposite signs so that $\lambda_1 > 0$ and $\lambda_2 < 0$. From (5.4.7), the magnitude of $(\lambda_2 - a)x - by$ increases with time whereas that of $(\lambda_1 - a)x - by$ diminishes. The origin can never be reached unless $C_1 = 0$ when the trajectory is a straight line. The trajectory is also a straight line when $C_2 = 0$, but now the origin is departed from. The behaviour of the trajectories is displayed in Figure 5.4.3. The critical point is called a **saddle-point**. Clearly a point (x, y) started near the origin cannot stay near it in general and there is no stability.

So far we have assumed that $b \neq 0$. If $b = 0$ we see at once from (5.4.1) that $x = C_1 e^{at}$ and then

$$y = C_2 e^{dt} + cC_1 e^{at}/(a - d).$$

We remark that $a \neq d$ because $(a - d)^2$ must be positive when $b = 0$. In this case, the dividing lines are $x = 0$ and $(a - d)y = cx$. Apart from this change the pictures are practically unaltered. There is a node if a and d have the same sign (stable if $a < 0$, unstable if $a > 0$) and a saddle-point if a and d

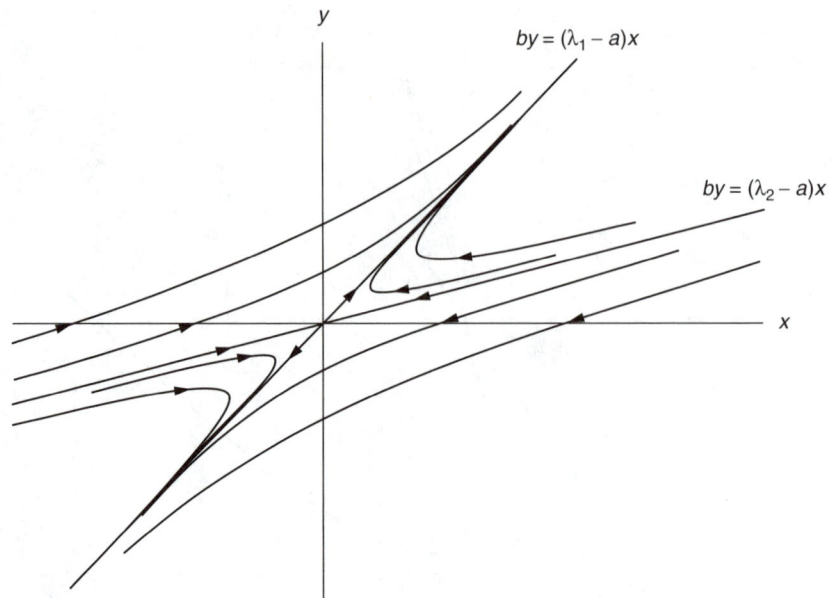

FIGURE 5.4.3:   The saddle-point.

have opposite signs.

$$\textbf{(b) } (a - d)^2 + 4bc < 0$$

This possibility can occur only when $bc < 0$; so neither b nor c vanishes and they have opposite signs. The roots λ_1 and λ_2 are still distinct but they are now complex conjugates. Write $\lambda_1 = \frac{1}{2}(a+d)+i\omega$, $\lambda_2 == \frac{1}{2}(a+d)-i\omega$ where

$$\omega^2 = -\tfrac{1}{4}(a-d)^2 - bc.$$

Then

$$x = Ae^{\frac{1}{2}(a+d)t}\cos(\omega t - \alpha) \tag{5.4.9}$$

where A and α are arbitrary constants. From (5.4.1)

$$y = (A/b)e^{\frac{1}{2}(a+d)t}\left\{\tfrac{1}{2}(d-a)\cos(\omega t - \alpha) - \omega\sin(\omega t - \alpha)\right\}. \tag{5.4.10}$$

The formulae (5.4.9) and (5.4.10) can be combined to give

$$cx^2 + (d-a)xy - by^2 = -(\omega^2 A^2/b)e^{(a+d)t}. \tag{5.4.11}$$

Suppose that $a+d = 0$ so that λ_1 and λ_2 are pure imaginary. The equation of a trajectory is given by (5.4.11) as

$$cx^2 + (d-a)xy - by^2 = -\omega^2 A^2/b.$$

Rotate the axes by means of the transformation

$$x = X \cos \theta - Y \sin \theta, \quad y = X \sin \theta + Y \cos \theta$$

where

$$\tan 2\theta = \frac{d - a}{b + c}.$$

The equation of the curve goes over to

$$A'X^2 + C'Y^2 = -\omega^2 A^2 / b$$

where

$$A' = \tfrac{1}{2}(c - b) + \tfrac{1}{2}\{(b + c)^2 + (d - a)^2\}^{1/2},$$
$$C' = \tfrac{1}{2}(c - b) - \tfrac{1}{2}\{(b + c)^2 + (d - a)^2\}^{1/2}.$$

Since $A'C' = \omega^2$, A' and C' have the same sign. Also $A' + C' = c - b$ so that if $b < 0$, which implies $c > 0$, A' and C' are positive, whereas if $b > 0$, which makes $c < 0$, A' and C' are negative. Thus A' and C' have the opposite sign to b and the trajectory is an ellipse with semi-axes $\omega|A|/(-bA')^{1/2}$ and $\omega|A|/(-bC')^{1/2}$. Typical trajectories are drawn in Figure 5.4.4; the critical point is known as a **centre**. With regard to the direction of motion on a

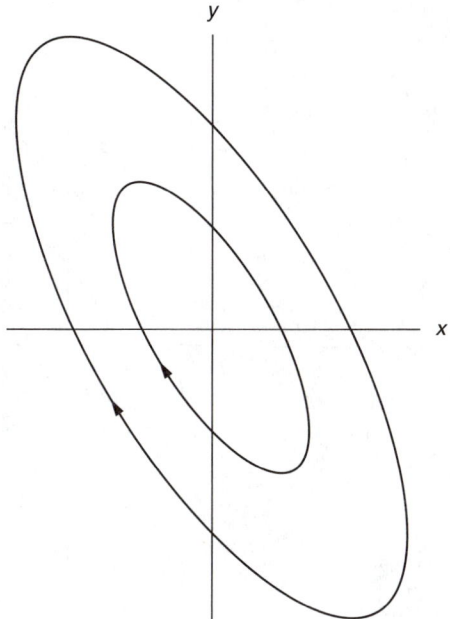

FIGURE 5.4.4: The centre.

trajectory, we see from (5.4.1) that when $x = 0$, $\dot{x} = by$. Hence, when $b > 0$, x must be increasing at positive y and so the direction is as shown in Figure 5.4.4; if $b < 0$ the arrows have to be reversed.

The equations (5.4.9) and (5.4.10) make it evident that x and y vary harmonically when $a + d = 0$. The point (x, y) therefore makes continual circuits round the origin and is forever retracing its path. A trajectory started from near the origin never leaves the neighbourhood but never swings into the origin. Therefore, there is stability in the sense that (x, y) remains in the vicinity of the critical point, if it is initially near there, but it never attains the critical point.

It should be observed that any closed trajectory implies periodic motion because it entails there being a fixed T such that $x(t + T) = x(t)$, $y(t + T) = y(t)$ for all t. The motion need not, however, have the simple harmonic character mentioned above when the system is more general than (5.4.1).

Turning now to the case when $a + d \neq 0$, we note that the only difference is the exponential factor in (5.4.11). The trajectory may be thought of instantaneously as an ellipse whose axes are changing exponentially. The trajectory therefore spirals about the origin as shown in Figure 5.4.5. If $a + d < 0$, the point (x, y) must approach the origin as $t \rightarrow \infty$. The direction of motion along a trajectory is then that of Figure 5.4.5 and the critical point is called a **stable focus**. When $a + d > 0$, (x, y) departs from the origin, the arrows are reversed and we have an **unstable focus**.

$$\textbf{(c) } (\mathbf{a} - \mathbf{d})^2 + 4\mathbf{bc} = \mathbf{0}$$

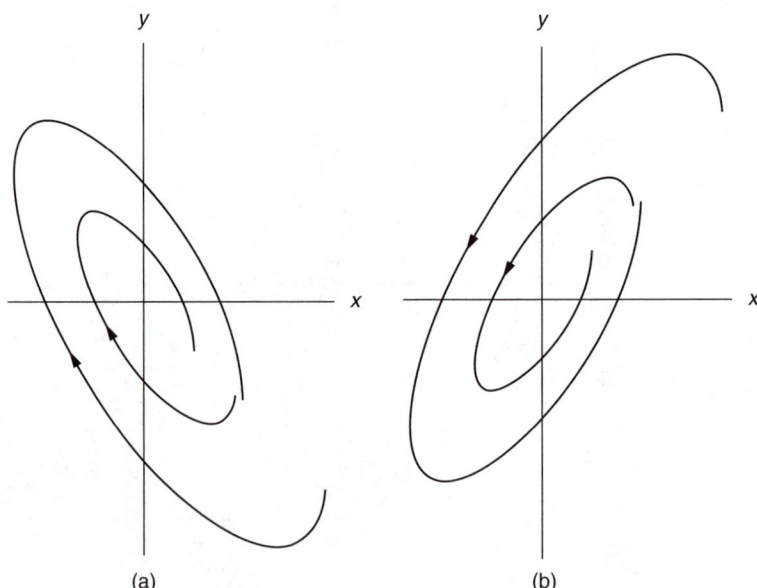

(a)   (b)

FIGURE 5.4.5:   The focus when $a + d < 0$; (a) $b > 0$, (b) $b < 0$.

In this case $\lambda_1 = \lambda_2 = \frac{1}{2}(a+d)$. However,

$$(a+d)^2 = (a-d)^2 + 4ad = -4bc + 4ad \neq 0$$

by (5.4.2) so λ_1 and λ_2 are non-zero.

If $b \neq 0$, (5.4.3) and (5.4.4) supply only the single solution $\alpha = b$, $\beta = \frac{1}{2}(d-a)$. To find a second solution we try, according to Section 3.4, $x = (\gamma + \alpha t)e^{\frac{1}{2}(a+d)t}$, $y = (\delta + \beta t)e^{\frac{1}{2}(a+d)t}$ with the result that $\gamma = 0$, $\delta = 1$. The technique of Section 3.5 leads to the same answer. Consequently

$$x = b(C_1 + C_2 t)e^{\frac{1}{2}(a+d)t},$$
$$y = \{C_2 + \frac{1}{2}(d-a)(C_1 + C_2 t)\}e^{\frac{1}{2}(a+d)t}.$$

The line $by = \frac{1}{2}(d-a)x$ cannot be crossed and the structure of the trajectories is similar to that of Figure 5.4.2 when the dividing lines coalesce. This critical point is therefore also termed a node; it is stable if $a+d < 0$ and unstable if $a+d > 0$.

If $b = 0$, then $a = d$ and $x = C_1 e^{at}$, $y = (C_2 + cC_1 t)e^{at}$ and the trajectories are not much changed in shape if $c \neq 0$. If, in addition, $c = 0$ the trajectories are the straight lines $y/x = $ constant (see Figure 5.4.6). The critical point is still designated a node, stable if $a < 0$ and unstable if $a > 0$.

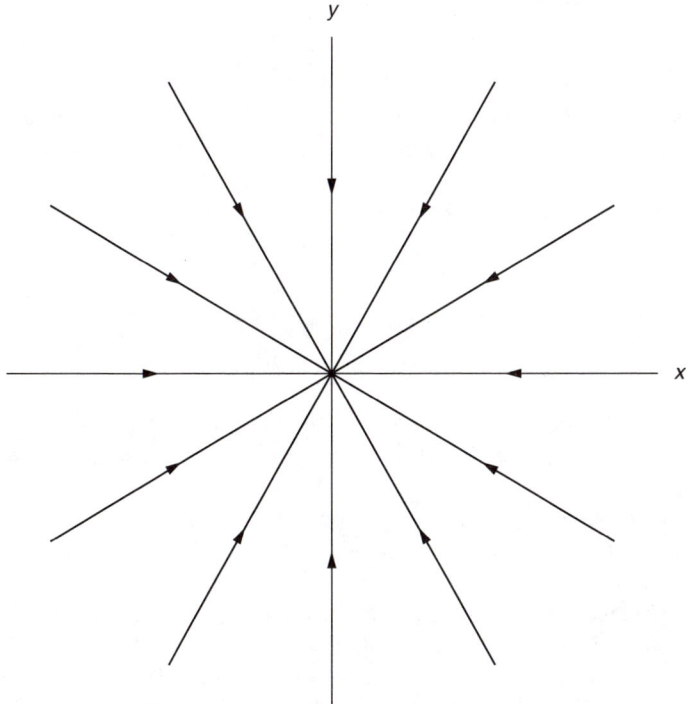

FIGURE 5.4.6: The case $b = c = 0$, $a < 0$.

It is convenient to summarise these results as follows: *if $ad \neq bc$ the trajectories of (5.4.1) form, when λ_1 and λ_2 are*

(a) *real, (i) a node if λ_1 and λ_2 have the same sign, (ii) a saddle-point if λ_1 and λ_2 have opposite signs;*

(b) *purely imaginary, a centre;*

(c) *complex but not purely imaginary, a focus.*

The reader will note that there can be no oscillation if λ_1 and λ_2 are real; in particular, the system will not be oscillatory if $bc > 0$.

5.5 Stability

We now want to investigate what general conclusions can be drawn about the behaviour of solutions to (5.3.1) on the basis of the model of (5.3.4), (5.3.5) and the trajectories determined in Section 5.4. To fix ideas, we consider a somewhat generalised model of epidemics in which the number of susceptibles x and of infected y satisfy

$$\dot{x} = h(x, y)x, \qquad \dot{y} = k(x, y)y.$$

The behaviour near an equilibrium point in which neither x nor y is zero is of interest. So x_0 and y_0 are taken to satisfy

$$h(x_0, y_0) = 0, \qquad k(x_0, y_0) = 0.$$

It then follows that, in the notation of Section 5.3,

$$(\partial f/\partial x)_0 = h_1 x_0, \qquad (\partial f/\partial y)_0 = h_2 x_0,$$
$$(\partial g/\partial x)_0 = k_1 y_0, \qquad (\partial g/\partial y)_0 = k_2 y_0$$

where h_1, h_2, k_1 and k_2 are the values of $\partial h/\partial x, \partial h/\partial y, \partial k/\partial x$ and $\partial k/\partial y$, respectively, at (x_0, y_0). Thus, in the theory of Section 5.4, $a = h_1 x_0, b = h_2 x_0$, $c = k_1 y_0, d = k_2 y_0$.

Since the presence of the infected tends to reduce the number of susceptibles by infection, we expect $h_2 < 0$. As the number of infected increases there will be less opportunity to affect the susceptibles and so $k_2 < 0, k_1 > 0$. If there is a birth rate of susceptibles, we can suppose $h_1 > 0$, though h_1 will be rather small in comparison with other partial derivatives in most epidemics, because they tend to spread much faster than susceptibles are created.

Since $ad - bc = (h_1 k_2 - h_2 k_1)x_0 y_0$ we can be sure that $ad \neq bc$ when h_1 is small, as suggested above, and the theory of Section 5.4 can be applied.

In order that there can be any kind of oscillation, we must have case (b) of Section 5.4, i.e.,

$$(h_1 x_0 - k_2 y_0)^2 + 4 h_2 k_1 x_0 y_0 < 0.$$

The second term on the right-hand side is negative; so the inequality is feasible if the first term is not too large. Since h_1 is small, this will be true if k_2 is not too large. The oscillations are likely to remain near equilibrium because $a + d = h_1 x_0 + k_2 y_0$ is negative on account of the smallness of h_1. The critical point will be a stable focus or, possibly, a centre.

In the absence of oscillations, λ_1 and λ_2 will both be negative because $ad - bc$ is positive and the equilibrium will be a stable node.

Quite a lot of qualitative information about the behaviour of the solution has been obtained without too specific assumptions about h and k. Of course, further conclusions could be drawn if more was known about h and k. In other problems, the signs of the partial derivatives might be different and the behaviour near equilibrium changed thereby. Some of the unstable critical points might occur. However, such instability merely means departure from equilibrium and, once this exceeds a certain amount, the model of (5.3.4) and (5.3.5) loses its validity because it assumed motion near equilibrium. We then enter the arena of global stability, with the possibility of some kind of stable behaviour away from equilibrium, a matter to be discussed in the next section.

In the foregoing the presence of an equilibrium point has been assumed and it will not be amiss to say a word or two about how the existence of an equilibrium point is verified. Often it will be done most simply by graphical means but sometimes an analytical argument is helpful. With $\partial k / \partial x > 0, k(x, y) = 0$ can be solved for x to give a unique function $x(y)$ of y. Because

$$\frac{dx}{dy} = -\frac{\partial k / \partial y}{\partial k / \partial x}$$

we have $dx/dy > 0$ when $\partial k / \partial y < 0$. There can be no infected if there is no population, and so $x(y) > 0$ but less than some bound. Thus $h(x(y), y)$ is such that

$$\frac{dh}{dy} = \frac{\partial h}{\partial x} \frac{dx}{dy} + \frac{\partial h}{\partial y},$$

which will be negative when both $\partial h / \partial x$ and $\partial h / \partial y$ are. Then $h(x(y), y)$ decreases as y increases so that if $h(x(y), y)$ is positive for small y and negative for large y there will be one and only one equilibrium point.

As a final illustration we consider a case in which the trajectories can be traced completely, namely

$$\dot{x} = y, \qquad \dot{y} = \tfrac{1}{2}(1 - x^2)$$

or, equivalently,

$$\ddot{x} + \tfrac{1}{2}(x^2 - 1) = 0.$$

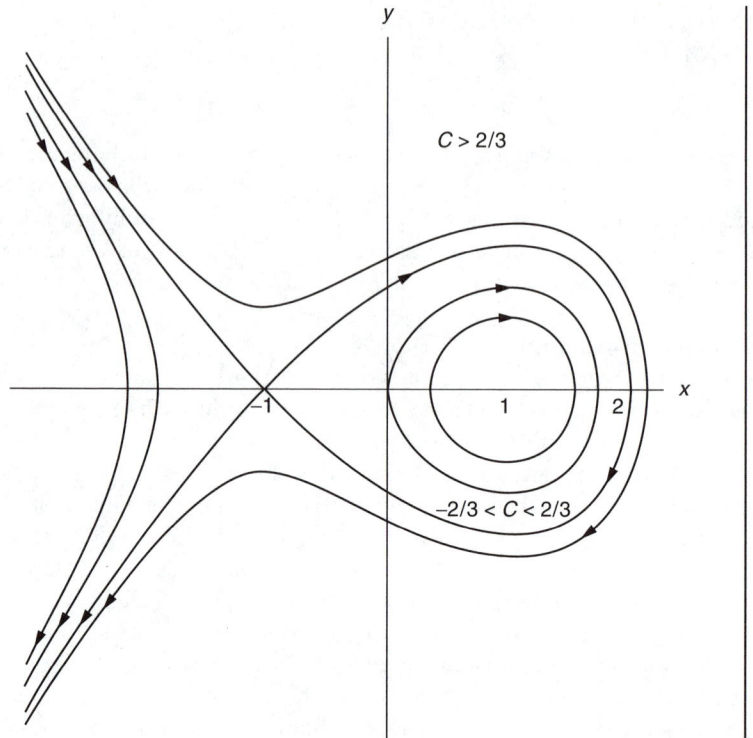

FIGURE 5.5.1:   Non-linear conservative system.

We have

$$y\frac{dy}{dx} = \tfrac{1}{2}(1 - x^2)$$

which gives, on integration,

$$y^2 = x - \tfrac{1}{3}x^3 + C.$$

These trajectories are displayed graphically in Figure 5.5.1. The critical point $x = 1$, $y = 0$ is a centre and $x = -1$, $y = 0$ is a saddle-point. If $-\tfrac{2}{3} < C < \tfrac{2}{3}$ the closed curves surrounding $x = 1$, $y = 0$ show that periodic motion is possible with proper initial conditions. If $C = \tfrac{2}{3}$ no oscillations are allowed but $x = -1$, $y = 0$ can be tended to if the initial conditions are appropriate. If $C > \tfrac{2}{3}$, x and y always approach negative infinity as $t \to \infty$.

A quick idea of the shapes of the trajectories can be obtained as follows. Put $v(x) = \tfrac{1}{3}x^3 - x$ so that the equation of the trajectories is

$$y^2 = C - v(x).$$

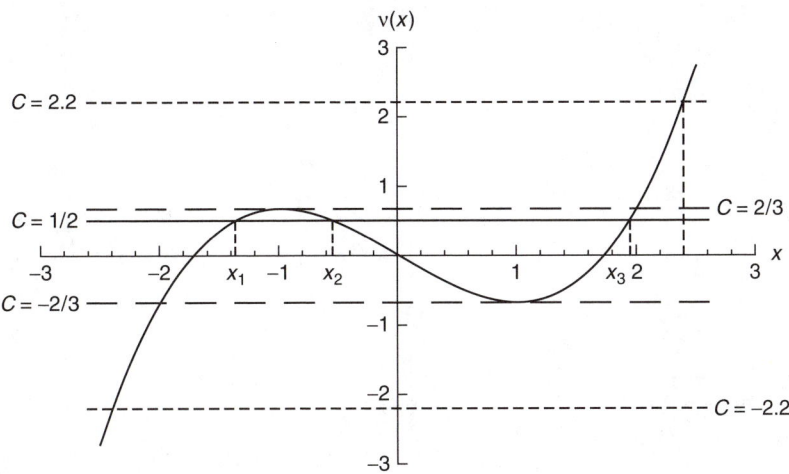

FIGURE 5.5.2: Qualitative determination of trajectories.

Since y^2 cannot be negative the only values of x that can occur are those that satisfy $v(x) \leq C$. Such values can be seen easily from a graph of $v(x)$. In Figure 5.5.2 a graph is displayed together with various possibilities for C. The line on which $C = 2.2$ intersects the curve at $x = 2.4$ approximately and is above the curve for $x < 2.4$. Hence $v(x)$ is below 2.2 for $x < 2.4$. Thus the trajectory starts at $x = -\infty$ with y positive (because $\dot{x}$ must be positive at the start) and moves to the right until it crosses the x-axis at $x = 2.4$ approximately. Thereafter it returns to $x = -\infty$ with y negative. Likewise, when $C = -2.2$, x is restricted to $x < -2.4$; the trajectory goes from $-\infty$ and back again, crossing the x-axis at $x = -2.4$ approximately. On the other hand, when $C = \frac{1}{2}$, there are three intersections with the curve at x_1, x_2 and x_3. Two trajectories are possible now. On one $x \leq x_1$ and it is similar to the one when $C = -2.2$. On the other trajectory x is confined to the interval $x_2 \leq x \leq x_3$ so that the trajectory is a closed curve. It is clear from Figure 5.5.2 that trajectories that are closed curves are possible only for $-\frac{2}{3} < C < \frac{2}{3}$.

This is an example of a *non-linear conservative system*. A more general version is

$$\dot{x} = y, \qquad \dot{y} = -\frac{d}{dx}V(x)$$

or

$$\ddot{x} + \frac{dV(x)}{dx} = 0.$$

The trajectories can be determined as above and are

$$\tfrac{1}{2}y^2 + V(x) = C.$$

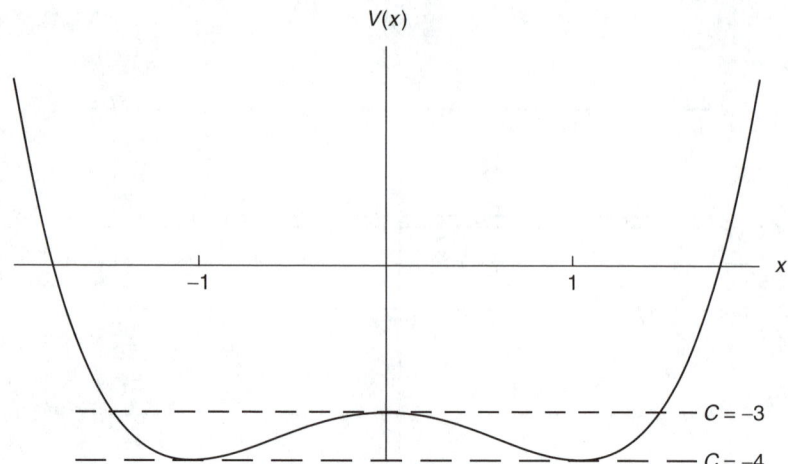

FIGURE 5.5.3:   Graph of $x^4 - 2x^2 - 3$.

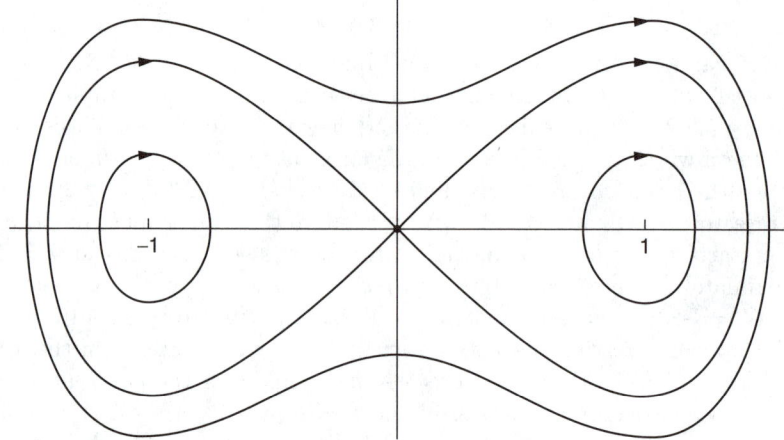

FIGURE 5.5.4:   The trajectories corresponding to Figure 5.5.3.

The graphical method will give an indication of the shapes of the trajectories. As an illustration take

$$V(x) = x^4 - 2x^2 - 3.$$

Its graph is drawn in Figure 5.5.3. Evidently, when $C > -3$ the trajectory consists of a single closed curve. When $-3 > C > -4$, however, the trajectory has two parts. One is a closed curve surrounding $x = -1$ while the other is a closed curve around $x = 1$. There is no trajectory for C below -4. Since the equilibrium points are $x = 0, \pm 1$ the trajectories are as shown in Figure 5.5.4. The points $x = 1$ and $x = -1$ are centres whereas $x = 0$ is a saddle-point.

5.6   Limit cycles

As a beginning to the discussion of global stability, let us examine the system

$$\dot{x} = y, \qquad \dot{y} = \epsilon(1 - x^2)y - x$$

where ϵ is a non-negative constant. Eliminating y we obtain

$$\ddot{x} - \epsilon(1 - x^2)\dot{x} + x = 0 \qquad (5.6.1)$$

which is known as **van der Pol's equation**.

Assume that ϵ is small. As a first approximation one would ignore the term involving ϵ. The general solution is then

$$x = C \cos(t - \alpha)$$

where C and α are arbitrary. Let us study how the presence of the term in ϵ affects the solution $x = C \cos t$. Since this is expected to be a reasonable first approximation, we try $x = A \cos \omega t$ where ω is a constant that is nearly 1. Substitution in (5.6.1) leads to

$$A(1 - \omega^2)\cos \omega t = -\epsilon(1 - A^2 \cos^2 \omega t)\omega A \sin \omega t$$

$$= \epsilon \omega A \left\{ \frac{A^2}{4} \sin 3\omega t - \left(1 - \frac{1}{4}A^2\right) \sin \omega t \right\}.$$

This equation can be satisfied for all t only if the coefficients of different sinusoidal terms vanish. The term in $\cos \omega t$ disappears if $\omega = 1$, and the term in $\sin \omega t$ is removed by taking $A = 2$. Thus we choose

$$\omega = 1, \qquad A = 2.$$

That still leaves the term containing $\sin 3\omega t$ unaccounted for because we have now fixed ω and A. To get rid of the extra term we need to try a higher approximation such as

$$x = A \cos \omega t + B_1 \cos 2\omega t + C_1 \sin 2\omega t$$

and repeat the process. We shall expect ω to be nearly 1, A to be nearly 2 and B_1, C_1 small. Even then, there will still be extra terms in the equation which we cannot dispose of. To tackle these we could contemplate adding further terms to the expression for x by introducing sinusoidal functions of $3\omega t, 4\omega t, \ldots$. However, we shall not go into this complication but stay with our first approximation in the belief that the extra terms will represent a small correction. According to our first approximation, there is a periodic motion

$$x = 2 \cos t$$

in which the amplitude 2 is in error by order ϵ and the error in the argument of the cosine is of order ϵ^2.

The influence of a small non-linearity has been radical. Instead of a simple harmonic motion in which the amplitude C can take any value the periodic motion has been restricted to the single amplitude 2.

Of course, we do not know whether this periodic state can ever be reached and the theory of local stability is no aid because the motion is nowhere near the critical point at $x = 0$, $y = 0$. Nevertheless, some progress can be made. Multiply (5.6.1) by $\dot{x}$ and integrate with respect to t from $t = \tau$ to $t = \tau + 2\pi$. Then

$$\tfrac{1}{2}\left[x^2 + y^2\right]_{\tau}^{\tau+2\pi} = \int_{\tau}^{\tau+2\pi} \epsilon(1 - x^2)\dot{x}^2 dt.$$

Although x is not known precisely, it can be expected to be substantially of the form $B\cos t$ for a time interval of 2π. The error in calculating the integral by this formula should not be more than order ϵ^2. Now

$$\int_{\tau}^{\tau+2\pi} \epsilon(1 - x^2)\dot{x}^2 dt = \int_{\tau}^{\tau+2\pi} \epsilon(1 - B^2 \cos^2 t)B^2 \sin^2 t\, dt$$
$$= \tfrac{1}{4}\pi\epsilon B^2(4 - B^2).$$

Thus, if $B > 2$, $x^2 + y^2$ is reduced after one period. Repeating the argument for each consecutive period we conclude that if the motion starts with $B > 2$ it must eventually arrive at the position where $B = 2$. Similarly, for a motion that begins with $B < 2$, $x^2 + y^2$ increases after each period and continues to do so until $B = 2$. Consequently, the system ends up in the specified periodic motion whatever the initial point.

A typical trajectory can be seen in Figure 5.6.1. Any closed trajectory that is eventually reached by a system is called a **limit cycle**. For van der Pol's equation $x^2 + y^2 = 4$ is a limit cycle, according to our first approximation.

The theory can be arranged to cover the more general

$$\ddot{x} + x = -\epsilon g(x, \dot{x}) \tag{5.6.2}$$

and the corresponding system. Put $x = A\cos \omega t$ and let

$$g(A\cos \omega t, -A\omega \sin \omega t) = a_1 \cos \omega t + b_1 \sin \omega t + a_2 \cos 2\omega t + b_2 \sin 2\omega t + \cdots.$$

Taking advantage of the orthogonality of the trigonometric functions and Section 3.7 we have

$$\int_{0}^{2\pi/\omega} g(A\cos \omega t, -A\omega \sin \omega t) \cos \omega t\, dt = \pi a_1/\omega,$$
$$\int_{0}^{2\pi/\omega} g(A\cos \omega t, -A\omega \sin \omega t) \sin \omega t\, dt = \pi b_1/\omega.$$

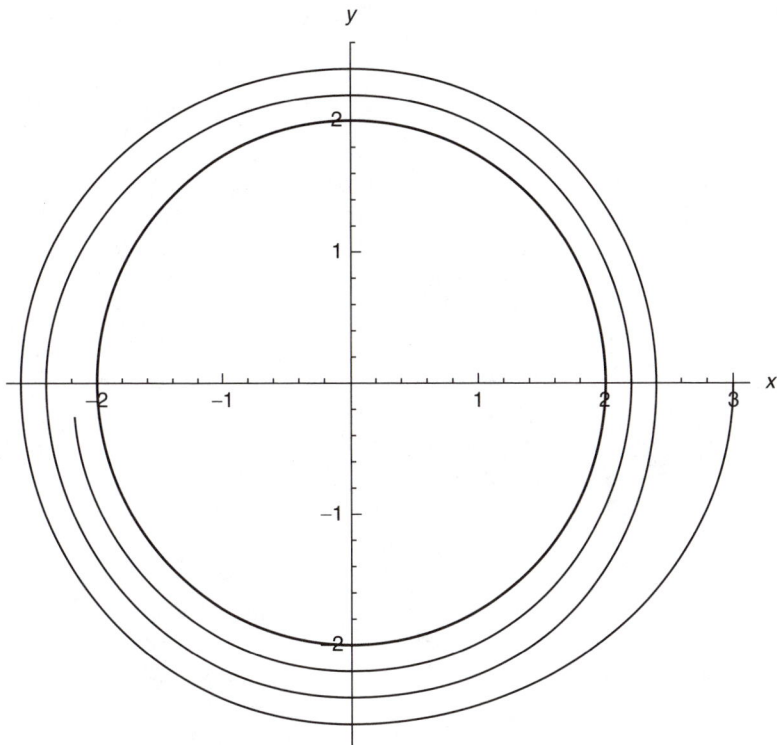

FIGURE 5.6.1: The limit cycle.

Since the left-hand side of (5.6.2) is $A(1 - \omega^2)\cos \omega t$ the coefficients of $\cos \omega t$ and $\sin \omega t$ can be made to agree by requiring that

$$1 - \omega^2 = -\omega a_1/A = -\frac{\epsilon \omega}{\pi A} \int_0^{2\pi/\omega} g(A\cos \omega t, -A\omega \sin \omega t)\cos \omega t\, dt, \quad (5.6.3)$$

$$0 = b_1 \pi/\omega = \int_0^{2\pi/\omega} g(A\cos \omega t, -A\omega \sin \omega t)\sin \omega t\, dt. \quad (5.6.4)$$

Equation (5.6.3) tells us that $\omega = 1 + O(\epsilon)$ and, taking advantage of this in (5.6.4), we obtain

$$\int_0^{2\pi} g(A\cos t, -A\sin t)\sin t\, dt = 0 \quad (5.6.5)$$

to determine A. Substitution back into (5.6.3) then leads to a more accurate determination of ω. Knowing ω and A enables us to locate the positions of any limit cycles.

To establish whether the system tends to a limit cycle we note that when $x = B \cos t$

$$\frac{1}{2}\left[x^2 + y^2\right]_0^{2\pi} = \epsilon B \int_0^{2\pi} g(B \cos t, -B \sin t) \sin t \, dt. \qquad (5.6.6)$$

It is too much to expect to be able to estimate the sign of the right-hand side of (5.6.6) in general, but it is possible to form an opinion of what happens near a limit cycle, i.e., when $B \approx A$ where A satisfies (5.6.5). We want the left-hand side of (5.6.6) to decrease if $B > A$ and to increase when $B < A$ if the trajectory is to approach the limit cycle from either side, i.e., we need the right-hand side of (5.6.6) to go from positive to negative values as B increases through A. This demands that the derivative of the right-hand side of (5.6.6) with respect to B shall be negative when $B = A$, i.e.,

$$\epsilon \int_0^{2\pi} \{g + [g_x]A \cos t - [g_y]A \sin t\} \sin t \, dt < 0$$

where

$$[g_x] = \left[\frac{\partial}{\partial x} g(x, y)\right], \qquad [g_y] = \left[\frac{\partial}{\partial y} g(x, y)\right]$$

and the substitution $x = A \cos t$, $y = -A \sin t$ is made after the derivatives have been performed. The first integrand gives a zero contribution by virtue of (5.6.5) but is retained because, by integration by parts,

$$\int_0^{2\pi} g \sin t \, dt = \int_0^{2\pi} \frac{\partial g}{\partial t} \cos t \, dt$$

$$= -\int_0^{2\pi} A \cos t \{[g_x] \sin t + [g_y] \cos t\} dt.$$

Hence the inequality becomes

$$\int_0^{2\pi} [g_y] dt > 0 \qquad (5.6.7)$$

since ϵ and A are both positive. The inequality (5.6.7) is the condition under which the system tends to go into the periodic motion of the limit cycle, i.e., the limit cycle is **stable**. If the inequality in (5.6.7) is reversed, i.e., the left-hand side is negative, the system would tend to depart from the limit cycle, which would then be **unstable**.

The number of possible limit cycles is fixed by the number of distinct positive values of A which satisfy (5.6.5). Their stability can be assessed by means of (5.6.7). One can thereby obtain an idea of which limit cycle (if any) will be attained under given initial conditions.

The theory expounded above has been based on ϵ being small and is plausible rather than rigorous. Notwithstanding this, it indicates what can happen.

To deal with cases when ϵ is not small, much more elaborate analysis is necessary and we shall confine ourselves to quoting two theorems.

LIMIT CYCLE CRITERION

In (5.6.2) let $g(x, \dot{x}) = \dot{x}G(x)$ where $G(x)$ is an even function of x such that $G(x) < 0$ for $|x| < 1$ and $G(x) > 0$ for $|x| > 1$. Suppose further that for some finite x_0 there is $G_0 > 0$ such that $G(x) \geq G_0$ for $|x| \geq x_0$. Then, for any $\epsilon > 0$, (5.6.2) has just one limit cycle and this limit cycle is stable.

POINCARÉ-BENDIXSON THEOREM

If there is a bounded region D in the (x, y)-plane such that any solution of the system

$$\dot{x} = f(x, y), \qquad \dot{y} = g(x, y)$$

that starts in D remains in D, then D contains either a stable critical point or a limit cycle.

The Poincaré-Bendixson theorem deals with a more general situation than the Limit Cycle criterion but provides less specific information. The reader should not, however, attempt to apply the Poincaré-Bendixson theorem to a system of more than two differential equations.

Limit cycles make their presence felt only with non-linear differential equations; they do not occur when the governing equations are linear. Caution should therefore be exercised in introducing a linear model for a natural phenomenon where the behaviour is essentially non-linear. At best it will describe local stability characteristics but it may give no clue as to what takes place globally or, worse, may suggest misleading conclusions.

The Poincaré-Bendixson theorem does not distinguish between the cases of a critical point and a limit cycle. It is sometimes helpful in deciding between them to observe that *a limit cycle must contain a critical point.* For example, take the limit cycle as the boundary of D in the Poincaré-Bendixson theorem. Then there must be a limit cycle or critical point inside. If there is no limit cycle the assertion is proved and if there is one we repeat the argument for that one.

5.7 Forced oscillations

The investigation of the effect of an external oscillatory disturbance on a non-linear system is very difficult. If the non-linearity is small some insight can be gained by the method of Section 5.6. To illustrate the technique and to display the new features that can be present we shall consider van der Pol's equation with a forcing term, namely

$$\ddot{x} - \epsilon(1 - x^2)\dot{x} + x = E \sin \Omega t \tag{5.7.1}$$

where E and Ω are constants, and ϵ is small.

When $E = 0$, previous theory indicates a stable limit cycle $x = 2\cos t$ and the system can be expected to go into a **self-excited oscillation**. If $\epsilon = 0$ but $E \neq 0$, there is a forced oscillation with harmonic time variation of argument Ωt (Section 2.6). When neither ϵ nor E is zero, both types of oscillation have to be allowed for.

Therefore, try

$$x = A\cos(\omega t + \gamma) + C\sin(\Omega t + \theta) \tag{5.7.2}$$

where the constants γ and θ have been incorporated to cover the influence of the self-excited and forced oscillations on each other. It is to be expected that ω is near 1 and, if Ω^2 is not near 1, that C is not far from $E/(1 - \Omega^2)$ with θ small. No estimate of A and γ can be made at this stage.

The substitution of (5.7.2) in (5.7.1) leads to

$$\begin{aligned}
A(1 - \omega^2)&\cos(\omega t + \gamma) + C(1 - \Omega^2)\sin(\Omega t + \theta) - E\sin\Omega t \\
= \epsilon\big[&\Omega C\big(1 - \tfrac{1}{2}A^2 - \tfrac{1}{4}C^2\big)\cos(\Omega t + \theta) \\
&+ \big(\tfrac{1}{4}A^2 + \tfrac{1}{2}C^2 - 1\big)\omega A\sin(\omega t + \gamma) \\
&- \tfrac{1}{4}A^2C(\Omega + 2\omega)\cos\{(\Omega + 2\omega)t + 2\gamma + \theta\} + \tfrac{1}{4}A^3\omega\sin 3(\omega t + \gamma) \\
&+ \tfrac{1}{4}A^2C(2\omega - \Omega)\cos\{(\Omega - 2\omega)t - 2\gamma + \theta\} + \tfrac{1}{4}\Omega C^3\cos 3(\Omega t + \theta) \\
&- \tfrac{1}{4}AC^2(\omega + 2\Omega)\sin\{(\omega + 2\Omega)t + \gamma + 2\theta\} \\
&+ \tfrac{1}{4}AC^2(2\Omega - \omega)\sin\{(\omega - 2\Omega)t + \gamma - 2\theta\}\big].
\end{aligned} \tag{5.7.3}$$

All the sinusoidal terms on the right-hand side of (5.7.3) have different time variations unless $\Omega = 0$, $\tfrac{1}{3}\omega$, ω or 3ω. The case $\Omega = 0$ has already been disposed of (being the same as $E = 0$) and, for the moment, the exceptional cases in which $\Omega = \tfrac{1}{3}\omega$, ω or 3ω will be ignored. On the two sides of (5.7.3) equate the coefficients of $\cos(\omega t + \gamma)$, $\sin(\omega t + \gamma)$, $\cos\Omega t$ and $\sin\Omega t$ to obtain

$$A(1 - \omega^2) = 0, \tag{5.7.4}$$

$$\big(\tfrac{1}{4}A^2 + \tfrac{1}{2}C^2 - 1\big)\omega A = 0, \tag{5.7.5}$$

$$C(1 - \Omega^2)\sin\theta - \epsilon\Omega C\big(1 - \tfrac{1}{2}A^2 - \tfrac{1}{4}C^2\big)\cos\theta = 0, \tag{5.7.6}$$

$$C(1 - \Omega^2)\cos\theta + \epsilon\Omega C\big(1 - \tfrac{1}{2}A^2 - \tfrac{1}{4}C^2\big)\sin\theta = E. \tag{5.7.7}$$

It is evident from (5.7.6) and (5.7.7) that

$$C = E\cos\theta/(1 - \Omega^2), \qquad \tan\theta = \epsilon\Omega\big(1 - \tfrac{1}{2}A^2 - \tfrac{1}{4}C^2\big)/(1 - \Omega^2).$$

Thus, so long as Ω^2 is not near $1, \theta$ is small and $\cos\theta$ may be replaced by unity. In other words, *the non-linearity has very little effect on the forced oscillation.* On the other hand, (5.7.4) and (5.7.5) imply that

$$\omega = 1, \qquad A = (4 - 2C^2)^{1/2}.$$

As C^2 increases from 0, A decreases from 2 to 0 and finally becomes imaginary for $C^2 > 2$. Imaginary values of A are not permitted and so, when $C^2 > 2$, the only possible solution to (5.7.4) and (5.7.5) is $A = 0$. The influence of the forcing term, therefore, is to reduce the amplitude of the self-excited oscillation when $C^2 < 2$ and *to extinguish it completely* when $C^2 \geq 2$, i.e., $E^2 \geq 2(1 - \Omega^2)^2$.

Thus, even in the unexceptional case, a strong enough external vibration can obliterate totally the self-excited oscillation.

Turn now to the exceptional case in which $\Omega = 3\omega$. Since ω is near 1 this can occur only when $\Omega = 3 + \delta$ and $\omega = 1 + \frac{1}{3}\delta$ with $|\delta| \ll 1$. The terms involving $\sin 3(\omega t + \gamma)$ and $\cos\{(\Omega - 2\omega)t - 2\gamma + \theta\}$ in (5.7.3) cannot now be neglected. Equate the coefficients of $\cos(\omega t + \gamma)$, $\sin(\omega t + \gamma)$, $\sin 3\omega t$ and $\cos 3\omega t$ on the two sides of (5.7.3); then

$$A(1 - \omega^2) + \tfrac{1}{4}\epsilon A^2 C\omega \cos(3\gamma - \theta) = 0, \tag{5.7.8}$$
$$\left(\tfrac{1}{4}A^2 + \tfrac{1}{2}C^2 - 1\right)\omega A - \tfrac{1}{4}A^2 C\omega \sin(3\gamma - \theta) = 0, \tag{5.7.9}$$
$$C(1 - 9\omega^2)\cos\theta - E + 3\epsilon\omega C\left(1 - \tfrac{1}{2}A^2 - \tfrac{1}{4}C^2\right)\sin\theta = \tfrac{1}{4}\epsilon A^3\omega \cos 3\gamma, \tag{5.7.10}$$
$$C(1 - 9\omega^2)\sin\theta - 3\epsilon\omega C\left(1 - \tfrac{1}{2}A^2 - \tfrac{1}{4}C^2\right)\cos\theta = \tfrac{1}{4}\epsilon A^3\omega \sin 3\gamma. \tag{5.7.11}$$

From (5.7.10) and (5.7.11), $C = -E/8$ to the first order and $\tan\theta$ is small so that the forced oscillation is virtually the same as in the absence of the non-linearity. In (5.7.8) and (5.7.9) put $\omega = 1 + \frac{1}{3}\delta$, put $A\cos(3\gamma - \theta) = \xi$ and put $A\sin(3\gamma - \theta) = \eta$. Then, if only dominant terms are retained,

$$-\tfrac{2}{3}\delta + \tfrac{1}{4}\epsilon C\xi = 0,$$
$$\xi^2 + \left(\eta - \tfrac{1}{2}C\right)^2 = 4 - 7C^2/4.$$

Consequently, ξ and η (and thereby A and γ) are determined by the intersection of a straight line and a circle. The circle is imaginary if $C^2 > 16/7$ and the line does not intersect the circle if $4 - 7C^2/4 < (8\delta/3\epsilon C)^2$. Therefore A is non-zero only if $4 - 7C^2/4 > (8\delta/3\epsilon C)^2$; this inequality cannot be satisfied unless $(\delta/\epsilon)^2 < 9/28$. Hence, if $(\delta/\epsilon)^2 < 9/28$ and $4 - 7C^2/4 > (8\delta/3\epsilon C)^2$,

$$A^2 = 4 - \tfrac{3}{2}C^2 \pm \left\{4C^2 - \tfrac{7}{4}C^4 - (8\delta/3\epsilon)^2\right\}^{1/2}. \tag{5.7.12}$$

Having found A, we can proceed to determine γ. However, γ occurs only in the form 3γ so that if γ_0 is a possible value so are $\gamma_0 + 2\pi/3$ and $\gamma_0 + 4\pi/3$. Thus, to a given amplitude of the self-excited oscillation there correspond three possible distinct phases.

According to this approximation when the magnitude of the forcing term is small, C^2 is not large enough for A to exist. Thus application of a small forcing term of three times the natural frequency extinguishes the self-excited oscillation. As the magnitude of the forcing term and C^2 grow there comes a point where A is non-zero and self-excited oscillations can occur. Further increase of C^2 will eventually reach a point when A disappears again so that the self-excited oscillation is absent when the magnitude of the forcing term is large enough. For the upper sign in (5.7.12), A will increase with C^2 to a maximum before dropping back to zero. The possibility of A increasing and of surpassing the value 2 are new aspects. For instance, $A \geq 2$ if $0 \leq C^2 \leq 1$ when $\delta = 0$. In general, $A \geq 2$ if $(8\delta/3\epsilon)^2 < 1$ and

$$-\{1 - (8\delta/3\epsilon)^2\}^{1/2} \leq 2C^2 - 1 \leq \{1 - (8\delta/3\epsilon)^2\}^{1/2}.$$

This magnification of the self-excited oscillation by the application of a vibration, which is an integer multiple of the self-excited, is known as **subharmonic resonance**.

Our investigation does not permit us to say which of the many oscillations that have been uncovered can be reached by the system starting from given initial conditions. A more refined analysis reveals that the negative square root in (5.7.12) corresponds to an unstable state, whereas the upper sign provides a stable state. Therefore, under conditions in which A is non-zero, the system tends to adopt the larger value of A; the phase will depend on the initial conditions.

The exceptional case $\Omega = \frac{1}{3}\omega$ may be discussed in a similar manner. As regards the special case $\Omega = \omega$, there is no necessity to have both terms in (5.7.2) and C can be placed equal to zero.

5.8   Appendix: existence theory

(a)   Single first-order equation

The aim of existence theory is to specify conditions under which one can be sure that there is a solution to a differential equation such as

$$\dot{y} = f(t, y). \tag{5.8.1}$$

There is no point in wasting analytical and computational effort on trying to find a solution when there is not one. Basically, there are two ways of demonstrating existence, **non-constructive** in which no attempt is made to show how one might arrive at a solution, and **constructive** in which a method for building up the solution is described. We shall consider only a constructive approach, one that lays the foundation for a numerical attack when that is desired.

The initial value problem for (5.8.1) seeks a solution such that $y = y_0$ at $t = t_0$ and this is the problem that we shall discuss in some detail when y_0 is a prescribed constant.

The purpose of the analysis is to show that, under specified conditions, when t does not stray too far from t_0 there is a solution y which does not differ from y_0 by more than a certain amount. So we consider what happens

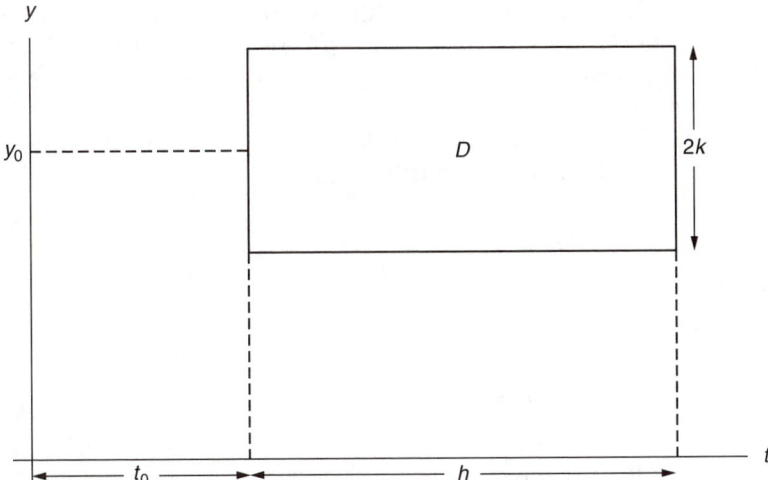

FIGURE 5.8.1: The domain of existence.

as t ranges from t_0 to $t_0 + h$ where h is positive (similar considerations apply when h is negative). In this range we are prepared to consider deviations of y from y_0 of magnitude k, i.e., we expect y to lie between $y_0 - k$ and $y_0 + k$. The points that originate values of $f(t, y)$ are then in the domain D of Figure 5.8.1. Suppose that f is bounded in D, say $|f| < M$; then we shall impose the restriction $h < k/M$. This can always be arranged by reducing h if necessary. The constraint is an expression of our expectation that the more t departs from t_0 the more y will deviate from y_0. Further conditions are placed on f in the following theorem.

EXISTENCE THEOREM I
Let $f(t, y)$ be a single-valued continuous function of t and y in D such that

(a) $|f(t, y)| < M$ *in* D,
(b) (**Lipschitz condition**)
$$|f(t, y) - f(t, y')| < K|y - y'|,$$

K being a finite constant, for any pair of points (t, y) and (t, y') in D. Then, for $h < k/M$, the differential equation (5.8.1) possesses one and only one continuous solution $y(t)$ in $t_0 \leq t \leq t_0 + h$ such that $y(t_0) = y_0$.

PROOF The proof starts by observing that the problem is equivalent to showing that

$$y(t) = y_0 + \int_{t_0}^{t} f(u, y(u))du \qquad (5.8.2)$$

has a solution. For, since f is bounded, the integral exists and tends to zero as $t \to t_0$ with the consequence that $y(t_0) = y_0$. Also, a derivative with respect to t of (5.8.2) returns, because of the assumed continuity of f, to (5.8.1). So it is sufficient to discuss (5.8.2).

Now solve (5.8.2) by iteration by making a series of approximations. First put $y(u) = y_0$ in the integral to generate $y_1(t)$ given by

$$y_1(t) = y_0 + \int_{t_0}^{t} f(u, y_0) du. \tag{5.8.3}$$

Now produce the sequence $y_n(t)$ defined by

$$y_n(t) = y_0 + \int_{t_0}^{t} f(u, y_{n-1}(u)) du. \tag{5.8.4}$$

In each approximation the right-hand side can be calculated and a practical mechanism of solution has been erected, provided that the iteration converges to a solution.

Note firstly that (5.8.3) implies that $y_1(t)$ is a continuous function of t and, since (u, y_0) is in D for $t < t_0 + h$,

$$|y_1(t) - y_0| < M(t - t_0) < Mh < k;$$

thus $(t, y_1(t))$ is in D for $t < t_0 + h$. The reasoning can now be repeated to show that $y_2(t)$ is a continuous function of t such that $(t, y_2(t))$ is in D for $t < t_0 + h$. It is then clear from (5.8.4) that $y_n(t)$ is a continuous function of t such that $(t, y_n(t))$ is in D for every n while $t < t_0 + h$.

Suppose now that

$$|y_n(t) - y_{n-1}(t)| < MK^{n-1}(t - t_0)^n/n! \tag{5.8.5}$$

for $t_0 \leq t \leq t_0 + h$, a result already proved for $n = 1$. Then, from (5.8.4),

$$|y_{n+1}(t) - y_n(t)| \leq \int_{t_0}^{t} |f(u, y_n(u)) - f(u, y_{n-1}(u))| du$$
$$< \int_{t_0}^{t} K|y_n(u) - y_{n-1}(u)| du$$

by the Lipschitz condition. Invoking our hypothesis, we have

$$|y_{n+1}(t) - y_n(t)| < MK^n \int_{t_0}^{t} (u - t_0)^n du/n!$$
$$< MK^n(t - t_0)^{n+1}/(n+1)!$$

which is the same as (5.8.5) except that n is replaced by $n+1$. Since it is true for $n = 1$ it follows by induction that (5.8.5) holds for every n.

Accordingly

$$\left| y_0 + \sum_{r=1}^{\infty} \{y_r(t) - y_{r-1}(t)\} \right| \leq |y_0| + \sum_{r=1}^{\infty} M K^{r-1} (t - t_0)^r / r!$$

$$\leq |y_0| + M e^{K(t-t_0)} / K$$

$$\leq |y_0| + M e^{Kh} / K$$

which reveals that the series on the left is absolutely and uniformly convergent (by the Weierstrass M-test) in $t_0 \leq t \leq t_0 + h$. But the sum of a uniformly convergent series of continuous functions is itself continuous and, since

$$y_0 + \sum_{r=1}^{\infty} \{y_r(t) - y_{r-1}(t)\} = y_n(t),$$

it follows that $\lim_{n \to \infty} y_n(t)$ exists and is a continuous function $\hat{y}(t)$ in $t_0 \leq t \leq t_0 + h$.

It remains to identify $\hat{y}$ as a solution of (5.8.2). Now

$$\hat{y}(t) = \lim_{n \to \infty} y_n(t) = y_0 + \lim_{n \to \infty} \int_{t_0}^{t} f(u, y_{n-1}(u)) du$$

$$= y_0 + \int_{t_0}^{t} f(u, \hat{y}(u)) du + \lim_{n \to \infty} \int_{t_0}^{t} \{f(u, y_{n-1}(u)) - f(u, \hat{y}(u))\} du.$$

By the Lipschitz condition, the magnitude of the last integral does not exceed

$$K \int_{t_0}^{t} |y_{n-1}(u) - \hat{y}(u)| du < K(t - t_0) \max_u |y_{n-1}(u) - \hat{y}(u)|$$

which tends to zero as $n \to \infty$. Equation (5.8.2) has been recovered.

The existence of a continuous solution has now been verified and to complete the theorem it is necessary to show there is no other. Suppose, in fact, there were another solution $Y(t)$ such that $Y(t_0) = y_0$, which is continuous in $t_0 \leq t \leq t_0 + H$ with $H \leq h$ and $|Y(t) - y_0| < k$. Then

$$Y(t) = y_0 + \int_{t_0}^{t} f(u, Y(u)) du.$$

Hence, if

$$|Y(t) - y_{n-1}(t)| < K^{n-1} k (t - t_0)^{n-1} / (n-1)!,$$

$$|Y(t) - y_n(t)| \leq \int_{t_0}^{t} |f(u, Y(u)) - f(u, y_{n-1}(u))| du$$

$$\leq k K^n (t - t_0)^n / n!.$$

This inequality is valid by induction if it is true for $n = 1$. But

$$|Y(t) - y_1(t)| \leq K \int_{t_0}^t |Y(u) - y_0| du \leq Kk(t - t_0)$$

and so the result for $n = 1$ holds. Letting $n \to \infty$ in the inequality we have $Y(t) = \lim_{n \to \infty} y_n(t) = \hat{y}(t)$ and uniqueness has been established. ☐

(b) System of first-order equations

The theory of the preceding section generalises to the system

$$\dot{y}_m = f_m(t, y_1, y_2, \ldots, y_n) \quad (m = 1, \ldots, n) \tag{5.8.6}$$

under the initial conditions $y_m = y_{m0}$ at $t = t_0$. The region D is not so simple to depict since it is a rectangular parallelepiped in space of $n + 1$ dimensions, because each y_m may change by a different amount from its initial value as t moves from t_0. So D is defined by $t_0 \leq t \leq t_0 + h$, $|y_m - y_{m0}| \leq k_m$ ($m = 1, \ldots, n$). Analogous to the conditions for a single equation the restrictions $|f_m| < M$ and $h < k_m/M$ for $m = 1, 2, \ldots$ are imposed.

EXISTENCE THEOREM II

Let $f_m(t, y_1, \ldots, y_n)$ be single-valued continuous functions of t, $y_1, \ldots, y_n$ in D such that for $m = 1, \ldots, n$

(a) *$|f_m(t, y_1, \ldots, y_n)| < M$ in D,*

(b) **(Lipschitz condition)**

$|f(t, y_1, \ldots, y_n) - f(t, y_1', \ldots, y_n')| < K_1|y_1 - y_1'| + \cdots + K_n|y_n - y_n'|$,

$K_1, \ldots, K_n$ being finite constants, for any $(t, y_1, \ldots, y_n)$ and $(t, y_1', \ldots, y_n')$ in D. Then, for $h < k_m/M$ ($m = 1, \ldots, n$) the system (5.8.6) possesses one and only one set of continuous solutions $y_1(t), \ldots, y_n(t)$ in $t_0 \leq t \leq t_0 + h$ such that $y_m(t_0) = y_{m0}(m = 1, \ldots, n)$.

PROOF The method of proof runs parallel to that for a single equation. It begins with

$$y_m(t) = y_{m0} + \int_{t_0}^t f_m(u, y_1(u), \ldots, y_n(u)) du \quad (m = 1, \ldots, n)$$

and an iteration is performed according to

$$y_{mr}(t) = y_{m0} + \int_{t_0}^t f_m(u, y_{1,r-1}(u), \ldots, y_{n,r-1}(u)) du.$$

In view of the similarity to the single equation, details will be omitted. ☐

(c) Differential equation of order n

The above theory can be applied to the n differential equation

$$\frac{d^n y}{dt^n} = f\left(t, y, \frac{dy}{dt}, \ldots, \frac{d^{n-1}y}{dt^{n-1}}\right) \tag{5.8.7}$$

with the initial conditions

$$y = y_{10}, \quad dy/dt = y_{20}, \quad \ldots, \quad d^{n-1}y/dt^{n-1} = y_{n0}$$

at $t = t_0$. Make the substitutions

$$y(t) = y_1(t), \quad \frac{dy}{dt} = y_2(t), \quad \ldots, \quad \frac{d^{n-1}y}{dt^{n-1}} = y_n(t).$$

The system

$$\dot{y}_1 = y_2, \quad \dot{y}_2 = y_3, \quad \ldots, \quad \dot{y}_{n-1} = y_n,$$
$$\dot{y}_n = f(t, y_1, y_2, \ldots, y_n)$$

is obtained. When this is compared with (5.8.6) we see that

$$f_m(t, y_1, \ldots, y_n) = \begin{cases} y_{m+1} & (m = 1, \ldots, n-1) \\ f(t, y_1, \ldots, y_n) & (m = n). \end{cases}$$

The f_m for $m = 1, \ldots, n-1$ obviously satisfy the conditions stated in Existence Theorem II. Therefore, if we make f_n comply with these conditions, that theorem is available for (5.8.7). Accordingly, we have the next theorem.

EXISTENCE THEOREM III

If $f(t, y_1, \ldots, y_n)$ is continuous and

$$|f(t, y_1, \ldots, y_n) - f(t, y_1', \ldots, y_n')| < K_1|y_1 - y_1'| + \cdots + K_n|y_n - y_n'|$$

the differential equation (5.8.7) has one and only one continuous solution $y(t)$ such that $dy/dt, \ldots, d^{n-1}y/dt^{n-1}$ are continuous for $t_0 \le t \le t_0 + h$ and such that $y, dy/dt, \ldots, d^{n-1}y/dt^{n-1}$ take given values at $t = t_0$.

It may be remarked that, if (5.8.7) is linear,

$$f(t, y_1, \ldots, y_n) = g(t) - a_0(t)y_1 - a_1(t)y_2 - \cdots - a_{n-1}(t)y_n.$$

Since this is a polynomial in $y_1, \ldots, y_n$ the conditions of Existence Theorem III are met except at those values of t where one or more of $g, a_0, \ldots, a_{n-1}$ are not continuous. Hence, a linear differential equation has a unique continuous solution to the initial value problem, provided that t_0 is not a point where there is lack of continuity on the part of $g, a_0, \ldots, a_{n-1}$.

5.9   Appendix: computing trajectories

A useful idea of the performance of trajectories can be obtained from two instructions in MATHEMATICA. Suppose that we are interested in the trajectory starting at $(x0, y0)$ in the phase plane and watching how it evolves as t goes from 0 to $t0$. For the differential equations associated with Figure 5.5.1 issue the instructions

```
Sol=NDSolve [{x'[t]==y[t],y'[t]==(1-x[t]2̂)/2,
x[0]==x0,y[0]==y0},{x,y},{t,t0}]

ParametricPlot[Evaluate[{x[t],y[t]}/.sol],{t,0,t0},
PlotRange->All]
```

A trace of the trajectory should appear then on your screen with axes superimposed (unless you have altered the default setting for axes). You may need to experiment with the value of $t0$. Above some value that depends on the differential equations and initial conditions, the computer will report

```
NDSolve::mxst:

Maximum number of steps reached at the point t=t1
```

where t1 is some numerical value in the interval $(0, t0)$. Sufficient information for your purpose may be available by reducing $t0$ until the warning message disappears. If that simple solution is not satisfactory there are two options open to you, assuming that the differential equation is such that NDSolve is giving accurate results. One is to start the trajectory from a point that has been found already; for instance, you could determine such a point by

```
{x[t2],y[t2]}/.sol
```

with t2 equal to t1 or a bit less. Two pieces of a trajectory can always be combined into a single figure by means of Show. The alternative is to permit NDSolve to take more steps. This can be done by adding an option such as

```
MaxSteps->2000
```

to NDSolve, though you may have to keep adjusting the number until there is no warning message.

Exercises

5.1 At what points are the conditions of Existence Theorem I not satisfied for

(a) $\dot{y} = y$,

(b) $\dot{y} = |y|^{1/2}$,

(c) $t^3 \dot{y} = y$,

(d) $t\dot{y} = y$,

(e) $\frac{1}{2}t\dot{y} = t^2 + y$?

Are the points where failure occurs singular points?

5.2 Sketch the trajectories in the phase plane of

(a) $\dot{x} = 5x + 2y$,
$\dot{y} = 2x + 2y$,

(b) $\dot{x} = 6x + 12y$,
$\dot{y} = 3x + y$,

(c) $\dot{x} = 4x + 5y$,
$\dot{y} = -5x - 4y$.

(d) $\dot{x} = y - 2x$,
$\dot{y} = 4x - 5y$,

(e) $\dot{x} = 5x - 5y$,
$\dot{y} = 5x - 3y$,

(f) $\dot{x} = 5x + 4y$,
$\dot{y} = 9x$,

(g) $\dot{x} = 4x + 13y$,
$\dot{y} = -13x - 6y$,

(h) $\dot{x} = 5x - 13y$,
$\dot{y} = 13x - 5y$.

5.3 Draw the trajectories in the phase plane of

$$\frac{d^2x}{dt^2} + 3\frac{dx}{dt} + 2x = 0.$$

5.4 Examine what critical points arise in the phase plane for

$$\frac{d^2x}{dt^2} + 2b\frac{dx}{dt} + ax = 0$$

where a and b are constants $(a \neq 0)$.

5.5 Discuss the trajectories of

$$\ddot{x} - 4\dot{x} + 40x = 0$$

in the phase plane by making the substitution $x = \rho\cos\phi$, $y = \rho\sin\phi$.

5.6 Examine the possibility of periodic solutions of

$$c\ddot{x} + (2 + 3ax + 4bx^2)x = 0$$

where a, b and c are constants, c being positive.

5.7 Sketch the trajectories in the phase plane of

$$\ddot{x} + \sin x = 0.$$

5.8 In the differential equation

$$(1 + a^2x^2)\ddot{x} + (b + a^2\dot{x}^2)x = 0$$

a and b are constants. Discuss the behaviour of the solution.

5.9 Discuss the trajectories of

$$\ddot{\theta} = (\cos\theta - \mu)\sin\theta$$

for $-\pi \le \theta \le \pi$, the constant μ ($\neq 1$) being positive.

5.10 In the differential equation

$$\ddot{x} + \lambda\dot{x}^2 + x = 0$$

the constant λ is positive. Obtain dy/dx where $y = \dot{x}$ and hence derive the differential equation satisfied by $w = y^2$. Find the trajectories and determine when they are closed curves.

5.11 Prove that

$$\ddot{x} - \epsilon(1 - x^4)\dot{x} + x = 0$$

has a stable limit cycle and that its amplitude is $2^{3/4}$ (≈ 1.68) when $0 < \epsilon \ll 1$.

5.12 Show that

$$\ddot{z} + f(\dot{z}) + z = 0$$

becomes

$$\ddot{x} + f'(x)\dot{x} + x = 0$$

on putting $\dot{z} = x$. Hence obtain information about the limit cycle of **Rayleigh's equation**

$$\ddot{z} + \epsilon(\tfrac{1}{3}\dot{z}^2 - \dot{z}) + z = 0$$

when $0 < \epsilon \ll 1$.

5.13 In **Duffing's equation**

$$\ddot{x} + D\dot{x} + x + \beta x^3 = E \sin \Omega t$$

$\Omega = 1+\delta$ and D, β, E and δ are all small. Show that, if an approximate solution is $x = A\sin\{(1+\delta)t + \theta\}$, A satisfies

$$\left(\tfrac{3}{4}\beta A^3 - 2\delta A\right)^2 + D^2 A^2 = E^2.$$

5.14 After the substitution $x = \rho\cos\phi$, $y = \rho\sin\phi$ the differential equations of a system are

$$\dot{\rho} = (\rho - 1)(a + \sin^2 \phi), \quad \dot{\phi} = 1.$$

Show that there are stable limit cycles when $a < -\tfrac{1}{2}$. What happens when $a > -\tfrac{1}{2}$?

Chapter 6

Mathematics of Heart Physiology

6.1 The local model

We begin by recalling the three basic features of the heart beat cycle upon which a mathematical model is to be developed. These are:

(a) the model must exhibit an equilibrium state corresponding to diastole;

(b) it should also contain a threshold for triggering the electrochemical wave emanating from the pacemaker causing the heart to contract into systole;

(c) it should reflect the rapid return to the equilibrium state.

As in Chapter 4, we suppose the important quantities that model these features to be x, a typical muscle fibre length, which will necessarily depend on the time t, and b, an electrical control variable which governs the electrochemical wave, and which also depends on t.

In order to make any reasonable progress in modelling the heart beat cycle, we must assume that the mathematical equations must be drawn from a particular class of equations which have in them the ability to describe at least the main features of heart physiology. The class we choose here is taken from the class of autonomous dynamical systems in two degrees of freedom. In other words we seek a mathematical model of the form

$$\frac{dx}{dt} = f(x, b),$$

$$\frac{db}{dt} = g(x, b). \tag{6.1.1}$$

For this model to satisfy the first important quality (a) above, we ask that it has a unique **stable rest state**. Suppose this occurs at the critical point (b_0, x_0) of (6.1.1). That is, (b_0, x_0) satisfy the equations

$$f(x_0, b_0) = g(x_0, b_0) = 0. \tag{6.1.2}$$

If the system (6.1.1) is linearised, as in Chapter 5, about the rest state (b_0, x_0)

we have

$$\frac{dx}{dt} = f(x_0, b_0) + (x - x_0)\frac{\partial f(x_0, b_0)}{\partial x} + (b - b_0)\frac{\partial f(x_0, b_0)}{\partial b}$$
$$+ \ higher \ order \ terms,$$
$$\frac{db}{dt} = g(x_0, b_0) + (x - x_0)\frac{\partial g(x_0, b_0)}{\partial x} + (b - b_0)\frac{\partial g(x_0, b_0)}{\partial b}$$
$$+ \ higher \ order \ terms.$$

That is,

$$\frac{dx}{dt} = a_{11}(x - x_0) + a_{12}(b - b_0) + higher \ order \ terms,$$
$$\frac{db}{dt} = a_{21}(x - x_0) + a_{22}(b - b_0) + higher \ order \ terms, \qquad (6.1.3)$$

where

$$a_{11} = \frac{\partial f}{\partial x}, \quad a_{12} = \frac{\partial f}{\partial b}, \quad a_{21} = \frac{\partial g}{\partial x}, \quad a_{22} = \frac{\partial g}{\partial b},$$

each evaluated at $x = x_0, b = b_0$.

From Chapter 5 we know that the local stability of the system (6.1.3) in the neighbourhood of (b_0, x_0) is governed by the roots λ_1, λ_2 of the quadratic equation

$$\lambda^2 - \lambda(a_{11} + a_{22}) + a_{11}a_{22} - a_{12}a_{21} = 0. \qquad (6.1.4)$$

Furthermore we know that our system is stable near (b_0, x_0) if the real parts of λ_1, λ_2 are negative. We shall in fact assume a little more and suppose λ_1, λ_2 to be real and negative, thus eliminating any undesirable oscillatory behaviour in our mathematical model. This requires that

$$a_{11} + a_{22} < 0,$$
$$a_{11}a_{22} - a_{12}a_{21} > 0. \qquad (6.1.5)$$

The inequalities (6.1.5) provide an infinite number of ways in which to choose the constants $a_{ij}, i, j = 1, 2$. Thus, to make further progress towards a possible model, we make some hypotheses, which, of course, ask for experimental confirmation. The hypotheses we make are the following:

H1: The rate of change of muscle fibre contraction depends, at any particular instant, on the tension of the fibre and on the chemical control.

H2: The chemical control changes at a rate directly proportional to muscle fibre extension.

For the moment we shall not specify under hypothesis H1 precisely how tension should enter into our model but focus on the implication of hypothesis H2. Mathematically, H2 is simply stated through the equation

$$\frac{db}{dt} = x - x_0. \qquad (6.1.6)$$

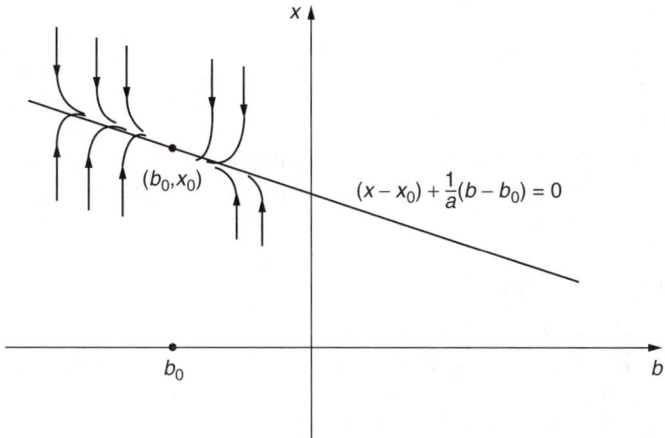

FIGURE 6.1.1: Phase plane for the local model (6.1.9).

In other words the function $g(x, b)$ in (6.1.5) is linear and independent of b and furthermore $a_{21} = 1, a_{22} = 0$. Consequently the inequalities (6.1.5) simplify to

$$a_{11} < 0, \quad a_{12} < 0. \tag{6.1.7}$$

The third of the features required by our mathematical model is that it should reflect the rapid return to the equilibrium state. This quality suggests that a_{11} be large and negative and since $\frac{db}{dt}$, by hypothesis H2, is proportional to $x - x_0$ we expect a_{12} to be large and negative as well.

With this information we try

$$a_{11} = -\frac{a}{\epsilon}, \quad a_{12} = -\frac{1}{\epsilon}, \tag{6.1.8}$$

where a and ϵ are positive constants with ϵ small.

Putting these remarks and conclusions together, we arrive at the "local linearised model"

$$\epsilon \frac{dx}{dt} = -a(x - x_0) - (b - b_0),$$
$$\frac{db}{dt} = x - x_0. \tag{6.1.9}$$

This model is depicted in the phase plane (Figure 6.1.1) and is constructed as follows. Take $b_0 < 0, x_0 > 0$, then along the line

$$(x - x_0) + \frac{1}{a}(b - b_0) = 0, \quad \frac{dx}{dt} = 0.$$

For

$$(x - x_0) + \frac{1}{a}(b - b_0) > 0, \quad \frac{dx}{dt} < 0,$$

that is, x is decreasing and for $x > x_0, \frac{db}{dt} > 0$ and so b is increasing.

Furthermore, away from the line

$$(x - x_0) + \frac{1}{a}(b - b_0) = 0,$$

$|\frac{dx}{dt}|$ is large because of the presence of the small factor ϵ and consequently the trajectories are largely vertical. Notice also that as the line

$$(x - x_0) + \frac{1}{a}(b - b_0) = 0$$

is approached from below along a trajectory with $b < b_0$, the trajectory crosses the line and is horizontal at the point of crossing. A similar phenomenon occurs when $b > b_0$ and we approach the line from above. In fact (b_0, x_0) is a stable node. Figure 6.1.1 summarises these remarks.

6.2 The threshold effect

In this section we seek to modify the local mathematical model (6.1.9) so as to incorporate the feature (b) of Section 6.1, namely that the model should contain a threshold for triggering the electrochemical wave. Of course we must specify what is meant by the term **threshold**.

To help define these terms and so improve on the model (6.1.9), we recall from Chapter 4 the following facts.

During the heart beat cycle there are two equilibrium states, namely diastole and systole. The diastolic state is included in our model (6.1.9) whereas the systolic state is not, and so some modifications to include this are required. Furthermore, we observe that the pacemaker triggers off an electrochemical wave, which spreads slowly over the atria causing the muscle fibres to contract fairly slowly. The wave then spreads rapidly causing the whole ventricle to contract into systole.

This discussion suggests that, during the first part of the heart beat cycle, the muscle fibre x contracts slowly at first and then at a certain point rapidly contracts further until the systolic equilibrium state is achieved. We shall call the point at which the rapid contraction occurs the **threshold**. While this contraction is going on, the chemical control variable b will be rising to a value b_1 corresponding to systole.

The remarks we have made so far are summarised in Figure 6.2.1.

Following contraction into systole, the muscle fibres rapidly relax and return the heart to diastole and thus complete the cycle. The return is depicted schematically by the dotted line in Figure 6.2.1.

The problem then is to seek to modify the equations (6.1.9) so as to incorporate the general features shown in Figure 6.2.1. Such a modification is

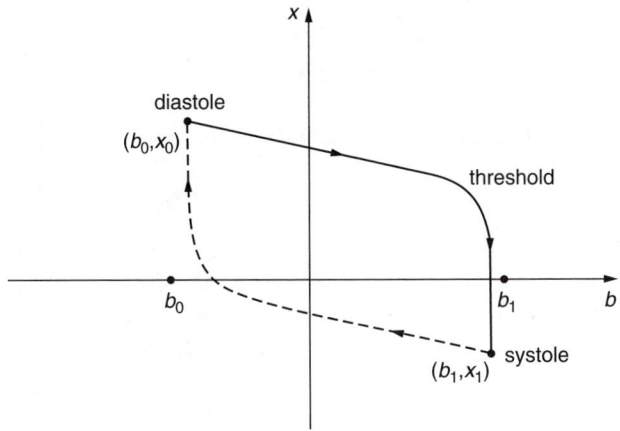

FIGURE 6.2.1:   The heart beat cycle.

largely developed by trial. We propose that the model (6.1.9) is modified to the non-linear form

$$\epsilon \frac{dx}{dt} = -a(x - x_0) - (b - b_0) - (x - x_0)^3 - 3x_0(x - x_0)^2,$$
$$\frac{db}{dt} = x - x_0. \tag{6.2.1}$$

In order to assist with the development, we wish to write (6.2.1) in the form

$$\epsilon \frac{dx}{dt} = -(x^3 - Tx + b), \quad T > 0,$$
$$\frac{db}{dt} = x - x_0. \tag{6.2.2}$$

If we use Taylor's theorem to expand the right-hand side of the first equation appearing in (6.2.2) about the point (b_0, x_0), we can compare the system (6.2.2) with the system (6.2.1). Thus from (6.2.2)

$$\epsilon \frac{dx}{dt} = -\big[x_0^3 - Tx_0 + b_0 + \big(3x_0^2 - T\big)(x - x_0) + 3x_0(x - x_0)^2 \\ + (x - x_0)^3 + (b - b_0)\big],$$

and so we deduce that

$$x_0^3 - Tx_0 + b_0 = 0$$

and

$$3x_0^2 - T = a,$$

from which

$$T = 3x_0^2 - a$$

and the control b_0 is expressed in terms of x_0 and a via the relation

$$b_0 = 2x_0^3 - ax_0.$$

When tension has been appropriately identified, we shall see in the following section that the system (6.2.2) can be considered as contributing to a "local" model for the heart beat cycle in that hypotheses H1 and H2 are included in the system and the qualities depicted in Figure 6.2.1 are almost accounted for.

6.3   The phase plane analysis and the heart beat model

In order to develop the final form of our model of the heart beat cycle, we begin by describing the phase plane portrait associated with the system (6.2.2).

To begin with consider the curve

$$x^3 - Tx + b = 0 \qquad (6.3.1)$$

shown by the solid line $ABCD$ in Figure 6.3.1.

On this curve $dx/dt = 0$ and the flow is parallel to the b axis. Near the equilibrium (b_0, x_0) the configuration has that which is depicted in Figure 6.1.1. Furthermore, the direction of flow is determined by the second member of the system (6.2.2).

If we are above the cubic curve (6.3.1), i.e., where

$$x^3 - Tx + b > 0,$$

then we see from (6.2.2) that dx/dt is large and negative and so the flow is largely vertically downwards, whereas if we are below the curve (6.3.1) the flow is vertically upwards. Thus, in general, the phase portrait consists of vertical trajectories except in the neighbourhood of the cubic curve.

We notice also that the trajectories always flow towards the portions AB and CD of (6.3.1) but always away from the portion BC. It is natural therefore to refer to the segments AB, CD as attractors for the flow and the segment BC as a repellor for the flow. The points B, C are important in that they can be associated with the threshold phenomenon discussed in the previous section.

If we now compare Figures 6.3.1 and 6.2.1, we see that the lower attractor CD gives rise to trajectories which follow a path back to (b_0, x_0) similar to the dotted line of Figure 6.2.1. Therefore the systolic state could be represented by a point (b_1, x_1) on the attractor CD.

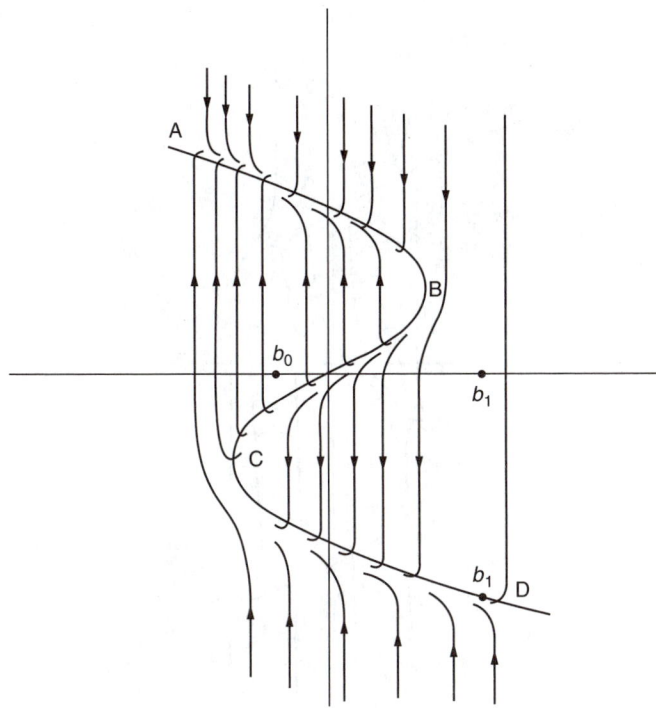

FIGURE 6.3.1:   Phase portrait for the system (6.2.2).

However, the flow on the upper attractor AB is not in accordance with the general feature depicted by the solid line in Figure 6.2.1 in that there is no mechanism for providing a trajectory corresponding to a smooth change of the chemical control from b_0 to b_1. The mechanism for doing this could be thought of as due to the pacemaker and thus is not present in the model (6.2.2). If we switch the equilibrium state from (b_0, x_0) to (b_1, x_1), then the flow on the upper attractor AB would provide trajectories similar to the solid line in Figure 6.2.1. This is achieved for the alternative model

$$\epsilon \frac{dx}{dt} = -(x^3 - Tx + b),$$

$$\frac{db}{dt} = x - x_1, \tag{6.3.2}$$

the phase portrait of which is shown in Figure 6.3.2.

In this figure we see that the flow along the upper attractor AB does conform to the general behaviour of the solid line in Figure 6.2.1 and the point B can be identified with the threshold. However, the flow along the lower attractor CD cannot now be identified with the systolic state of the heart beat cycle.

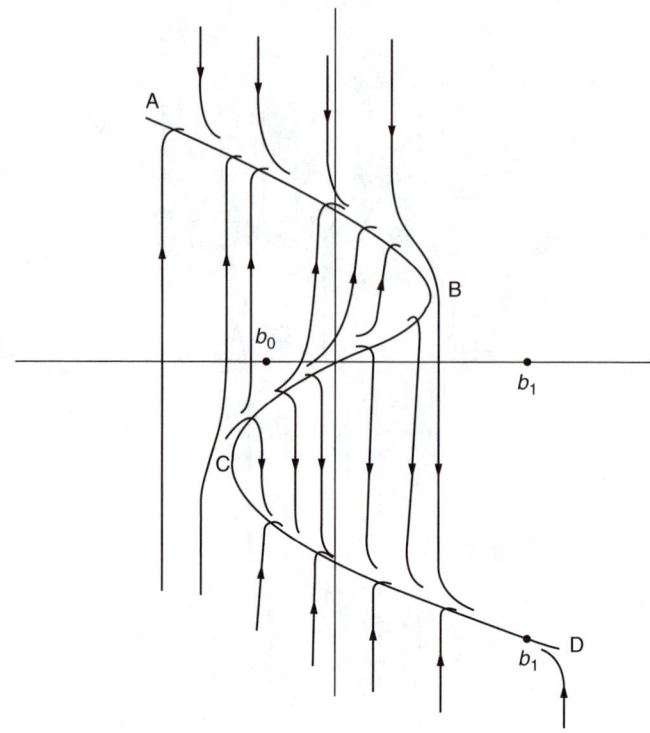

FIGURE 6.3.2: Phase portrait for the system (6.3.2).

What is required is a modification of either (6.2.2) or (6.3.2), which incorporates both the desirable features of Figures 6.3.1 and 6.3.2, but excludes those features that do not conform to the known physiological behaviour of the heart beat cycle. A model that does satisfy the above criteria is the following:

$$\epsilon \frac{dx}{dt} = -(x^3 - Tx + b),$$

$$\frac{db}{dt} = (x - x_0) + (x_0 - x_1)u, \tag{6.3.3}$$

where u is a control variable associated with the pacemaker and is defined as follows:

$u = 1$ for (a) $b_0 \leq b \leq b_1$ and for those values of x for which $x^3 - Tx + b > 0$ and for (b) $b > b_1$ and all values of x.

$u = 0$ otherwise.

The system (6.3.3) will be called the **heart beat equations** and the corresponding phase portrait is shown in Figure 6.3.3, where the dashed line indicates the heart beat cycle.

We should not leave the discussion leading up to the model (6.3.3) without drawing the reader's attention to the mathematical problem of proving that

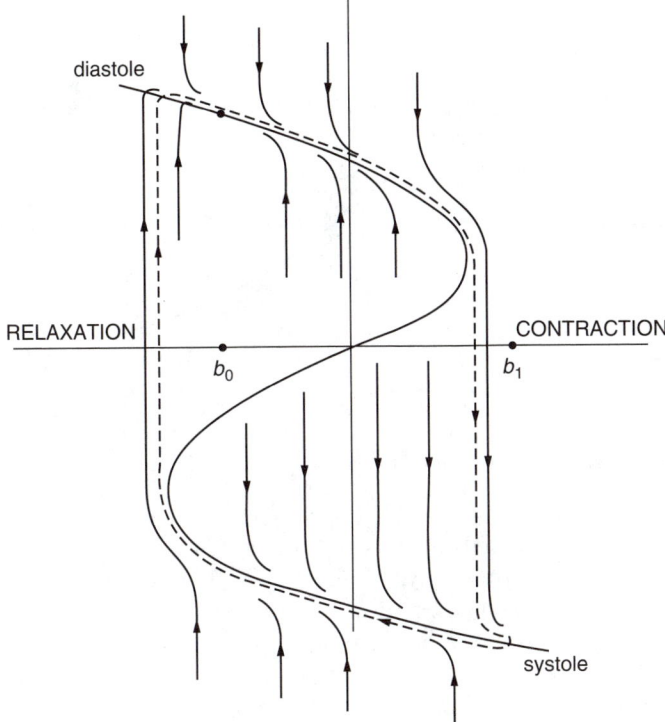

FIGURE 6.3.3: Phase portrait for the heart beat equations.

there is a closed trajectory, shown dotted in Figure 6.3.3, corresponding to a complete heart beat cycle. This is a difficult problem and will not be pursued here.

Finally we have to identify the contribution of tension to our model. To help us here, imagine that the muscle fibres are not under tension so that contraction into systole is slow and rather sluggish. In other words, we would not expect the sharp downward trajectories shown in Figure 6.3.3 but rather the slow behaviour shown in Figure 6.3.4.

Referring back to our model (6.3.3) we see that a portrait corresponding to Figure 6.3.4 is obtained if we set $T = 0$. We therefore identify T as tension. In the following section we shall consider this further when we discuss the predictions of the model (6.3.3) compared with known physiological facts.

6.4 Physiological considerations of the heart beat cycle

In this section we expand upon some of the physiological aspects of the heart beat outlined in Chapter 4 and see how they may be interpreted in the context of the heart beat equations (6.3.3).

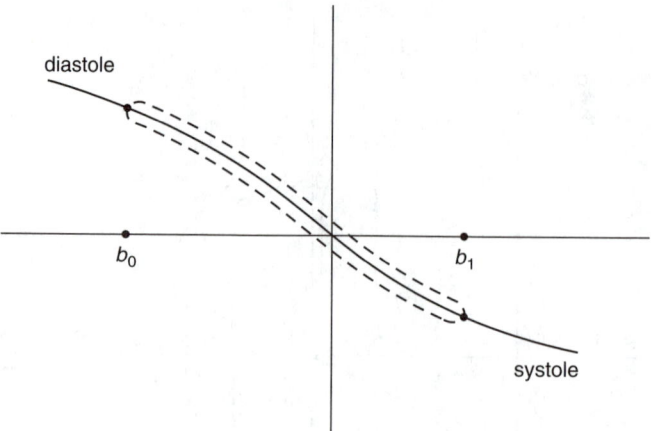

FIGURE 6.3.4: Low-tension heart beat.

Rybak in 1957 originated the following experiment. If the heart of a frog is taken out, then, not surprisingly, it ceases to beat. However, if it is then cut open into a flat membrane and subject to slight tension, it begins to beat once more and continues to do so for some hours. Alternatively, if the pacemaker is removed, then again beating stops. Rybak's experiment is analogous to setting $T = 0$ in (6.3.3), leading to the sluggish cycle shown in Figure 6.3.4. As T increases slightly we again obtain a figure similar to Figure 6.3.3, but this time most of the work is done by the pacemaker wave in moving the control b from b_0 to b_1 and providing the large amount of squeezing necessary to contract the heart into systole. In fact, this low-tension heart beat corresponds to the small atrial beat described in Chapter 4. Another relevant feature is known as Starling's law. This says that the more the muscle fibres are stretched before beating the more forcible is the beat. Therefore suppose excitement of one form or another causes adrenalin to be injected into the blood stream; the adrenalin then causes the arteries to contract and the pulse rate increases, which in turn causes the blood pressure to rise and the atria to push more blood into the ventricles. Starling's law describes how the stretched ventricles give a larger beat, overcoming the increased arterial back-pressure and circulating the blood faster. Starling's law is present in the model (6.3.3) if T is large, but not too large.

Finally, if the ventricles are overstretched beyond a certain point, as can happen, for example, when someone with high blood pressure receives a sudden shock, then the heart may fail to beat, or only beat feebly and cardiac failure may result. This particular aspect can be realised in our mathematical model if we increase the tension so much that the threshold extends beyond the systolic equilibrium point b_1, that is, when

$$T > \left(27b_1^2/4\right)^{1/2}. \tag{6.4.1}$$

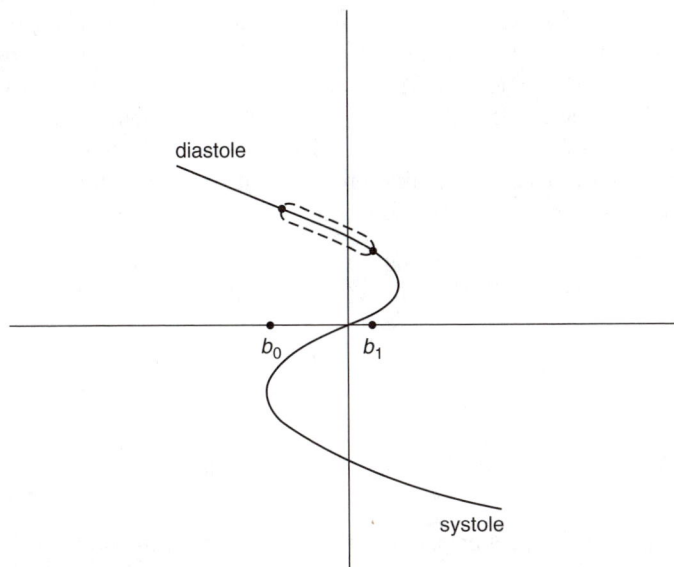

FIGURE 6.4.1: The overstretched heart leading to cardiac failure.

This condition is such as to prevent the trigger from reaching threshold and so the muscle fibres remain in diastole (see Figure 6.4.1). In other words the heart does not beat and cardiac failure has occurred.

6.5 A model of the cardiac pacemaker

The heart beat originates in the sino-atrial node, a region of cells which have the capability of depolarising spontaneously towards the threshold, firing and then recovering. However, the mechansm underlying the pacemaker wave is not fully understood, even though it is the basis of the field of electrocardiography.

In this section we propose a tentative mathematical model of the pacemaker firing mechanism that incorporates the control variable u introduced in the previous section.

Let us suppose that the pacemaker is characterised by a state $y, 0 \le y \le 1$, which satisfies the ordinary differential equation

$$\frac{dy}{dt} = -\gamma y + u, \tag{6.5.1}$$

and that, when $y = 1$, the pacemaker fires and jumps back to $y = 0$. We regard γ as a small positive number less than $1/4$ and intrinsic to the pacemaker.

Furthermore, it is natural to suppose that the motion of y is periodic, of period T say. Thus if $\{t_n\}$ denotes the set of firing times we suppose $u = 1$ for $t_n < t \leq t_n + T/2$ and $u = 0$ for $t_n + T/2 < t \leq t_{n+1}$ where $t_{n+1} - t_n = T$. In other words, the times $\{t_n\}$ correspond to the heart being in diastole while the times $\{t_n + T/2\}$ correspond to the heart being in systole.

Equation (6.5.1) is a simple differential equation of integrating factor type and is readily solved to give the solution

$$y(t) = \int_{t_n}^{t} e^{-\gamma(t-t')} u(t') dt', \tag{6.5.2}$$

where $y(t_n) = 0$ and $t_n \leq t \leq t_{n+1}$. The equation for the firing times is therefore

$$1 = \int_{t_n}^{t_{n+1}} e^{-\gamma(t_{n+1}-t')} u(t') dt',$$

which because of the properties of $u(t)$ can be written in the form

$$1 = \int_{t_n}^{t_n+T/2} e^{-\gamma(t_{n+1}-t')} dt',$$

i.e.,

$$\gamma = e^{-\gamma(t_{n+1}-t_n-T/2)} - e^{-\gamma(t_{n+1}-t_n)}$$

or

$$\gamma = e^{-\gamma T/2} - e^{-\gamma T}. \tag{6.5.3}$$

From (6.5.3) we can compute the period T as follows: setting $z = e^{-\gamma T/2}$ we can write (6.5.3) as the quadratic equation

$$z^2 - z + \gamma = 0,$$

the solutions of which are

$$z = \frac{1}{2}(1 \pm \sqrt{(1 - 4\gamma)}),$$

i.e.,

$$2e^{-\gamma T/2} = (1 \pm \sqrt{(1 - 4\gamma)}). \tag{6.5.4}$$

Since we expect γT to be small, we choose the positive root in (6.5.4), and if we neglect terms of order higher than γ^2 we have

$$2e^{-\gamma T/2} \approx 2 - 2\gamma - 2\gamma^2,$$

which further approximates to

$$2(1 - \gamma T/2) \approx 2 - 2\gamma - 2\gamma^2$$

or

$$T \approx 2 + 2\gamma.$$

Thus if T is known, this equation can be used to estimate the parameter γ, or conversely if γ is known from experiment then T can be estimated.

6.6 Notes

This chapter has been largely motivated by the work of E.C. Zeeman. Section 6.5 was inspired by the paper of B.W. Knight, Dynamics of encoding in a population of neurons, *J. Gen. Physiol.*, **59**, 734–766, 1972.

Exercises

6.1 Verify the phase portrait (Figure 6.1.1) for the "local model" (6.1.9) by solving the system for x and b as functions of t.

6.2 Provide a full phase plane analysis for the model (6.2.2).

6.3 Provide a full phase plane analysis for the heart beat equations (6.3.3).

6.4 Verify the statements in Section 6.4 regarding Rybak's experiment and Starling's law by analysing the behaviour of the phase plane trajectories for the heart beat equations as the tension T varies between $T = 0$ and $T = (27b_1^2/4)^{1/2}$.

6.5 If the "cubic" term in the first member of the heart beat equations (6.3.3) is replaced by a piecewise linear expression to that depicted in Figure 4.4.4, a simplified model is obtained. Specifically, consider the system

$$\epsilon \frac{dx}{dt} = -F(x, b),$$
$$\frac{db}{dt} = (x - x_0) + (x_0 - x_1)u,$$

where

$$F(x,b) = x + b + \sqrt{T}, \quad x < -\frac{1}{2}\sqrt{T},$$

$$= x - b, \quad -\frac{1}{2}\sqrt{T} \leq x \leq \frac{1}{2}\sqrt{T},$$

$$= x + b - \sqrt{T}, \quad x > \frac{1}{2}\sqrt{T}.$$

Provide a complete analysis of this model and compare the results with those of the system (6.3.3).

Chapter 7

Mathematics of Nerve Impulse Transmission

7.1 Excitability and repetitive firing

In this chapter we make a study of some of the principle properties of the simplified model of nerve impulse transmission (4.4.7) due to FitzHugh and Nagumo.

For the space clamped case the model is

$$\frac{du}{dt} = u(1-u)(u-a) - w + I(t),$$

$$\frac{dw}{dt} = bu - \gamma w, \qquad (7.1.1)$$

where $0 < a < 1, b > 0, \gamma \geq 0$ and $I(t)$ is the total membrane current, which may be an arbitrary function of time. If the space clamp is removed then the model becomes

$$\frac{\partial u}{\partial t} = \frac{\partial^2 u}{\partial x^2} + u(1-u)(u-a) - w,$$

$$\frac{\partial w}{\partial t} = bu - \gamma w. \qquad (7.1.2)$$

To begin with we wish to determine whether the model (7.1.1) exhibits the important **threshold** property mentioned in Chapter 4. In mathematical terms this property is the same as asking whether the system (7.1.1) is **excitable**.

Consider the ordinary differential equation,

$$\frac{dy}{dt} = y(1-y)(y-a), \qquad (7.1.3)$$

where $y = 0, a, 1$ are rest states. This equation has stable rest states at $y = 0, 1$ and an unstable rest state at $y = a$. These statements are easily checked by observing that if initially $y(0) < a$ then $\frac{dy}{dt} < 0$ and so $y(t) \to 0$, whereas if $y(0) > a$ then $\frac{dy}{dt} > 0$ and $y(t) \to 1$. The implication of this is that for "small" (i.e., $y(0) < a$) initial data the solution is attracted to the rest state

$y = 0$, whereas if the initial data is "large" (i.e., $y(0) > a$) then the solution is attracted to $y = 1$. We call the parameter a the **threshold** and call the equation (7.1.3) **excitable**.

Let us now see if a similar property is present in the system (7.1.1). For simplicity we only consider the case where $I(t) = 0$. The case $I(t) \neq 0$ will be discussed later. With $I(t) = 0$ the system (7.1.1) can be studied using the techniques developed in Chapter 5. First of all it is an important requirement that (7.1.1) has a unique rest state. That is, on setting

$$\frac{du}{dt} = \frac{dw}{dt} = 0,$$

we require the pair (i.e., the null clines)

$$w = u(1 - u)(u - a),$$
$$bu = \gamma w \qquad (7.1.4)$$

to have the unique solution $(u, w) = (0, 0)$. That is, the equation

$$u(1 - u)(u - a) = \frac{b}{\gamma}u$$

must have the single solution $u = 0$. For this to be the case the quadratic equation,

$$u^2 - (1 + a)u + a + \frac{b}{\gamma} = 0,$$

can only have complex roots. That is, the parameters a, b, γ must be restricted so that

$$(1 - a)^2 < 4\frac{b}{\gamma}, \quad \gamma > 0. \qquad (7.1.5)$$

Notice that if $\gamma = 0$ then (7.1.1) has the unique rest state $(0, 0)$ without restriction on the parameters a and b.

Linearising (7.1.1) about $(0, 0)$ results in the system

$$\frac{du}{dt} = -au - w,$$
$$\frac{dw}{dt} = bu - \gamma w. \qquad (7.1.6)$$

As in Chapter 5 we look for solutions of the form $u = \alpha \exp \lambda t, w = \beta \exp \lambda t$. Substituting these into equations (7.1.6) leads to the requirement that

$$(a + \lambda)\alpha + \beta = 0,$$
$$(\gamma + \lambda)\beta - b\alpha = 0,$$

which has non-trivial solutions α and β only if

$$\lambda^2 + (a + \gamma)\lambda + b + a\gamma = 0. \qquad (7.1.7)$$

Since $(b + a\gamma)$ and $(a + \gamma)$ are positive, it follows that the roots of (7.1.7) have negative real parts and so we conclude that the rest state $(0, 0)$ is locally stable. Such an analysis does not help to determine global properties of the trajectories. Nevertheless let us try to use the ideas followed in treating the simple problem (7.1.3). We consider the system

$$\frac{du}{dt} = u(1 - u)(u - a) - w,$$

$$\frac{dw}{dt} = bu - \gamma w, \qquad (7.1.8)$$

subject to super threshold initial data

$$u(0) = u_0, \quad a < u_0 < 1, \quad w(0) = 0.$$

It follows from this that initially $\frac{du}{dt} > 0$, $\frac{dw}{dt} > 0$ and so (u, w) moves upwards in the positive quadrant of the phase plane and further away from $(0, 0)$. Notice that as u increases $\frac{du}{dt}$ will decrease until it reaches the null-cline $w = u(1 - u)(u - a)$ where $\frac{du}{dt} = 0$ but $\frac{dw}{dt}$ is still increasing. Beyond this point $\frac{du}{dt} < 0$ but $\frac{dw}{dt}$ increases until the trajectory meets the null-cline $\gamma w = bu$, where $\frac{dw}{dt} = 0$ and $\frac{du}{dt} < 0$. Continuing this argument we see that the trajectory returns to again meet the null-cline $w = u(1 - u)(u - a)$ at D where again $\frac{du}{dt} = 0$ and $\frac{dw}{dt} < 0$. From here, either the trajectory progresses directly to $(0, 0)$, if the solutions of (7.1.7) are real and negative, or spirals towards $(0, 0)$, if the solutions of (7.1.7) are complex with negative real parts as shown in Figure 7.1.1.

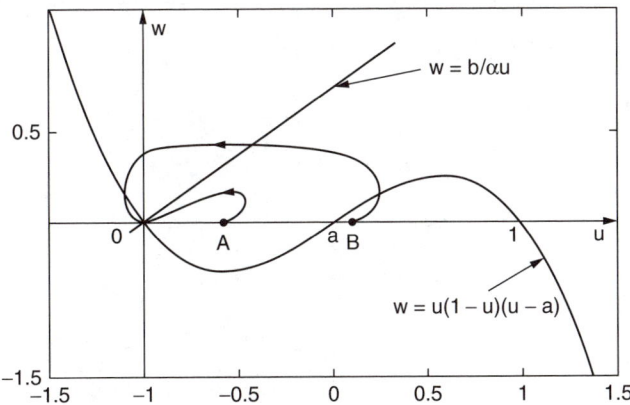

FIGURE 7.1.1: Global phase portrait for the system (7.1.8).

The same sequence of arguments is used if the initial data are chosen so that

$$u(0) = u_0 > 1, \qquad w(0) = 0.$$

Although this is not a conclusive proof, it is enough to convince one that the system (7.1.1) is excitable and that a is the threshold parameter.

Let us now consider the system in which the current $I(t)$ is not zero but set at some non-zero value I, which may be positive or negative. Whether $I < 0$ or $I > 0$, the resulting system is still excitable. The only difference being that the unique rest state is a solution to

$$u(1 - u)(u - a) - w + I = 0,$$
$$bu = \gamma w, \qquad (7.1.9)$$

and is no longer at $(0,0)$. That is, for $I > 0$ the null-clines are as shown in Figure 7.1.2(a) whereas for $I < 0$ the null-clines are as shown in Figure 7.1.2(b).

Of particular interest here is the case when $I > 0$. In their prize-winning work, Hodgkin and Huxley observed that on applying a constant current to the axon, repetitive firing of the action potential was observed. It is therefore of interest to see whether a similar phenomena is present in the FitzHugh-Nagumo system. Mathematically we ask whether the sysytem (7.1.1) has periodic orbits or limit cycles. A natural tool for exploring this is to use the Poincaré-Bendixson theorem discussed in Chapter 5, Section 5.6. To illustrate the ideas involved we consider the system

$$\frac{du}{dt} = u(1 - u^2) - w,$$
$$\frac{dw}{dt} = u. \qquad (7.1.10)$$

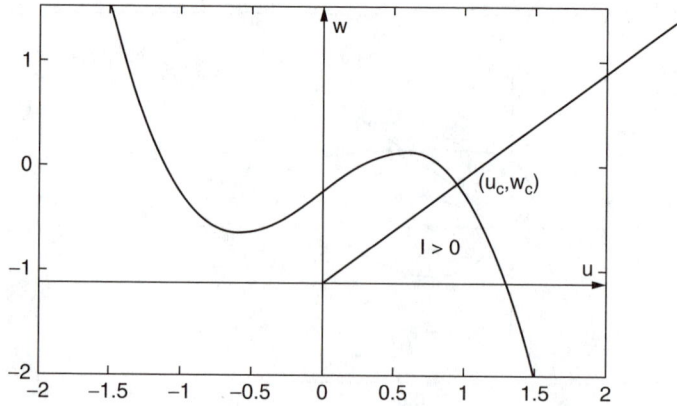

FIGURE 7.1.2(a):   Null-clines for $I > 0$.

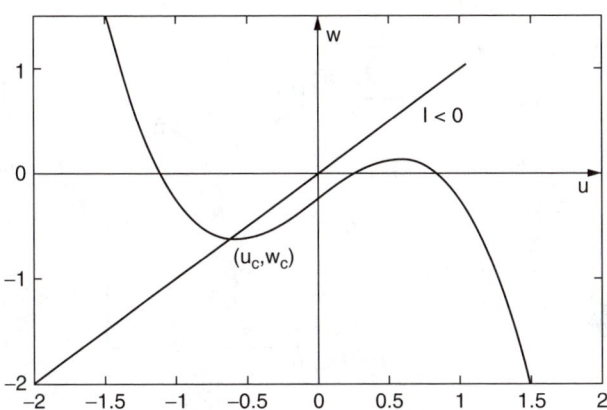

FIGURE 7.1.2(b): Null-clines for $I < 0$.

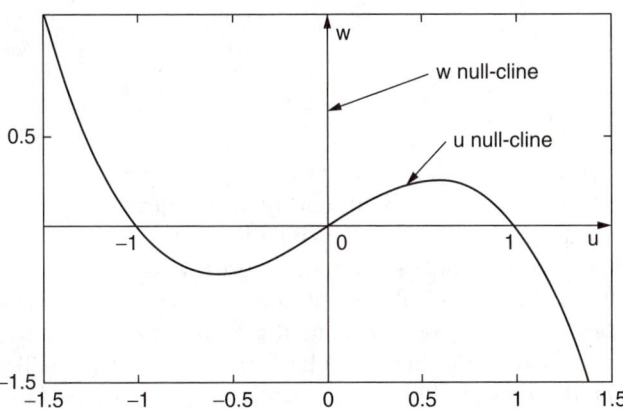

FIGURE 7.1.3: Null-clines of the system 7.1.10.

This system has the unique rest state $(u, w) = (0, 0)$ and the null-clines are shown in Figure 7.1.3.

Linearising (7.1.10) about $(0, 0)$ we have the system

$$\frac{du}{dt} = u - w,$$
$$\frac{dw}{dt} = u \qquad (7.1.11)$$

from which the characteristic determinant is

$$\begin{vmatrix} 1 - \lambda & -1 \\ 1 & -\lambda \end{vmatrix} = 0,$$

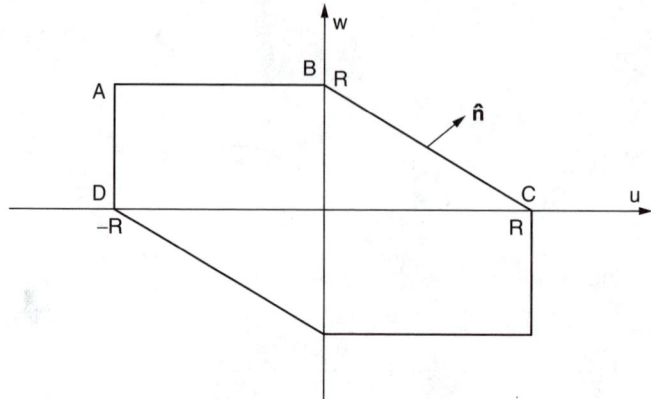

FIGURE 7.1.4:   Region Ω for the system 7.1.10.

or

$$\lambda^2 - \lambda + 1 = 0,$$

from which we conclude that $(0,0)$ is an unstable focus. This means that trajectories starting in a neighbourhood of $(0,0)$ move away from $(0,0)$. To be able to make use of the Poincaré-Bendixson theorem we need to show that there is a larger region Ω surrounding $(0,0)$ within which all trajectories starting in Ω remain in Ω for all time. Choosing such a region is often an art. A suitable choice for the problem at hand is that shown in Figure 7.1.4.

Notice that since on replacing (u, w) by $(-u, -w)$ leaves (7.1.10) unchanged, it is sufficient to consider Figure 7.1.4 with $u \geq 0$. On the face AB, $w = R, u \leq 0$ and $\frac{dw}{dt} = u \leq 0$, with zero only occurring at $u = 0$. That is, for any R, trajectories always enter Ω along AB. Similarly on the face $AD, u = -R, 0 \leq w \leq R$ and $\frac{du}{dt} = -R(1 - R^2) - w > 0$ for R sufficiently large. Thus we conclude that trajectories cross AD in the positive u direction for R large enough.

On the face $w = R - u$,

$$\frac{du}{dt} = u(1 - u^2) - R + u,$$
$$\frac{dw}{dt} = u$$

and so

$$\frac{dw}{du} = \frac{u}{u(1 - u^2) + u - R}.$$

If $\frac{dw}{du} > -1$ for $0 \le u \le R$, for R sufficiently large, then trajectories will cross in Ω along BC. Hence we need

$$\frac{dw}{du} = \frac{u}{u(1-u^2) + u - R} > -1,$$

for all $0 \le u \le R$. This means that $u < -u(1-u^2) - u + R$, or

$$H(u) \equiv 3u - u^3 - R < 0,$$

for $0 \le u \le R$.

Now $H(u)$ can be rewritten in the form

$$H(u) = -(u+2)(u-1)^2 + 2 - R < 0,$$

if $R \ge 2$. So we conclude that if R is sufficiently large then all trajectories of (7.1.10) remain in Ω. It now follows from the Poincaré-Bendixson theorem that all trajectories converge to a limit cycle contained in Ω.

An alternative way of considering the flow along the segment BC is to argue as follows:

Since $\mathbf{n} = \frac{1}{\sqrt{2}}(1,1)$ is the outward unit normal vector to BC and $(\frac{du}{dt}, \frac{dw}{dt})$ is tangential to all trajectories, then where a trajectory crosses the boundary BC, the flow will be inwards if

$$\mathbf{n} \cdot \left(\frac{du}{dt}, \frac{dw}{dt}\right) < 0, \quad w = R - u. \tag{7.1.12}$$

That is, we require

$$(1,1) \cdot (u(1-u^2) - R + u, u) = 3u - u^3 - R,$$
$$= H(u) < 0,$$

for $0 \le u \le R$ as above.

We now return to the system (7.1.1) with $I(t) = I > 0$ and constant. The system has the unique rest point (u_c, w_c) given by the solution to (7.1.8). For excitability we again wish to find an annular region Ω in the (u, w)-plane that does not contain the rest point (u_c, w_c) and for which the Poincaré-Bendixson theorem can be applied. Suppose we construct a large circle of radius R in the (u, w)-plane. We want to show that for R sufficiently large the flow is always directed inwards. To do this set

$$u = r \cos\theta, \qquad w = r \sin\theta. \tag{7.1.13}$$

Then

$$\frac{du}{dt} = \frac{dr}{dt}\cos\theta - r\frac{d\theta}{dt}\sin\theta,$$
$$= u(1-u)(u-a) - w + I,$$
$$\frac{dw}{dt} = \frac{dr}{dt}\sin\theta + r\frac{d\theta}{dt}\cos\theta,$$
$$= bu - \gamma w,$$

and so

$$\frac{dr}{dt} = u(1-u)(u-a)\cos\theta - w\cos\theta + I\cos\theta + bu\sin\theta - \gamma w\sin\theta,$$
$$= r\cos^2\theta(1 - r\cos\theta)(r\cos\theta - a) - r\sin\theta\cos\theta + I\cos\theta$$
$$+ br\sin\theta\cos\theta - \gamma r\sin^2\theta. \tag{7.1.14}$$

Now let $r \to \infty$. Due to the presence of the cubic term in u on the right-hand side of (7.1.14) we have $\frac{dr}{dt} \leq 0$, for all θ. From the Poincaré-Bendixson theorem we claim that there will be a limit cycle in the phase plane if the rest point (u_c, w_c) is an unstable node or focus. To investigate this we linearise (7.1.1) about (u_c, w_c).

Set

$$u = u_c + \xi,$$
$$w = w_c + \eta, \tag{7.1.15}$$

and expand $f(u) \equiv u(1-u)(u-a)$ as a Taylor series about (u_c, w_c) giving

$$f(u) = f(u_c) + f'(u_c)\xi + O(\xi^2). \tag{7.1.16}$$

The linearised system is

$$\begin{pmatrix} \dfrac{d\xi}{dt} \\ \dfrac{d\eta}{dt} \end{pmatrix} = \begin{pmatrix} f'(u_c) - \lambda & -1 \\ b & -\gamma - \lambda \end{pmatrix} \begin{pmatrix} \xi \\ \eta \end{pmatrix}. \tag{7.1.17}$$

The characteristic matrix is therefore

$$\begin{vmatrix} f'(u_c) - \lambda & -1 \\ b & -\gamma - \lambda \end{vmatrix} = 0,$$

or

$$\lambda^2 + (\gamma - f'(u_c))\lambda + b - f'(u_c)\gamma = 0. \tag{7.1.18}$$

For (u_c, w_c) to be unstable we need

$$\gamma - f'(u_c) < 0 \quad b - f'(u_c)\gamma > 0,$$
$$\gamma < f'(u_c) < b/\gamma. \tag{7.1.19}$$

An interpretation of the inequality (7.1.19) is that the slope of the u null-cline at (u_c, w_c) must be less than the slope of the w null-cline at that point. Finally we conclude from (7.1.19) and the Poincaré-Bendixson theorem that there is a limit cycle. That is, the FitzHugh-Nagumo system exhibits repetitive firing as originally observed by Hodgkin and Huxley.

7.2   Travelling waves

We now investigate travelling wave solutions to the system (7.1.2). Such solutions are of the form

$$u = \phi(x + ct), \qquad w = \psi(x + ct), \tag{7.2.1}$$

subject to the requirement that, with $\xi = x + ct$,

$$\lim_{|\xi| \to \infty} \phi(\xi) = \lim_{|\xi| \to \infty} \phi''(\xi) = \lim_{|\xi| \to \infty} \psi(\xi) = 0. \tag{7.2.2}$$

Among the many problems to be investigated in relation to travelling waves are the following:

(a) provide analytic evidence to support the conjecture that the potential $\phi(x + ct)$ has a form similar to that depicted in Figure 4.4.2;

(b) obtain estimates for the wave speed c.

In the course of investigating these problems, we shall see that the parameters a, b and γ entering in the system (7.1.2) cannot be chosen arbitrarily but must satisfy some simple constraints.

As a first step to analysing the above problems we substitute the supposed solutions (7.2.1) into (7.1.2) to obtain the coupled system of ordinary differential equations

$$\phi'' = c\phi' - \phi(1 - \phi)(\phi - a) + \psi,$$
$$c\psi' = b\phi - \gamma\psi, \tag{7.2.3}$$

where the primes denote differentiation with respect to ξ. If we set $\theta = \phi'$ then (7.2.3) can be written as the system

$$\theta' = c\theta - \phi(1 - \phi)(\phi - a) + \psi,$$
$$\phi' = \theta,$$
$$\psi' = \frac{b}{c}\phi - \frac{\gamma}{c}\psi. \tag{7.2.4}$$

The critical points of this system corresponding to "rest states" of the system (7.2.4) are given by

$$\left(0, \phi_i, \frac{b}{\gamma}\phi_i\right), \quad i = 1, 2, 3, \tag{7.2.5}$$

provided $\gamma > 0$ and ϕ_i is a root of the equation

$$x\left[(x - a)(1 - x) - \frac{b}{\gamma}\right] = 0. \tag{7.2.6}$$

An important requirement of the system (7.2.4) is that it has a unique "rest state" and this can only be achieved if the cubic equation (7.2.6) has only the real root $x = 0$. In other words the quadratic equation

$$x^2 - (1 + a)x + a + \frac{b}{\gamma} = 0$$

must have complex roots. As argued in Section 7.1 we know that this leads to the restriction

$$(1 - a)^2 < 4\frac{b}{\gamma}, \quad \gamma > 0. \tag{7.2.7}$$

Again if $\gamma = 0$ then (7.2.4) has the unique rest state $(0,0)$ without any restrictions on the parameters a, b.

A complete "phase plane" analysis of the system (7.2.4) is quite difficult due to the fact that it is a third order system rather than a second order one. The methods we employ in this situation will, as shown in the next section, be somewhat different from those discussed in Chapter 5. As a prelude to this consider the special case in which $b = 0$. Here the system (7.2.4) partially decouples in that ψ now satisfies the equation

$$\psi' = -\frac{\gamma}{c}\psi,$$

from which it follows that

$$\psi = A \exp\left(\frac{\gamma\xi}{c}\right).$$

For such a solution to satisfy the conditions (7.2.2) it is clear that $A = 0$. In other words $\psi \equiv 0$ and the system (7.2.4) simplifies to the second order system

$$\theta' = c\theta - \phi(1 - \phi)(\phi - a),$$
$$\phi' = \theta. \tag{7.2.8}$$

In the (ϕ, θ) phase plane there are three finite singular points, namely

$$(0,0), \quad (a,0), \quad (1,0). \tag{7.2.9}$$

Following the treatment of Chapter 5, we see that $(0,0)$ and $(1,0)$ are saddle-points whereas $(a,0)$ is a centre if $c = 0$, a repulsive spiral for $0 < c < 2\sqrt{[a(1-a)]}$ or a repulsive node if $c \geq 2\sqrt{[a(1-a)]}$ (see Figure 7.2.1). To provide a complete, that is global, analysis of the (ϕ, θ) phase plane is quite a formidable task and will not be pursued here. However, a glance at Figure 7.2.1 leads one to ask whether there is a trajectory which leaves $(0,0)$ and enters $(1,0)$. The answer to this is certainly "yes" and one such trajectory

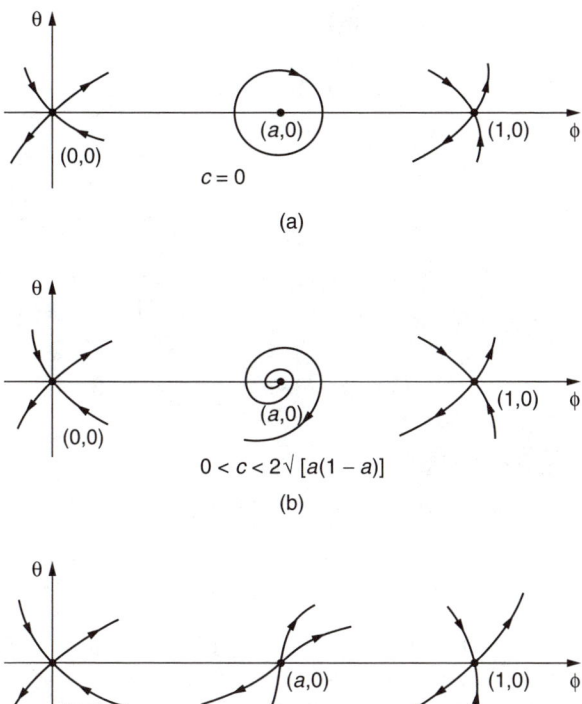

FIGURE 7.2.1: Local behaviour of the phase plane.

is provided by the answer to Exercise 4.8. That is, if $c = \sqrt{2}(\frac{1}{2} - a)$, $0 < a < \frac{1}{2}$, we have the solution

$$\phi(\xi) = \frac{1}{1 + \exp -\xi/\sqrt{2}}, \qquad \theta(\xi) = \frac{\exp -\xi/\sqrt{2}}{\sqrt{2}(1 + \exp -\xi/\sqrt{2})^2}. \qquad (7.2.10)$$

In fact this is the only trajectory that connects the points $(0,0)$, $(1,0)$.

7.3 Qualitative behaviour of travelling waves

If we look at Figure 4.4.2 we observe that the graph of the potential crosses the ξ axis at one point. Evidence that supports the same type of behaviour for ϕ can be demonstrated as follows. Integrate (7.2.3) over the range

$-\infty < \xi < \infty$ and use the conditions (7.2.2) to obtain

$$\int_{-\infty}^{\infty} \phi(1-\phi)(\phi-a)d\xi = \int_{-\infty}^{\infty} \psi \, d\xi$$

and

$$\int_{-\infty}^{\infty} \psi \, d\xi = \frac{b}{\gamma} \int_{-\infty}^{\infty} \phi \, d\xi$$

provided $\gamma \neq 0$. Eliminating ψ between these identities shows that

$$\int_{-\infty}^{\infty} \left(\phi(1-\phi)(\phi-a) - \frac{b}{\gamma}\phi \right) d\xi = 0.$$

This means that the function $\phi(1-\phi)(\phi-a) - \frac{b}{\gamma}\phi$ must change signs and, since we require the inequality (7.2.7) to hold, it follows that the equation

$$\phi(1-\phi)(\phi-a) - \frac{b}{\gamma}\phi = 0$$

has only the root $\phi = 0$. In other words $\phi = 0$ for at least one value of ξ. This argument of course does not preclude the possibility that $\phi = 0$ for more than one value of ξ and so does not completely confirm Figure 4.4.2.

In what follows we assume that ϕ and ψ are sufficiently well behaved for all integrals that occur to converge.

Multiply the first of equations (7.2.3) by ϕ and the second by ψ, and then integrate the resulting expressions with respect to ξ from $-\infty$ to ∞. This together with the conditions (7.2.2) gives

$$\int_{-\infty}^{\infty} \phi^2(1-\phi)(\phi-a)d\xi = \int_{-\infty}^{\infty} (\phi')^2 d\xi + b \int_{-\infty}^{\infty} \phi\psi \, d\xi,$$

$$\gamma \int_{-\infty}^{\infty} \psi^2 d\xi = \int_{-\infty}^{\infty} \phi\psi \, d\xi. \tag{7.3.1}$$

Next we repeat the process but this time we multiply by ϕ'' and ψ, respectively. The result is

$$c \int_{-\infty}^{\infty} (\phi)'^2 d\xi = -b \int_{-\infty}^{\infty} \phi'\psi \, d\xi,$$

$$c \int_{-\infty}^{\infty} (\psi)'^2 d\xi = \int_{-\infty}^{\infty} \phi\psi' \, d\xi. \tag{7.3.2}$$

But

$$\int_{-\infty}^{\infty} \phi'\psi \, d\xi = - \int_{-\infty}^{\infty} \phi\psi' d\xi$$

and so from (7.3.2) we find that

$$\int_{-\infty}^{\infty} (\phi)'^2 d\xi = b \int_{-\infty}^{\infty} (\psi)'^2 \, d\xi. \qquad (7.3.3)$$

Finally, multiply the second of equations (7.2.3) by ϕ and integrate, using (7.3.1), to get

$$\int_{-\infty}^{\infty} (\phi)^2 d\xi = c \int_{-\infty}^{\infty} \phi\psi' d\xi + \gamma^2 \int_{-\infty}^{\infty} (\psi)^2 \, d\xi. \qquad (7.3.4)$$

If we now eliminate $\int_{-\infty}^{\infty} (\phi)'^2 d\xi, \int_{-\infty}^{\infty} (\psi)^2 d\xi, \int_{-\infty}^{\infty} \phi\psi' d\xi$ and $\int_{-\infty}^{\infty} \phi\psi \, d\xi$ between equations (7.3.1)–(7.3.4) we obtain

$$\int_{-\infty}^{\infty} \left(\frac{b}{\gamma} \phi^2 - \phi^2 (1-\phi)(\phi - a) \right) d\xi = \frac{b}{\gamma} (c^2 - \gamma) \int_{-\infty}^{\infty} (\psi)'^2 d\xi. \qquad (7.3.5)$$

If, instead of eliminating $\int_{-\infty}^{\infty} \psi^2 d\xi$, we eliminate $\int_{-\infty}^{\infty} (\psi')^2 d\xi$ then we obtain the identity

$$\int_{-\infty}^{\infty} \left(\frac{b}{c^2} \phi^2 - \phi^2 (1-\phi)(\phi - a) \right) d\xi = \frac{b\gamma}{c^2} (\gamma - c^2) \int_{-\infty}^{\infty} \psi^2 \, d\xi. \qquad (7.3.6)$$

Now consider the integrand on the left-hand side of (7.3.5). We have

$$\phi^2 \left(\frac{b}{\gamma} + (a - \phi)(1 - \phi) \right) = \phi^2 \left[\left(\phi - \frac{(1+a)^2}{2} \right)^2 + \frac{b}{\gamma} - \frac{(1-a)^2}{4} \right] \geq 0,$$

on noting the inequality (7.2.7). Thus since the left-hand side of (7.3.5) is positive, the same must be true of the right-hand side and this means that

$$c^2 > \gamma, \qquad (7.3.7)$$

which provides a lower bound on the wave speed c. A better lower bound can be achieved as follows. Using the inequality (7.3.7) we deduce from (7.3.6) that

$$\int_{-\infty}^{\infty} \left(\frac{b}{c^2} \phi^2 - \phi^2 (1-\phi)(\phi - a) \right) d\xi < 0$$

and so the expression $\frac{b}{c^2} - (1 - \phi)(\phi - a)$ must take negative values. This is possible if and only if

$$(1 + a)^2 > 4 \left(\frac{b}{c^2} + a \right),$$

i.e.,

$$(1 - a)^2 > 4 \frac{b}{c^2}$$

and so

$$c^2 > 4b(1-a)^2 > \gamma. \tag{7.3.8}$$

In establishing this lower bound on c we can obtain further insight into the role of a as a threshold parameter. Since the expression $\frac{b}{c^2} - (1-\phi)(\phi - a)$ must take negative values, it follows that the maximum value ϕ_{max} of ϕ must exceed the smallest root of the equation

$$\frac{b}{c^2} - (1-\phi)(\phi - a) = 0.$$

That is,

$$\phi_{max} > \frac{1+a}{2} - \frac{1}{2}\left((1-a)^2 - \frac{4b}{c^2}\right)^{1/2}.$$

But more than this, it is easy to see that

$$a < \frac{1+a}{2} - \frac{1}{2}\left((1-a)^2 - \frac{4b}{c^2}\right)^{1/2},$$

and so

$$\phi_{max} > a.$$

From this fact we again see that a plays the role of a threshold parameter.

To conclude our discussion of the qualitative behaviour of the travelling wave $\phi(\xi)$ we remark that there are a number of problems remaining:

(a) Obtain an upper bound for the wave speed c.

(b) Is the travelling wave ϕ stable with respect to small perturbations?

(c) Are there other wave-like solutions?

These and other problems have been discussed in the literature but we shall not pursue them further here.

7.4 Notes

A detailed discussion of the phase plane depicted in Figure 7.2.1 together with the global behaviour of the system (7.2.8) is contained in the paper by H.P. McKean, Nagumo's equation, *Adv. Math.*, **4** 209–223, 1970.

Exercises

7.1 Consider the space clamp model (7.1.1) where the induced current I is assumed constant. Determine the critical points of this system when the current takes the values $I = \frac{b}{\gamma}, I = \frac{ab}{\gamma}$.

7.2 Prove that the system (7.1.1) has a single unstable critical point only when it lies on that part of the u-null-cline which has a positive gradient.

7.3 Verify the local behaviour of the phase plane depicted in Figure 7.2.1 for the system (7.2.8).

7.4 Consider the general system:

$$\frac{\partial u}{\partial t} = \frac{\partial^2 u}{\partial x^2} + f(u) - w,$$
$$\frac{\partial w}{\partial t} = bu - \gamma w,$$

where $f(0) = 0$.
Derive the equations to be satisfied by a travelling wave $u(x,t) = \phi(x + ct)$, $w(x,t) = \psi(x + ct)$, and determine conditions on $f(\phi)$ so that the resulting system has a unique rest state.

7.5 Suppose the function $f(u)$ in question 7.4 is specialised to

$$f(u) = -u, \quad u \leq a/2,$$
$$= u - a, \quad a/2 \leq u \leq (1+a)/2,$$
$$= 1 - u, \quad (1+a)/2 \leq u.$$

Does this system exhibit the same types of behaviour as found in the original FitzHugh-Nagumo system?

7.6 Derive the identities (7.3.5) and (7.3.6).

Chapter 8

Chemical Reactions

8.1 Wavefronts for the Belousov-Zhabotinskii reaction

In this chapter we discuss a simplification of the Belousov-Zhabotinskii reaction model described in Chapter 4. In particular, we give a qualitative analysis of the front of certain travelling concentration waves which have been observed frequently in experiments.

To begin with, we assume that the wave front depends primarily on the concentrations of bromous acid $(HBrO_2)$, which we have denoted by X, and the bromide ion (Br^-) denoted by Y and to a lesser extent on the concentration of the oxidised state $Ce(IV)$ denoted by Z. We also assume that the diffusion coefficients D_X, D_Y are constant and that $D_X = D_Y = D$. Furthermore, for simplicity, we consider only one space dimension x.

Therefore, if we neglect the concentration Z and take note of the above assumptions, the system (4.5.9)–(4.5.11) reduces to

$$\frac{\partial X}{\partial t} = k_1 AY - k_2 XY + k_3 AX - 2k_4 X^2 + D\frac{\partial^2 X}{\partial x^2},$$

$$\frac{\partial Y}{\partial t} = -k_1 AY - k_2 XY + D\frac{\partial^2 Y}{\partial x^2}. \tag{8.1.1}$$

For later purposes it is convenient to non-dimensionalise (8.1.1) by setting

$$u = \frac{2k_4 X}{k_3 A}, \quad v = \frac{k_2 Y}{k_3 Ar},$$

$$x' = \left(\frac{k_3 A}{D}\right)^{1/2} x, \quad t' = k_3 At,$$

$$L = \frac{2k_4 k_1}{k_2 k_3}, \quad M = \frac{k_1}{k_3}, \quad b = \frac{k_2}{2k_4}, \tag{8.1.2}$$

where r is a suitable parameter. The reason for making these transformations is that the solutions u and v, which are of interest, lie in the interval $0 \leq u, v \leq 1$.

With the transformations (8.1.2), the system (8.1.1) takes the form

$$\frac{\partial u}{\partial t'} = Lrv + u(1 - u - rv) + \frac{\partial^2 u}{\partial x'^2},$$

$$\frac{\partial v}{\partial t'} = Mv - buv + \frac{\partial^2 v}{\partial x'^2}.$$

From experimental measurements it is found that r varies from between 10 to about 50 while

$$L \sim 8.4 \times 10^{-6}, \quad M \sim 2.1 \times 10^{-4}, \quad \text{and} \quad b \sim 2.5 \times 10.$$

Since u and v are of order 1, the values of L and M allow us to neglect the first terms on the right of the above system. With this further simplification and removing the primes from x', t', we arrive at the travelling front model:

$$\frac{\partial u}{\partial t} = u(1 - u - rv) + \frac{\partial^2 u}{\partial x^2},$$

$$\frac{\partial v}{\partial t} = -buv + \frac{\partial^2 v}{\partial x^2}, \tag{8.1.3}$$

where $x \in (-\infty, \infty)$ and $0 \leq u, v \leq 1$.

The solutions we are interested in are those that satisfy the conditions

$$\lim_{x \to -\infty} u(x, t) = 0 = \lim_{x \to \infty} v(x, t),$$

$$\lim_{x \to \infty} u(x, t) = 1 = \lim_{x \to -\infty} v(x, t). \tag{8.1.4}$$

One class of solutions that satisfy these conditions precisely are the desired travelling fronts. However, before discussing these further we observe that if $u = 1 - v$ and $b = 1 - r, r \leq 1$, the system (8.1.3) reduces to the single equation

$$\frac{\partial u}{\partial t} = bu(1 - u) + \frac{\partial^2 u}{\partial x^2}, \tag{8.1.5}$$

where $b = 1 - r$. This equation is called **Fisher's equation** and arises in the study of population genetics. It also occurs in many other problems and we shall come across it again in a different context in Chapter 11. It is therefore sufficiently important to warrant some attention before going on to the general case.

8.2   Phase plane analysis of Fisher's equation

We seek a solution u of (8.1.5) in the form of a travelling wave, viz.

$$u(x, t) = \phi(x + ct), \tag{8.2.1}$$

where $0 \leq \phi \leq 1$ and c (>0) is the wave speed.

Bearing in mind the conditions (8.1.4) we see that, as a function of $\xi = x + ct$, ϕ must satisfy

$$\lim_{\xi \to -\infty} \phi(\xi) = 0, \quad \lim_{\xi \to \infty} \phi(\xi) = 1. \tag{8.2.2}$$

On substituting (8.2.1) into (8.1.5) we see that ϕ satisfies the non-linear ordinary differential equation

$$\phi'' - c\phi' + b\phi(1 - \phi) = 0, \tag{8.2.3}$$

where the primes denote differentiation with respect to ξ.

By writing $\phi' = \psi$ (8.2.3) can be rewritten as the system

$$\psi' = c\psi - b\phi(1 - \phi),$$
$$\phi' = \psi. \tag{8.2.4}$$

The singular points of this system in the (ϕ, ψ) plane are $(0, 0)$ and $(1, 0)$, respectively.

If we analyse the character of these singular points we find that $(0, 0)$ is an unstable node if $c \geq 2\sqrt{b}$, a stable focus if $0 < c < 2\sqrt{b}$ and a centre if $c = 0$. The point $(1, 0)$ is a saddle-point for all $c \geq 0$.

Observe that since we are looking for the particular solution of (8.2.4) satisfying the condition $0 \leq \phi \leq 1$ and (8.2.2), the range $0 \leq c < 2\sqrt{b}$ is inadmissable, since in this situation ϕ would become negative near the singular point $(0, 0)$. Thus for the required solution we must have $c \geq 2\sqrt{b}$.

In fact it can be shown that there exists a unique trajectory leaving $(0, 0)$ and entering $(1, 0)$, which remains inside the strip $(0 \leq \phi \leq 1, \psi \geq 0)$, but we shall not prove this here.

8.3 Qualitative behaviour in the general case

Let us now look for wavefront solutions

$$u(x, t) = \phi(x + ct), \quad v(x, t) = \psi(x + ct), \tag{8.3.1}$$

to the system (8.1.3), (8.1.4). As before if we set $\xi = x + ct$, then ϕ, ψ satisfy the equations

$$\phi'' - c\phi' + \phi(1 - \phi - r\psi) = 0,$$
$$\psi'' - c\psi' - b\phi\psi = 0, \tag{8.3.2}$$

where

$$\lim_{\xi \to \infty} \phi(\xi) = 1 = \lim_{\xi \to -\infty} \psi(\xi),$$
$$\lim_{\xi \to -\infty} \phi(\xi) = 0 = \lim_{\xi \to \infty} \psi(\xi). \tag{8.3.3}$$

It is possible to write (8.3.2) as an equivalent system of first-order equations. However, the resulting system is of fourth order and a phase plane analysis would be extremely difficult to perform. Instead, we shall treat (8.3.2) by methods similar to those used to discuss the FitzHugh-Nagumo equations of Chapter 7.

Recall that we are looking for wavefronts ϕ, ψ such that $0 \leq \phi, \psi \leq 1$. If we suppose $\phi \geq 0$ then we shall show that ϕ and ψ have the following properties:

$$\text{(i)} \ \psi \geq 0, \quad \text{(ii)} \ \psi' \leq 0, \quad \text{(iii)} \ \phi' \geq 0. \tag{8.3.4}$$

In other words, we shall show that ϕ is monotonic increasing from 0 to 1 while ψ is monotonic decreasing from 1 to 0 as ξ varies from $-\infty$ to $+\infty$.

To prove (i) suppose $\psi < 0$ for some range of ξ. Then since $\psi(-\infty) = 1$ and $\psi(\infty) = 0$ there must be a negative minimum for which $\psi' = 0$ and $\psi'' > 0$. But from (8.3.2) $\psi'' = b\phi\psi < 0$ at such a point and this contradicts the above inequalities. Thus we must have $\psi \geq 0$ for all values of ξ.

To show that $\psi' \leq 0$ we argue as follows. Since $\phi \geq 0$ by assumption and we have just proved that $\psi \geq 0$, then from (8.3.2) we have

$$\psi'' - c\psi' = b\phi\psi \geq 0.$$

Now ψ cannot have a positive maximum since this would mean that for some point $\xi_0, \psi'(\xi_0) = 0, \psi'' < 0$, which is impossible. Thus either $\psi' \geq 0$ or $\psi' \leq 0$ for all ξ. Since $\psi \geq 0, \psi(-\infty) = 1, \psi(\infty) = 0$, we have $\psi' \leq 0$.

The statement (iii) that $\phi' \geq 0$ is a little more involved and has to be proved in steps depending on whether $0 < r < 1, r = 1$ or $r > 1$. To begin with we note that $1 - \phi - r\psi \neq 0$, since if it were not so (8.3.2) would yield

$$\phi'' - c\phi' = 0$$

giving $\phi = A + B \exp c\xi$, which cannot satisfy the conditions (8.3.3) for any choice of A and B. Set

$$F = 1 - \phi - r\psi \tag{8.3.5}$$

and combine equations (8.3.2) to give

$$F'' - cF' - \phi F = -br\phi\psi \leq 0, \tag{8.3.6}$$

from which it follows that F cannot have a negative minimum. From the boundary conditions (8.3.3) we have

$$\lim_{\xi \to -\infty} F(\xi) = 1 - r, \quad \lim_{\xi \to \infty} F(\xi) = 0. \tag{8.3.7}$$

There are now three cases, $0 < r < 1, r = 1, r > 1$, to consider.

Take the case $0 < r < 1$; then since $F(-\infty) > 0$ and the fact that F cannot have a negative minimum we must have $F \geq 0$ for all ξ. Furthermore, F cannot have a positive minimum followed by a positive maximum. Suppose this could happen; then there would be points ξ_{min}, ξ_{max} where $\xi_{min} < \xi_{max}$ such that

$$F' = 0, \quad F'' > 0$$

at $\xi = \xi_{min}$, i.e.,

$$F'' = \phi F - br\phi\psi > 0$$

or

$$F > br\psi$$

at $\xi = \xi_{min}$. Similarly, at $\xi = \xi_{max}$ we have

$$F < br\psi$$

at $\xi = \xi_{max}$.

Since we have supposed that $F(\xi_{max}) > F(\xi_{min})$, the above inequalities could only hold if $\psi(\xi_{max}) > \psi(\xi_{min})$, which is impossible because we have already proved that $\psi' \leq 0$. This contradiction proves the statement. It is possible, however, for F to have a positive maximum. In any case, we have proved that $F \geq 0$ for all ξ and so from (8.3.2)

$$\phi'' - c\phi' = -\phi F \leq 0.$$

Suppose ϕ has a positive maximum, then in order to satisfy the condition $\phi \to 1$ as $\xi \to \infty, \phi$ would have to approach 1 from above, which would mean that $F = 1 - \phi - r\psi < 0$ for some range of ξ. This is a contradiction and so $\phi' \geq 0$ for all ξ and $0 < r < 1$. The possibility $\phi' \leq 0$ is excluded by virtue of the boundary conditions (8.3.3).

Similar but more straightforward arguments are used to show that $\phi' \geq 0$ for all ξ when $r \geq 1$.

Let us now turn to the problem of estimating the wavefront speed c. Throughout remember that $0 \leq \phi, \psi \leq 1$. Integrating the equations (8.3.2) from $-\infty$ to ∞ gives

$$c = \int_{-\infty}^{\infty} \phi(1 - \phi - r\psi)d\xi, \tag{8.3.8}$$

and

$$c = b \int_{-\infty}^{\infty} \phi\psi \, d\xi. \tag{8.3.9}$$

Substituting for

$$\int_{-\infty}^{\infty} \phi\psi \, d\xi$$

then gives

$$c = \frac{b}{b+r} \int_{-\infty}^{\infty} \phi(1 - \phi)d\xi. \tag{8.3.10}$$

Now as $\xi \to -\infty$, the boundary conditions (8.3.3) require $\phi \to 0$ and $\psi \to 1$ and so (8.3.2) behave near $\xi = -\infty$ as the linearised equations

$$\phi'' - c\phi' + \phi(1 - r) = 0,$$
$$\psi'' - c\psi' - b\phi = 0,$$

the solutions of which are oscillatory unless

$$c^2 > 4(1 - r), \quad r \le 1. \tag{8.3.11}$$

Oscillatory solutions are not possible asymptotically when $r > 1$. Thus (8.3.11) gives a lower bound on the wave speed.

In order to obtain more useful information about the wave speed c and in particular on its variation with b, we proceed as follows. Multiply the first equation in (8.3.2) by ϕ and integrate to get

$$I \equiv \int_{-\infty}^{\infty} \phi'^2 d\xi = -\frac{c}{2} + \int_{-\infty}^{\infty} \phi^2(1 - \phi - r\psi)d\xi. \tag{8.3.12}$$

Similarly, repeating the process but multiplying first by ϕ' gives

$$cI = \int_{-\infty}^{\infty} \phi'\phi \, d\xi - \int_{-\infty}^{\infty} \phi'\phi^2 d\xi - r\int_{-\infty}^{\infty} \phi'\phi\psi \, d\xi$$
$$= \frac{1}{2} - \frac{1}{3} - r\int_{-\infty}^{\infty} \phi'\phi\psi \, d\xi.$$

i.e.,

$$I = \frac{1}{6c} - \frac{r}{c}\int_{-\infty}^{\infty} \phi'\phi\psi \, d\xi. \tag{8.3.13}$$

We now estimate the integral appearing on the right-hand side of equation (8.3.13) via **Schwarz's inequality** which states that if F and G are integrable then

$$\int_a^b FG \, d\xi \le \left(\int_a^b F^2 d\xi\right)^{1/2}\left(\int_a^b G^2 d\xi\right)^{1/2}.$$

In our case we take $a = -\infty, b = \infty, F = \phi', G = \phi\psi$ to obtain

$$\int_{-\infty}^{\infty} \phi'\phi\psi \, d\xi \le \left(\int_{-\infty}^{\infty} \phi'^2 d\xi\right)^{1/2}\left(\int_{-\infty}^{\infty} (\phi\psi)^2 d\xi\right)^{1/2}$$
$$= I^{1/2}\left(\int_{-\infty}^{\infty} (\phi\psi)^2 d\xi\right)^{1/2}$$
$$\le I^{1/2}\left(\int_{-\infty}^{\infty} (\phi\psi)d\xi\right)^{1/2}$$
$$= I^{1/2}\left(\frac{c}{b}\right). \tag{8.3.14}$$

Substituting this into (8.3.13) results in the inequality

$$I > \frac{1}{6c} - \frac{r}{c}\left(\frac{c}{b}\right) I^{1/2}$$

or

$$I + \frac{r}{(cb)^{1/2}} I^{1/2} - \frac{1}{6c} > 0,$$

i.e.,

$$\left(I^{1/2} + \frac{r}{2(cb)^{1/2}}\right)^2 > \frac{1}{6c} + \frac{r^2}{4cb}$$

$$= \frac{1}{2c}\left(\frac{1}{3} + \frac{r^2}{2b}\right)$$

and so

$$I^{1/2} > -\frac{r}{2(cb)^{1/2}} + \frac{1}{(2c)^{1/2}}\left(\frac{1}{3} + \frac{r^2}{2b}\right)^{1/2},$$

i.e.,

$$I > \frac{1}{4bc}\left[\left(r^2 + \frac{2}{3}b\right)^{1/2} - r\right]^2. \tag{8.3.15}$$

Next, since $0 \le \phi, \psi \le 1$ we have from (8.3.12)

$$I = -\frac{c}{2} + \int_{-\infty}^{\infty} \phi^2(1-\phi)d\xi - r\int_{-\infty}^{\infty} \phi^2\psi\,d\xi,$$

$$\le -\frac{c}{2} + \int_{-\infty}^{\infty} \phi(1-\phi)d\xi,$$

$$= -\frac{c}{2} + c\left(\frac{b+r}{b}\right),$$

i.e.,

$$I < \frac{c}{2b}(b + 2r). \tag{8.3.16}$$

Combining (8.3.16) and (8.3.15) to eliminate I we find that

$$\frac{1}{4bc}\left[\left(r^2 + \frac{2}{3}b\right)^{1/2} - r\right]^2 \le \frac{c}{2b}(b + 2r)$$

from which it follows that

$$c^2 \ge \frac{\left[\left(r^2 + \frac{2}{3}b\right)^{1/2} - r\right]^2}{2(b + 2r)}.$$

Numerical results show that $c^2 \leq 4$ and so we have the final result

$$\frac{\left[\left(r^2 + \frac{2}{3}b\right)^{1/2} - r\right]^2}{2(b + 2r)} \leq c^2 \leq 4. \tag{8.3.17}$$

Again from numerical results (8.3.11) is a sharper lower bound for c^2 when $0 < r < 1$.

8.4 Notes

For further information about the Belousov-Zhabotinskii reaction and many related topics see J. D. Murray, *Mathematical Biology*, Springer-Verlag, Heidelberg, 1993.

Exercises

8.1 Show that the transformations (8.1.2) reduce equations (8.1.1) to the equations (8.1.3).

8.2 Analyse the behaviour of trajectories for the system (8.2.4) in the neighbourhood of the singular points (0, 0), (1, 0) for all positive values of the wavefront speed c.

8.3 Consider the system (8.3.2), (8.3.3). Use the arguments of Section 8.3 to show that $\phi' \geq 0$ for $\xi \in (-\infty, \infty)$ when $r \geq 1$.

8.4 In the inequality (8.3.17) show that as $b \to 0$ or $r \to \infty$ it reduces to $c^2 \geq 0$ and that if $b \to \infty$ or $r \to 0$ the inequality reduces to $c^2 \geq 1/3$. In the latter case, compare the result with that predicted by the inequality (8.3.11).

Chapter 9

Predator and Prey

9.1 Catching fish

Often one species uses another as food. For example, man uses fish, lions consume gazelles, and some wasps eat the caterpillars of moths. The presence of one species can therefore have effects (sometimes irreversible) on another. The study of such dynamic interactions is the topic of this chapter. For simplicity, consideration will be limited to two species, though it must be understood that in nature the situation is frequently much more complicated, with several species involved and interacting in a complex manner. Nevertheless, useful conclusions can be drawn from even relatively simple models and the influence of various actions on the level of available resources ascertained.

Usually, the species regarded as food is called the **prey** and the consuming species the **predator**. The aim is to construct predator-prey models that are relevant, whether it is man killing deer, wasps eating caterpillars or man harvesting timber, though humans often behave differently from other organisms; some general aspects have already been treated in Section 4.6. To fix on a specific illustration for the simplest model, we shall discuss how fishing can affect the population of fish.

Let $N(t)$ be the number of fish in a designated zone at time t. It will be assumed that N varies continuously with t; in fact, N changes only by integers but the error introduced by our approximation should not be appreciable except possibly when N is small. In the absence of fishing, new fish can be expected to be born at a rate proportional to existing numbers, say $bN(t)$ where b is a constant. Similarly, the rate at which fish die will be taken as $dN(t)$ with d constant. Then

$$\frac{dN}{dt} = (b - d)N. \tag{9.1.1}$$

The solution of this differential equation has been found in Section 1.1 and is

$$N(t) = N(0)e^{(b-d)t} \tag{9.1.2}$$

where $N(0)$ is the population at time $t = 0$.

The solution (9.1.2) reveals that the fish population eventually disappears if the birth rate is less than the death rate, i.e., $b < d$, is in equilibrium if $b = d$

and increases exponentially if $b > d$, i.e., the reproductive rate is greater than the mortality rate. Exponential growth in which the population doubles in every interval $0.69/(b-d)$ of time is indeed exhibited by many species under ideal conditions when they have boundless space, food and the resources they need available. However, such growth cannot continue indefinitely in a finite world and there must come a time when shortages of supplies inhibit the exponential growth.

The preparation of models that allow for environmental constraints has followed a variety of routes. One of the most popular is to replace (9.1.1) by

$$\frac{dN}{dt} = a\left(1 - \frac{N}{N_0}\right)N \tag{9.1.3}$$

where a and N_0 are positive constants. According to Section 1.3, this leads to logistic growth and the population always ends up at N_0 no matter what level it started at. Thus N_0 can be regarded as the maximum population that can be sustained under (9.1.3). Equation (9.1.3) often appears in ecology in the form

$$\frac{dN}{dt} = r\left(1 - \frac{N}{K}\right)N.$$

Whether (9.1.3) is applicable to any particular population depends upon either producing a convincing argument for the presence of each term or comparing its predictions with actual observations of the numbers in a population. For instance, yeast grown in laboratory cultures follows the logistic curve admirably. On the other hand, the human population of the United States, while fitting the logistic curve well from 1790 to 1910, grew much more strongly than predicted from 1920 onwards and departed widely from the curve. In general, it seems that the more complex the life history of an organism the less likely its population is to sit on the logistic curve. This notwithstanding, there are often sufficient periods of time for which the logistic curve can be applied to justify retaining it as a model.

Another view is that environmental factors do not instantaneously change the population but take some time to permeate through it. In that case the rate of change depends upon the size of the population at an earlier time and a possible model is

$$\frac{dN(t)}{dt} = a\left(1 - \frac{N(t-T)}{N_0}\right)N(t) \tag{9.1.4}$$

where T is a constant representing the delay that transpires before effects are felt. Equation (9.1.4) is commonly solved on the computer but its solution bears little resemblance to the logistic curve in many circumstances and can display oscillatory behaviour.

Although (9.1.3) can be solved analytically it is desirable to see what can be deduced by employment of the phase plane since that method may be available

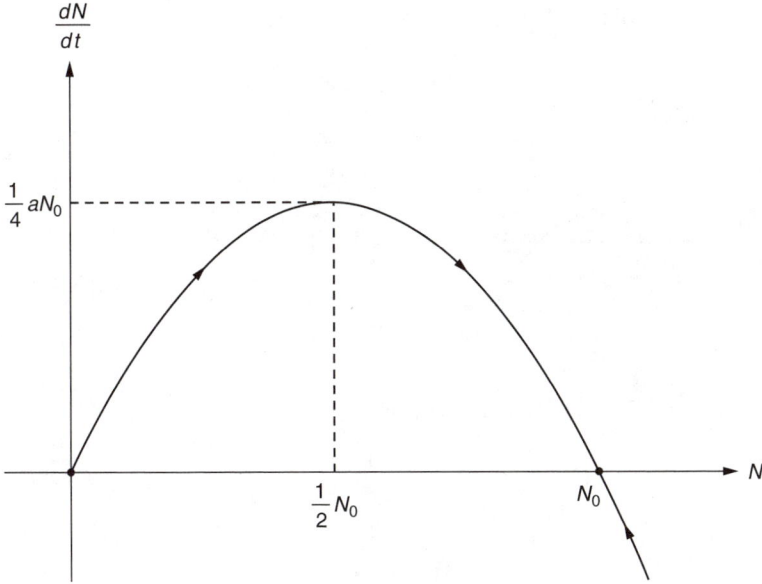

FIGURE 9.1.1: The trajectory for logistic growth.

when analytical solution is precluded. Note first that there is equilibrium when $dN/dt = 0$, i.e., when $N = 0$ or $N = N_0$. Secondly, if dN/dt is plotted against N in the phase plane the trajectory obtained is the parabola shown in Figure 9.1.1. The parabola has a maximum of $\frac{1}{4}aN_0$ at $N = \frac{1}{2}N_0$. Also dN/dt is positive when N is slightly less than N_0 and negative when N is a bit larger than N_0. Consequently the directions on the trajectory are those indicated by the arrows on the diagram. From these it is clear that N always approaches the equilibrium value N_0 whereas a slight disturbance will cause it to depart from the equilibrium at $N = 0$. Thus $N = N_0$ corresponds to stable equilibrium and $N = 0$ to unstable equilibrium. Evidently, the main features of the behaviour can be extracted from the phase plane as well as from the explicit solution.

9.2 The effect of fishing

A stable environment allows the fish population to adopt a pattern as set out in the previous section, but the disturbance caused by removing some of the fish for food may be profound. Some idea of how the population is affected and whether equilibrium still supervenes can be achieved by supposing the fish are caught at the constant rate c but the population is otherwise governed

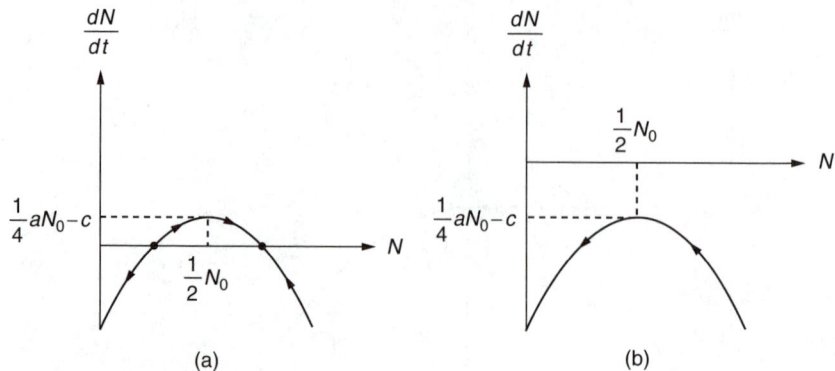

FIGURE 9.2.1: The trajectory for constant fishing rate: (a) $c < \frac{1}{4}aN_0$, (b) $c > \frac{1}{4}aN_0$.

by (9.1.3). Then

$$\frac{dN}{dt} = a\left(1 - \frac{N}{N_0}\right)N - c. \tag{9.2.1}$$

The differential equation (9.2.1) can be solved explicitly but, instead of finding the analytical solution, we shall consider it in the phase plane. The corresponding parabolic trajectory is displayed in Figure 9.2.1. If $c < \frac{1}{4}aN_0$ there are two points of equilibrium of which the lower is unstable and the higher stable. Therefore, if

$$N(0) > \frac{1}{2}N_0\{1 - (1 - 4c/aN_0)^{1/2}\},$$

the population will eventually arrive at the value of $\frac{1}{2}N_0\{1+(1-4c/aN_0)^{1/2}\}$. In contrast, if

$$N(0) < \frac{1}{2}N_0\{1 - (1 - 4c/aN_0)^{1/2}\},$$

the fish will steadily decrease in number until they disappear altogether.

When $c > \frac{1}{4}aN_0$ it is transparent from Figure 9.2.1(b) that there are no points of equilibrium so that the fish population steadily decays and dies out.

For $c = \frac{1}{4}aN_0$ the picture is similar to Figure 9.2.1(b) but the parabola just touches the N-axis at $N = \frac{1}{2}N_0$. So there is equilibrium at $N = \frac{1}{2}N_0$ but any perturbation which tends to reduce N will result in the fish becoming extinct.

We conclude that, if the fish population is not to be decimated, the catching rate must not exceed $\frac{1}{4}aN_0$ and the initial population must not be too small. If the catching rate is near the maximum the population will be reduced to about half the natural level of N_0 and will be subject to extinction if any other adverse factor enters the scene. It is therefore advisable to have the catching rate significantly below the maximum permitted.

The above simple model has led to a useful recommendation on the catching rate to be adopted if the fish population is to survive, but it must be recognised that the model becomes increasingly unrealistic as the population falls. The scarcer the fish become the more effort has to be expended in catching the same number of fish per unit time. Since there is a limit on the resources for fishing, the catch is likely to become proportional to the fish population when it drops below a certain value. Moreover, at low levels the population is likely to be in a state where exponential growth is pertinent if fishing were to cease. Therefore, when considering survival prospects, the appropriate model is

$$\frac{dN}{dt} = (b - d - f)N \qquad (9.2.2)$$

where f is the factor which accounts for the loss due to fishing. There is equilibrium if $f = b - d$ so that the population will not vary if the catch is restricted to $(b - d)N$. Fishing at a greater rate will lead to extinction. By putting $f = 0$ in (9.2.2) we can determine how long it will take for the population to attain an acceptable level when there is no fishing.

Before any model can be applied in a practical context, it is necessary to be able to measure the parameters that occur. This is a far from trivial task. For example, an estimate of b may require a consideration of the breeding cycle, the sex ratio at birth and environmental factors. Again, the mortality rate may be different for infants, adults and different sexes. Furthermore, there is a distinct possibility that b and d vary with N. We shall say no more than that adequate estimates are feasible for some species because we now want to examine the complicated matter of interaction between species.

9.3 The Volterra-Lotka model

Imagine that there is an island occupied by men whose sole source of food consists of fish. This is an example of a predator-prey problem and, to illustrate the ideas, we shall consider a very much simplified model in which humans do not evolve new methods of fishing as the fish population changes. Let $x(t)$ be the population of fish at time t and $y(t)$ that of humans. Assume that the fish, when undisturbed, grow exponentially and that, in the absence of fish, the starvation rate of the human population is greater than the birth rate so that it decays exponentially. The number of fish consumed will depend upon the frequency of encounter with humans. If we assume that, when they meet, there is a constant probability that the human will catch a fish and the time taken to eat it is negligible we obtain

$$\dot{x} = ax - bxy, \tag{9.3.1}$$
$$\dot{y} = -cy + dxy, \tag{9.3.2}$$

where a, b, c and d are all positive constants. Notice that only solutions in which $x \geq 0$ and $y \geq 0$ are of interest.

The system of (9.3.1) and (9.3.2) has critical points where

$$ax - bxy = 0, \qquad -cy + dxy = 0,$$

i.e., $x = 0$ and $y = 0$ or $x = c/d$ and $y = a/b$. By linearisation about $x = 0$ and $y = 0$ as in Section 5.3, we obtain the system

$$\dot{\xi} = a\xi, \qquad \dot{\eta} = -c\eta,$$

which shows that the origin is a saddle-point (Section 5.4). Linearisation near $(c/d, a/b)$ leads to

$$\dot{\xi} = -bc\eta/d, \qquad \dot{\eta} = da\xi/b$$

showing that this critical point is a centre. Also when $x = c/d$, (9.3.1) reveals that $\dot{x}$ is positive and x increasing when $y < a/b$. Therefore the trajectories have the structure of Figure 9.3.1.

The precise equations of the trajectories can be derived by observing that

$$\frac{dy}{dx} = \frac{(dx - c)y}{(a - by)x}$$

which is a separable first-order differential equation. Its solution is

$$a \ln y - by + c \ln x - dx = K \tag{9.3.3}$$

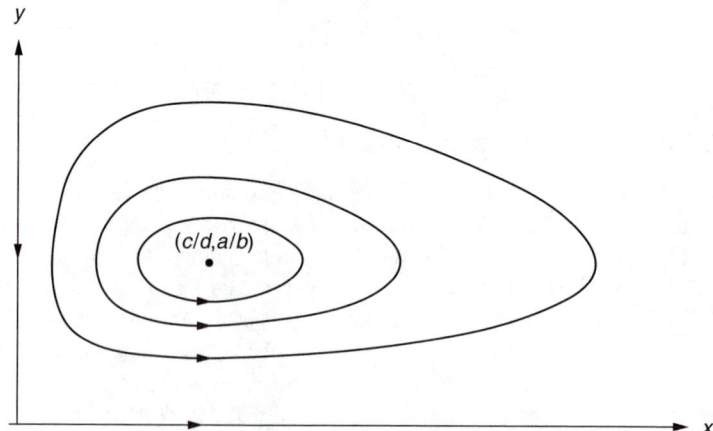

FIGURE 9.3.1:   Phase plane for a predator-prey problem.

where K is an arbitrary constant. The left-hand side has a minimum at $(c/d, a/b)$ and so the contours (9.3.3) must form closed curves in the neighbourhood of this point, in agreement with its being a centre and with Figure 9.3.1.

Since a typical trajectory is a closed curve, the variation of x and y with time must be periodic unless either x or y is initially zero, a possibility that will be excluded from further consideration. It is clear from Figure 9.3.1 that a maximum (minimum) in y occurs about a quarter period after a maximum (minimum) in x. An interpretation is that when the human population is at a maximum the fish population is declining and that decline induces a drop in the number of humans. The reduction in predators allows the fish to thrive followed by an increase in the humans and the cycle repeats itself. The actual levels attained depend upon which trajectory is traced. A change in the environment may switch the system from one trajectory to another but the periodic fluctuation continues with no tendency to an equilibrium state.

In place of exponential growth for the fish when there are no predators, it can be supposed that the growth is logistic so that

$$\dot{x} = ax - gx^2 - bxy, \tag{9.3.4}$$
$$\dot{y} = -cy + dxy, \tag{9.3.5}$$

which includes (9.3.1) as a special case when $g = 0$. The system (9.3.4) and (9.3.5) is said to constitute a **Volterra-Lotka model** for the predator-prey problem. It can be discussed in the phase plane in a similar fashion to that when $g = 0$.

Variants to the Volterra-Lotka model have been proposed in order to incorporate other facets of the interaction. In one version it is supposed that x_0 of the fish can find a refuge that makes it impossible for them to be caught. The modification to (9.3.4) and (9.3.5) is then

$$\dot{x} = ax - gx^2 - by(x - x_0), \tag{9.3.6}$$
$$\dot{y} = -cy + dy(x - x_0). \tag{9.3.7}$$

On the other hand, more general assumptions have been discussed. For example, the **Rosenzweig-MacArthur model** has

$$\dot{x} = f(x) - h(x, y), \tag{9.3.8}$$
$$\dot{y} = -cy + kh(x, y), \tag{9.3.9}$$

where f and h are appropriate functions. Again logistic growth might be preferred for y and (9.3.5) replaced by

$$\dot{y} = c'y - g'y^2 + dxy,$$

c' and g' being constants.

An indication of how information can be extracted from the phase plane can be given by examining

$$\dot{x} = \{\lambda(x) - \mu(x, y)\}x,$$
$$\dot{y} = \{v(x) + \rho(x, y)\}y.$$

Since predators consume prey we expect that $\partial\mu/\partial y > 0$. As the predators increase there will be less prey per predator and so

$$\frac{\partial}{\partial y}(v + \rho) < 0.$$

If the number of prey drops there will be less prey per predator and so

$$\partial\rho/\partial x > 0.$$

We shall also assume that

$$\partial(\lambda - \mu)/\partial x \leq 0$$

consistent with (9.3.1) and (9.3.4).

If there is a point of equilibrium (x_0, y_0) where neither species is extinct, linearisation gives

$$\dot{\xi} = a\xi + b\eta, \qquad \dot{\eta} = c\xi + d\eta$$

where

$$a = x_0\frac{\partial}{\partial x}(\lambda - \mu), \qquad b = -x_0\frac{\partial\mu}{\partial y},$$
$$c = y_0\frac{\partial\rho}{\partial x}, \qquad d = y_0\frac{\partial}{\partial y}(v + \rho),$$

the partial derivatives being evaluated at (x_0, y_0). Thus $a \leq 0, b > 0, c > 0$ and $d < 0$. Consequently, the point of equilibrium is stable, being a node or a focus. There are oscillations if, and only if,

$$\left(x_0\frac{\partial}{\partial x}(\lambda - \mu) - y_0\frac{\partial}{\partial y}(v + \rho)\right)^2 - 4x_0y_0\frac{\partial\mu}{\partial y}\frac{\partial\rho}{\partial x} < 0.$$

This can be met if the first term is not too large. If $d = 0$ and $a = 0$ were also an allowable case, the equilibrium point would become a centre and periodic fluctuations would be quite likely. Whether any of these states could be reached from specified initial conditions would depend on global considerations, which have been mentioned in Section 5.6.

The discussion of this section has centred on the phase plane but the reader should be aware that a change of variable can sometimes lead to a simpler situation. For instance, the obvious choice with

$$\ddot{x} - 12x\dot{x} + 16x^3 = 0$$

is

$$\dot{x} = y, \qquad \dot{y} = 12xy - 16x^3$$

but one might be better off with the choice

$$\dot{x} = 2(x^2 - y^2), \qquad \dot{y} = 4xy.$$

Moreover, the phase plane can be helpful in other ways. The differential equation

$$\frac{dy}{dx} = \frac{x + y}{x - y}$$

is of homogeneous type and can be solved as in Section 1.6. The solution is

$$y = x \tan\{\ln[K'(x^2 + y^2)^{1/2}]\},$$

$$K' = (x_0^2 + y_0^2)^{-1/2} \exp\{\tan^{-1}(y_0/x_0)\},$$

when $y = y_0$ at $x = x_0$. However, in the phase plane we have

$$\dot{x} = x - y, \qquad \dot{y} = x + y$$

with solution

$$x = e^t(x_0 \cos t - y_0 \sin t), \qquad y = e^t(y_0 \cos t + x_0 \sin t)$$

which is much easier to handle than the previous form.

Exercises

9.1 For whales, evidence suggests that b and d in (9.2.2) are given by $b = 0.14$ and $d = 0.09$ when t is measured in years. How long would the population take to increase from 800 to 8500 in the absence of fishing? If there were 4000 whales, what catch per year would you recommend? If $f = 0.75$, show that the population will take about 14 years to recover from a year's fishing.

9.2 For buffaloes $b = 0.081$ and $d = 0.05$ in (9.2.2) when time is calculated in years. If there were 40 million, what is the maximum number to be shot if the herd is to be sustained?

9.3 Discuss the Volterra-Lotka model of (9.3.4) and (9.3.5) in the phase plane when none of the constants is zero.

9.4 Discuss the system

$$\dot{x} = ax - gx^2 - bxy,$$
$$\dot{y} = c'y - g'y^2 + dxy$$

in the phase plane, all the constants being positive. Will the predators die out if there is no prey?

9.5 Show that the system

$$\dot{x} = -ax + b(1 - e^{-ky})x,$$
$$\dot{y} = -ay + c(1 - e^{-ky})x,$$

where a, b, c and k are positive constants with $b > a$, has equilibrium points at $(0,0)$ and $(\{-b/ck\}\ln\{(b-a)/b\}, -\{1/k\}\ln\{(b-a)/b\})$ of which the second is a saddle-point. Examine the trajectories in the phase plane.

9.6 Compare in the phase plane the systems

$$\dot{x} = ax - bxy,$$
$$\dot{y} = c'y - d'y^2/x$$

and

$$\dot{x} = ax^2 - bx^2y,$$
$$\dot{y} = c'xy - d'y^2.$$

9.7 Discuss the behaviour of the system

$$\dot{x} = -ax + gx^2 - bxy,$$
$$\dot{y} = -cy + dxy$$

where a, b, c, d and g are all positive constants.

9.8 Draw the trajectories of (9.3.1) and (9.3.2) when $a = 10, b = 0.2, c = 10^{-4}$ and $d = 0.2$.

9.9 A model in which the prey x and predator y are removed by an external mechanism as well as interacting is

$$\dot{x} = xf(x, y) - A,$$
$$\dot{y} = yg(x, y) - B$$

where A and B are non-negative constants (if they are negative, restocking is occurring). Assume that $\partial f/\partial x < 0$, $\partial g/\partial x > 0$, $\partial g/\partial y \leq 0$ for all $x > 0$ and $y > 0$. Show that, when $A = B = 0$, the prey can grow from a small initial population when $f(0,0) \geq 0$ provided that $f(0, y)$ is positive for $y > 0$; if $f(0, y)$ is positive for $0 < y < y_0$ and $f(0, y_0) = 0$ show that the prey can still establish itself so long as the initial predator population is less than y_0.

9.10 As a specific model for Exercise 9.9 take

$$f(x,y) = 2 - \frac{x}{30} - \frac{y}{x+10},$$

$$g(x,y) = \frac{x-20}{3(x+10)}.$$

Show that in the phase plane there can be two critical points, of which one is a saddle-point and the other, P, is not.

If $A = 13$ and $B = 2$, P is stable and there is a trajectory from the saddle-point to P; both species can coexist. The same is true when $A = 3$ and $B = 3$.

If $A = 3$ and $B = \frac{1}{2}$, P is unstable and there is a trajectory from P to the saddle-point; one of the species becomes extinct.

If $A = 3$ and $B = 2$ or $\frac{3}{2}$, P is unstable but is surrounded by a stable limit cycle; the populations can exhibit oscillatory behaviour.

Chapter 10

Partial Differential Equations

10.1 Characteristics for equations of the first order

In this section we set out a method of finding solutions of linear partial differential equations of the first order. We start by considering a particular example, namely

$$2\frac{\partial u}{\partial x} + 3\frac{\partial u}{\partial y} = 0. \tag{10.1.1}$$

Associated with this are certain curves in the (x, y)-plane specified by obliging them to have the slope which satisfies

$$\frac{dy}{dx} = \frac{3}{2}, \tag{10.1.2}$$

the right-hand side being the ratio of the coefficients of the two partial derivatives in (10.1.1). The general solution of (10.1.2) is $y = \frac{3}{2}x + C$ where C is a constant and the curves are, in fact, straight lines. They are drawn for various values of C in Figure 10.1.1. These special curves are known as the **characteristics** of the partial differential equation (10.1.1).

The idea in solving (10.1.1) is to introduce a new set of coordinates, called ξ and η, in which the characteristics can be expressed as $\xi = \text{constant}$. The other coordinate η must be selected so that to a given point (ξ, η) there corresponds only one point in the x- and y-coordinates. In other words, we want the curve $\eta = \text{constant}$ to intersect the curve $\xi = \text{constant}$ in the (x, y)-plane in one, and only one, point. Now, in Figure 10.1.1, a line parallel to the y-axis meets a characteristic in one, and only one, point. Therefore make the change of coordinates

$$\xi = y - \tfrac{3}{2}x, \qquad \eta = x. \tag{10.1.3}$$

Notice that (10.1.3) does supply just one (x, y) for a given (ξ, η); in fact $x = \eta$ and $y = \xi + \frac{3}{2}\eta$. Other choices of η are acceptable providing that they meet the criterion that the curve $\eta = \text{constant}$ intersects $\xi = \text{constant}$ once and once only; for example, $\eta = y$ and $\eta = 3y + 2x$ are other possibilities.

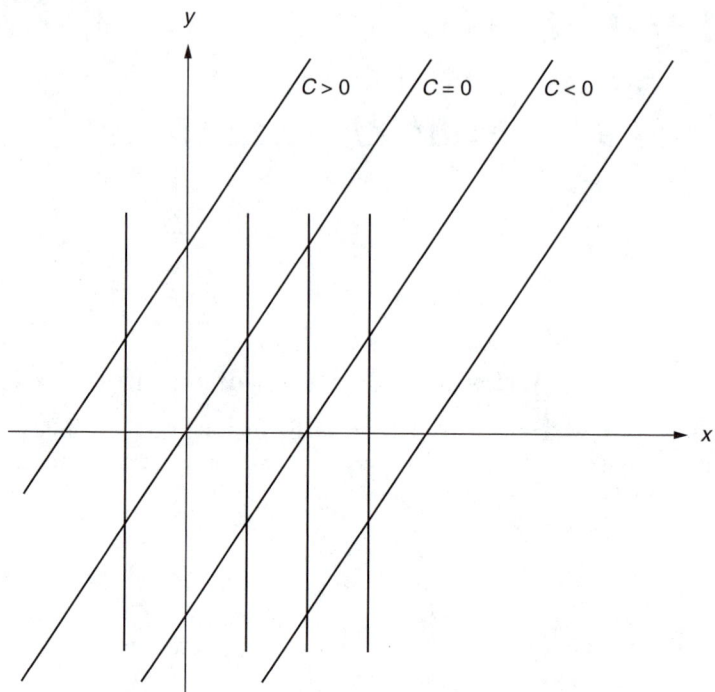

FIGURE 10.1.1: The characteristics of (10.1.1).

From (10.1.3), by the chain rule for partial derivatives,

$$\frac{\partial u}{\partial x} = \frac{\partial u}{\partial \xi}\frac{\partial \xi}{\partial x} + \frac{\partial u}{\partial \eta}\frac{\partial \eta}{\partial x} = -\frac{3}{2}\frac{\partial u}{\partial \xi} + \frac{\partial u}{\partial \eta},$$

$$\frac{\partial u}{\partial y} = \frac{\partial u}{\partial \xi}\frac{\partial \xi}{\partial y} + \frac{\partial u}{\partial \eta}\frac{\partial \eta}{\partial y} = \frac{\partial u}{\partial \xi}$$

since $\partial \xi/\partial x = -3/2$, $\partial \xi/\partial y = 1$, $\partial \eta/\partial x = 1$, $\partial \eta/\partial y = 0$. Substitution in (10.1.1) gives

$$2\frac{\partial u}{\partial \eta} = 0. \tag{10.1.4}$$

Thus the change of coordinates (10.1.3) has simplified the partial differential equation and, in fact, one partial derivative $\partial u/\partial \xi$ has been removed altogether. It was with the object of causing this partial derivative to disappear that the special choice of ξ was made.

Equation (10.1.4) tells us that u is independent of η and so it must be a function of ξ only, i.e.,

$$u = F(\xi) = F(y - \tfrac{3}{2}x) \tag{10.1.5}$$

on employing (10.1.3).

The formula (10.1.5) shows that u is a constant on a characteristic since $y - \frac{3}{2}x$ is constant on a characteristic. Hence, if u is known at one point of a characteristic it is known at every point of the characteristic. The constant can be different on different characteristics, but we cannot say how it will change without additional information.

Suppose, indeed, that it is required that $u = 3y$ on $x = 0$ for $y \geq 0$. Then, putting $x = 0$ in (10.1.5), we are obliged to have

$$F(y) = 3y$$

for $y \geq 0$. Using this expression for F we have $F(y - \frac{3}{2}x) = 3(y - \frac{3}{2}x)$ and so, from (10.1.5),

$$u = 3y - 9x/2$$

for $y \geq 3x/2$. Thus u has been found at every point on and above the line $y = 3x/2$. Nothing can be said about the behaviour of u below this line, other than that it is a function of $y - 3x/2$, because there is insufficient information. In other words, the solution is restricted to that part of the (x, y)-plane that is covered by the characteristics that intersect the arc on which the initial data are given.

We might have been asked to find u so that $u = 3y$ on the line $y = 3x/2$ for $y \geq 0$. Then (10.1.5) would require $F(0) = 3y$ for $y \geq 0$, which no F will satisfy because the left-hand side is constant whereas the right-hand side is not. Indeed, the only problem which can be solved when u is given on $y = 3x/2$ is for u to be specified as a constant there. A similar situation arises for the characteristic $y - 3x/2 = $ constant. Thus the initial data can be specified arbitrarily for u only when given on an arc that does not coincide with a characteristic. When initial data are prescribed on a characteristic they must take a special form if inconsistency is to be avoided.

Let us turn now to the more general linear partial differential equation

$$a(x, y)\frac{\partial u}{\partial x} + b(x, y)\frac{\partial u}{\partial y} = c(x, y)u + d(x, y). \tag{10.1.6}$$

We succeeded in solving (10.1.1) because we were able to discover a substitution which eliminated one partial derivative and led to (10.1.4). So we want to try to find a similar substitution for (10.1.6). Now, if $a \equiv 0$ or $b \equiv 0$ in the region under consideration, only one partial derivative occurs anyway so it is only when neither a nor b vanishes identically that further discussion is necessary. We then consider the curves whose slope satisfies

$$\frac{dy}{dx} = \frac{b(x, y)}{a(x, y)}. \tag{10.1.7}$$

They are known as **characteristics** and have no multiple points provided that a and b do not vanish at the same point, a possibility which will be excluded here. The characteristics need not be straight lines as they were for (10.1.1).

Let the equation for the characteristics be written in the form

$$\phi(x, y) = C \tag{10.1.8}$$

where C is a constant which gives different characteristics for different values. A derivative of (10.1.8) with respect to x supplies

$$\frac{\partial \phi}{\partial x} + \frac{\partial \phi}{\partial y} \frac{dy}{dx} = 0.$$

But, since the slope is forced to be the same as that of (10.1.7), we must have

$$a(x, y)\frac{\partial \phi}{\partial x} + b(x, y)\frac{\partial \phi}{\partial y} = 0. \tag{10.1.9}$$

Suppose now that there is a set of curves with equation

$$\psi(x, y) = C_1,$$

C_1 being constant, with the property that each curve meets each characteristic in one and only one point in the region of interest while ψ varies continuously along a characteristic. There are usually many possible ψ which are open to the solver and the one most convenient to the problem in hand should be chosen.

Make the substitution

$$\xi = \phi(x, y), \qquad \eta = \psi(x, y). \tag{10.1.10}$$

The conditions placed on ϕ and ψ ensure that a point may be fixed by either (x, y) or (ξ, η). Since

$$\frac{\partial u}{\partial x} = \frac{\partial u}{\partial \xi}\frac{\partial \phi}{\partial x} + \frac{\partial u}{\partial \eta}\frac{\partial \psi}{\partial x}, \qquad \frac{\partial u}{\partial y} = \frac{\partial u}{\partial \xi}\frac{\partial \phi}{\partial y} + \frac{\partial u}{\partial \eta}\frac{\partial \psi}{\partial y}$$

(10.1.6) transforms to

$$\left(a\frac{\partial \psi}{\partial x} + b\frac{\partial \psi}{\partial y}\right)\frac{\partial u}{\partial \eta} = cu + d \tag{10.1.11}$$

on account of (10.1.9). Equation (10.1.11) is often called the **characteristic form** of the partial differential equation.

If values for x, y in terms of ξ, η are inserted in (10.1.11) from (10.1.10), the coefficients in (10.1.11) become known functions of ξ and η. For fixed ξ (10.1.11) is an ordinary differential equation of the first order for u in terms of η. Therefore u can be determined but contains an arbitrary constant (which may be different for different values of ξ). Thus, if u is specified at a point of $\xi =$ constant it is known at all points of that curve. In other words, if u is given at one point of a characteristic it can be found at all points of the characteristic. In particular, when u is designated to have certain values on an arc that is met at most once by a characteristic it is determined in the region covered by the characteristics intersecting the arc. Again, as for (10.1.1), data cannot be prescribed arbitrarily all along a characteristic; they must comply with (10.1.11) if there is to be a solution.

The existence theorems of Chapter 5 can be employed to check the existence and uniqueness of solutions to (10.1.11).

Example 10.1.1

Find the solution of

$$2x\frac{\partial u}{\partial x} + (x+1)\frac{\partial u}{\partial y} = y$$

in $x > 0$ such that $u = 2y$ when $x = 1$.

The differential equation for the characteristics is, from (10.1.7),

$$\frac{dy}{dx} = \frac{\bar{x}+1}{2x} = \frac{1}{2}\left(1 + \frac{1}{x}\right).$$

Therefore the characteristics are

$$y = \tfrac{1}{2}(x + \ln x) + C. \tag{10.1.12}$$

The slope of the characteristics is positive in $x > 0$. Also $y \to -\infty$ as $x \to 0$ and $y \to \infty$ as $x \to \infty$. The shape of the characteristics is, consequently, that shown in Figure 10.1.2, the larger C the closer the characteristic is to the y-axis.

A set of curves that intersects characteristics once and once only is furnished by lines parallel to the x-axis; another set is provided by lines parallel to the y-axis. Choose those parallel to the y-axis since the initial values are intimated on $x = 1$. Then, according to (10.1.10),

$$\xi = y - \tfrac{1}{2}(x + \ln x), \qquad \eta = x. \tag{10.1.13}$$

Note that it is always worthwhile considering parallel straight lines first for the ψ curves since they lead to the simplest form for η. Now

$$\frac{\partial u}{\partial x} = \frac{\partial u}{\partial \xi}\left(-\frac{1}{2} - \frac{1}{2x}\right) + \frac{\partial u}{\partial \eta}, \qquad \frac{\partial u}{\partial y} = \frac{\partial u}{\partial \xi}$$

and the given partial differential equation is converted to

$$2x\frac{\partial u}{\partial \eta} = y.$$

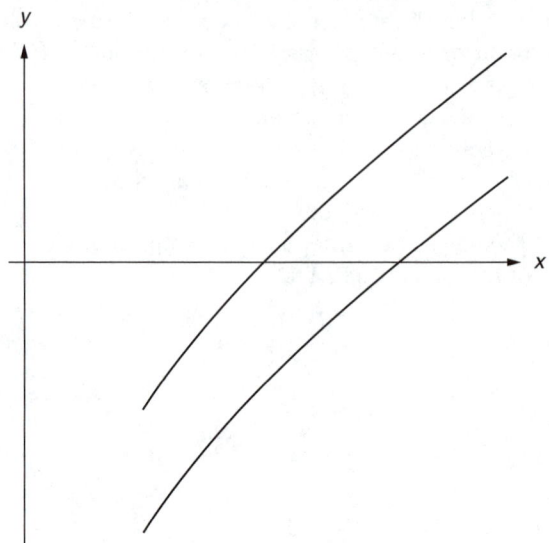

FIGURE 10.1.2:   The characteristics of (10.1.12).

From (10.1.13), $x = \eta$ and $y = \xi + \frac{1}{2}(\eta + \ln \eta)$ so that

$$2\eta \frac{\partial u}{\partial \eta} = \xi + \frac{1}{2}(\eta + \ln \eta)$$

has to be solved. From

$$\frac{\partial u}{\partial \eta} = \frac{\xi}{2\eta} + \frac{1}{4}\left(1 + \frac{1}{\eta}\ln \eta\right) \tag{10.1.14}$$

is deduced

$$u = \tfrac{1}{2}\xi \ln \eta + \tfrac{1}{4}\eta + \tfrac{1}{8}(\ln \eta)^2 + F(\xi).$$

The arbitrary function $F(\xi)$ of ξ is used rather than a constant because of the partial derivative in (10.1.14).

It is required that $u = 2y$ when $x = 1$, i.e., $u = 2\xi + \eta + \ln \eta$ when $\eta = 1$ or $u = 2\xi + 1$ when $\eta = 1$, and this is to hold for all ξ. Therefore

$$2\xi + 1 = \tfrac{1}{4} + F(\xi)$$

whence

$$u = \tfrac{1}{2}\xi \ln \eta + \tfrac{1}{4}\eta + \tfrac{1}{8}(\ln \eta)^2 + 2\xi + \tfrac{3}{4}.$$

Replacing ξ, η by x, y via (10.1.13) we obtain

$$u = \left(y - \tfrac{1}{2}x - \tfrac{1}{2}\ln x\right)\left(2 + \tfrac{1}{2}\ln x\right) + \tfrac{1}{4}x + \tfrac{3}{4} + \tfrac{1}{8}(\ln x)^2.$$

This solution is valid everywhere in $x > 0$, because all the characteristics intersect $x = 1$ and every point in $x > 0$ lies on some characteristic. ▯

Example 10.1.2
Find the solution of

$$-x\frac{\partial u}{\partial x} + y\frac{\partial u}{\partial y} = 1$$

in $0 < x < y$ such that $u = 2x$ on $y = 3x$.
   The characteristics satisfy

$$\frac{dy}{dx} = -\frac{y}{x}$$

whence their equation is $\ln y = -\ln x +$ constant, which may be simplified to

$$xy = \text{constant}.$$

Each characteristic in the first quadrant is therefore the branch of a hyperbola (Figure 10.1.3). The intersecting curves may again be chosen as straight lines and to illustrate the fact that they need not be parallel to the coordinate axes we select them to be parallel to $y = x$. (What would be the objection to making them parallel to $y = -x$?) Thus the substitution

$$\xi = xy, \qquad \eta = y - x \qquad (10.1.15)$$

is made.
   With this change of variable

$$\frac{\partial u}{\partial x} = \frac{\partial u}{\partial \xi}y - \frac{\partial u}{\partial \eta}, \qquad \frac{\partial u}{\partial y} = \frac{\partial u}{\partial \xi}x + \frac{\partial u}{\partial \eta}$$

and the partial differential equation is transformed to

$$(x + y)\frac{\partial u}{\partial \eta} = 1.$$

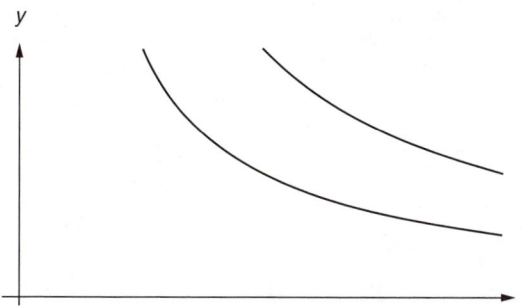

FIGURE 10.1.3:   The characteristic hyperbolae.

Now, if y is eliminated from (10.1.15),

$$x^3 + \eta x - \xi = 0$$

so that $x = \frac{1}{2}\{-\eta \pm (\eta^2 + 4\xi)^{1/2}\}$. However, x and ξ must both be positive in the first quadrant so that the upper sign must be selected, i.e.,

$$x = \tfrac{1}{2}\{-\eta + (\eta^2 + 4\xi)^{1/2}\},$$
$$y = \tfrac{1}{2}\{\eta + (\eta^2 + 4\xi)^{1/2}\}.$$

Hence we are led to

$$(\eta^2 + 4\xi)^{1/2}\frac{\partial u}{\partial \eta} = 1.$$

It follows that

$$\begin{aligned}u &= \ln\{\eta + (\eta^2 + 4\xi)^{1/2}\} + F(\xi) \\ &= \ln 2y + F(xy)\end{aligned}$$

on inserting (10.1.15) and the formulae for x and y just found.

Since $u = 2x$ when $y = 3x$,

$$2x = \ln 6x + F(3x^2)$$

which implies that

$$F(z) = \frac{2}{\sqrt{3}}z^{1/2} - \ln(6z^{1/2}/\sqrt{3})$$

when z is positive. Consequently

$$\begin{aligned}u &= \ln 2y + 2(xy)^{1/2}/\sqrt{3} - \ln\{6(xy)^{1/2}/\sqrt{3}\} \\ &= \ln(y/3x)^{1/2} + 2(xy)^{1/2}/\sqrt{3}.\end{aligned}$$

Since each characteristic intersects $y = 3x$ this solution is valid throughout the first quadrant. Nothing can be said about what happens outside the first quadrant because the initial data are given only on characteristics that are confined to the first quadrant.   ▯

10.2   Another view of characteristics

Suppose that u satisfies

$$a(x,y)\frac{\partial u}{\partial x} + b(x,y)\frac{\partial u}{\partial y} = c(x,y)u + d(x,y) \tag{10.2.1}$$

and we are given that $u = f(y)$ on $x = 0$. It is plausible to try to find a solution near $x = 0$ by making a series expansion in powers of x, i.e.,

$$u(x, y) = u(0, y) + x \left[\frac{\partial}{\partial x} u(x, y) \right]_{x=0} + \tfrac{1}{2} x^2 \left[\frac{\partial^2}{\partial x^2} u(x, y) \right]_{x=0} + \cdots .$$

The first term is known from the information given, namely $u(0, y) = f(y)$. Can we find $\partial u / \partial x$ at $x = 0$ from the governing partial differential equation? Clearly, $\partial u(0, y) / \partial y = f'(y)$ and so putting $x = 0$ in (10.2.1) furnishes

$$a(0, y) \left[\frac{\partial}{\partial x} u(x, y) \right]_{x=0} = c(0, y) f(y) + d(0, y) - b(0, y) f'(y). \qquad (10.2.2)$$

So long as $a(0, y) \neq 0$, this determines $\partial u / \partial x$ at $x = 0$ as a known function of y.

Now take a partial derivative with respect to x of (10.2.1). Then

$$a(x, y) \frac{\partial^2 u}{\partial x^2} = \text{terms in } u, \frac{\partial u}{\partial x}, \frac{\partial u}{\partial y}, \frac{\partial^2 u}{\partial x \partial y}.$$

All the terms on the right-hand side are available at $x = 0$ because

$$\frac{\partial}{\partial y} \left[\frac{\partial}{\partial x} u(x, y) \right]_{x=0}$$

can be deduced from (10.2.2). Therefore $\partial^2 u / \partial x^2$ can be found at $x = 0$ when $a(0, y) \neq 0$. Further partial derivatives with respect to x will supply higher terms in the expansion provided that a, b, c and d are sufficiently differentiable. Thus we can hope to find u off $x = 0$ when $a(0, y) \neq 0$.

If, however, $a(0, y) = 0$ the process fails immediately because the term involving $\partial u / \partial x$ disappears from (10.2.2). In this case u cannot be determined away from $x = 0$. In fact, f cannot be selected arbitrarily because (10.2.2) forces

$$c(0, y) f(y) + d(0, y) = b(0, y) f'(y) \qquad (10.2.3)$$

so that f is indeterminate only to the extent of a constant. To put it another way, the value of f can be prescribed at a single point of $x = 0$; at all other points its value is obtained from (10.2.3).

These properties coincide with those of a characteristic as discovered in the preceding section. Hence $a(0, y) = 0$ makes $x = 0$ a characteristic.

Let us now invoke this result after making a transformation of coordinates. Then we can say that if, after the change of variables $\xi = \phi(x, y)$, $\eta = \psi(x, y)$, the coefficient of $\partial u / \partial \xi$ vanishes when $\xi = 0$ then $\xi = 0$ is a characteristic. If follows that, if the coefficient vanishes whatever ξ, then $\xi = \text{constant}$ is a characteristic.

This way of looking at characteristics will be useful in discussing partial differential equations of the second order.

10.3 Linear partial differential equations of the second order

The linear partial differential equation of the second order that will be discussed in this section is

$$a(x,y)\frac{\partial^2 u}{\partial x^2} + 2b(x,y)\frac{\partial^2 u}{\partial x \partial y} + c(x,y)\frac{\partial^2 u}{\partial y^2}$$

$$+ e(x,y)\frac{\partial u}{\partial x} + f(x,y)\frac{\partial u}{\partial y} + g(x,y)u = 0 \tag{10.3.1}$$

where $a, \ldots, g$ are real. As a first step, let us examine the possibility of finding a solution by means of a power series, as in the preceding section, given initial data on $x = 0$. Two pieces of data will now be appropriate because the differential equation is of the second order. Suppose, indeed, that

$$u(0,y) = F(y), \qquad \left[\frac{\partial}{\partial x}u(x,y)\right]_{x=0} = G(y).$$

Then

$$\frac{\partial}{\partial y}u(0,y) = F'(y), \qquad \frac{\partial^2}{\partial y^2}u(0,y) = F''(y),$$

$$\frac{\partial}{\partial y}\left[\frac{\partial}{\partial x}u(x,y)\right]_{x=0} = G'(y)$$

so that, when $x = 0$, (10.3.1) becomes

$$a(0,y)\left[\frac{\partial^2 u}{\partial x^2}\right]_{x=0} = -\,2b(0,y)G'(y) - c(0,y)F''(y) - e(0,y)G(y)$$

$$- f(0,y)F'(y) - g(0,y)F(y). \tag{10.3.2}$$

Therefore, if $a(0,y) \neq 0$, we can find the second partial derivative of u with respect to x and higher derivatives can be determined from derivatives of (10.3.1) with $x = 0$. In this case u can be continued off $x = 0$.

If, however, $a(0,y) = 0$ the process fails; moreover, F and G cannot both be prescribed arbitrarily because (10.3.2) insists that

$$2b(0,y)G'(y) + c(0,y)F''(y) + e(0,y)G(y) + f(0,y)F'(y) + g(0,y)F(y) = 0.$$

By analogy with the first-order equation, we say that $x = 0$ is a characteristic.

It is now desirable to generalise the approach so that other curves may play the role of characteristics. Change variables by means of the transformation

$$\xi = \phi(x,y), \qquad \eta = \psi(x,y).$$

Then

$$\frac{\partial u}{\partial x} = \frac{\partial u}{\partial \xi}\frac{\partial \phi}{\partial x} + \frac{\partial u}{\partial \eta}\frac{\partial \psi}{\partial x}, \qquad \frac{\partial u}{\partial y} = \frac{\partial u}{\partial \xi}\frac{\partial \phi}{\partial y} + \frac{\partial u}{\partial \eta}\frac{\partial \psi}{\partial y},$$

$$\frac{\partial^2 u}{\partial x^2} = \frac{\partial^2 u}{\partial \xi^2}\left(\frac{\partial \phi}{\partial x}\right)^2 + 2\frac{\partial^2 u}{\partial \xi \partial \eta}\frac{\partial \phi}{\partial x}\frac{\partial \psi}{\partial x} + \frac{\partial^2 u}{\partial \eta^2}\left(\frac{\partial \psi}{\partial x}\right)^2 + \frac{\partial u}{\partial \xi}\frac{\partial^2 \phi}{\partial x^2} + \frac{\partial u}{\partial \eta}\frac{\partial^2 \psi}{\partial x^2},$$

$$\frac{\partial^2 u}{\partial y^2} = \frac{\partial^2 u}{\partial \xi^2}\left(\frac{\partial \phi}{\partial y}\right)^2 + 2\frac{\partial^2 u}{\partial \xi \partial \eta}\frac{\partial \phi}{\partial y}\frac{\partial \psi}{\partial y} + \frac{\partial^2 u}{\partial \eta^2}\left(\frac{\partial \psi}{\partial y}\right)^2 + \frac{\partial u}{\partial \xi}\frac{\partial^2 \phi}{\partial y^2} + \frac{\partial u}{\partial \eta}\frac{\partial^2 \psi}{\partial y^2},$$

$$\frac{\partial^2 u}{\partial x \partial y} = \frac{\partial^2 u}{\partial \xi^2}\frac{\partial \phi}{\partial x}\frac{\partial \phi}{\partial y} + \frac{\partial^2 u}{\partial \xi \partial \eta}\left(\frac{\partial \phi}{\partial x}\frac{\partial \psi}{\partial y} + \frac{\partial \phi}{\partial y}\frac{\partial \psi}{\partial x}\right) + \frac{\partial^2 u}{\partial \eta^2}\frac{\partial \psi}{\partial x}\frac{\partial \psi}{\partial y}$$
$$+ \frac{\partial u}{\partial \xi}\frac{\partial^2 \phi}{\partial x \partial y} + \frac{\partial u}{\partial \eta}\frac{\partial^2 \psi}{\partial x \partial y}.$$

Substitution of these formulae into (10.3.1) leads to

$$a_1 \frac{\partial^2 u}{\partial \xi^2} + 2b_1 \frac{\partial^2 u}{\partial \xi \partial \eta} + c_1 \frac{\partial^2 u}{\partial \eta^2} + e_1 \frac{\partial u}{\partial \xi} + f_1 \frac{\partial u}{\partial \eta} + g_1 u = 0 \qquad (10.3.3)$$

where

$$a_1 = a\left(\frac{\partial \phi}{\partial x}\right)^2 + 2b\frac{\partial \phi}{\partial x}\frac{\partial \phi}{\partial y} + c\left(\frac{\partial \phi}{\partial y}\right)^2, \qquad (10.3.4)$$

$$b_1 = a\frac{\partial \phi}{\partial x}\frac{\partial \psi}{\partial x} + b\left(\frac{\partial \phi}{\partial x}\frac{\partial \psi}{\partial y} + \frac{\partial \phi}{\partial y}\frac{\partial \psi}{\partial x}\right) + c\frac{\partial \phi}{\partial y}\frac{\partial \psi}{\partial y}, \qquad (10.3.5)$$

$$c_1 = a\left(\frac{\partial \psi}{\partial x}\right)^2 + 2b\frac{\partial \psi}{\partial x}\frac{\partial \psi}{\partial y} + c\left(\frac{\partial \psi}{\partial y}\right)^2, \qquad (10.3.6)$$

$$e_1 = a\frac{\partial^2 \phi}{\partial x^2} + 2b\frac{\partial^2 \phi}{\partial x \partial y} + c\frac{\partial^2 \phi}{\partial y^2} + e\frac{\partial \phi}{\partial x} + f\frac{\partial \phi}{\partial y}, \qquad (10.3.7)$$

$$f_1 = a\frac{\partial^2 \psi}{\partial x^2} + 2b\frac{\partial^2 \psi}{\partial x \partial y} + c\frac{\partial^2 \psi}{\partial y^2} + e\frac{\partial \psi}{\partial x} + f\frac{\partial \psi}{\partial y}, \qquad (10.3.8)$$

$$g_1 = g. \qquad (10.3.9)$$

From our earlier discussion, ξ = constant will be a characteristic if the coefficient of $\partial^2 u/\partial \xi^2$ vanishes for this value of ξ. This requires $a_1 = 0$ and so, from (10.3.4), $\phi(x,y)$ = constant is a characteristic when

$$a\left(\frac{\partial \phi}{\partial x}\right)^2 + 2b\frac{\partial \phi}{\partial x}\frac{\partial \phi}{\partial y} + c\left(\frac{\partial \phi}{\partial y}\right)^2 = 0. \qquad (10.3.10)$$

We may also seek to make ψ = constant a characteristic. Then $c_1 = 0$ and

$$a\left(\frac{\partial \psi}{\partial x}\right)^2 + 2b\frac{\partial \psi}{\partial x}\frac{\partial \psi}{\partial y} + c\left(\frac{\partial \psi}{\partial y}\right)^2 = 0. \qquad (10.3.11)$$

The slope of the curve $\phi = $ constant is given by

$$\frac{dy}{dx} = -\frac{\partial\phi/\partial x}{\partial\phi/\partial y}$$

and consequently, from (10.3.10), it satisfies

$$a\left(\frac{dy}{dx}\right)^2 - 2b\frac{dy}{dx} + c = 0. \qquad (10.3.12)$$

Comparison of the structures of (10.3.10) and (10.3.11) reveals that the slope of the characteristic $\psi = $ constant also complies with (10.3.12). Thus the slopes of both sets of characteristics are provided by the solution of (10.3.12), namely

$$\frac{dy}{dx} = \frac{1}{a}\{b \pm (b^2 - ac)^{1/2}\}. \qquad (10.3.13)$$

There are three possible cases to consider.

(I) If $ac > b^2$ in some region D there are two families of complex characteristics. The partial differential equation is then said to be **elliptic** in D. A typical example of an elliptic differential equation is Laplace's equation

$$\frac{\partial^2 u}{\partial x^2} + \frac{\partial^2 u}{\partial y^2} = 0$$

in which $a = 1$, $b = 0$ and $c = 1$.

(II) If $ac < b^2$ in some region D there are two families of real characteristics. The partial differential equation is called **hyperbolic** in D. A standard illustration of a hyperbolic differential equation is the equation of one-dimensional wave propagation

$$\frac{\partial^2 u}{\partial x^2} - \frac{1}{\alpha^2}\frac{\partial^2 u}{\partial y^2} = 0$$

in which $a = 1$, $b = 0$ and $c = -1/\alpha^2$.

(III) If $ac = b^2$ throughout some region D the two slopes coincide and there is one family of real characteristics. The partial differential equation is now **parabolic** in D. The equation of heat conduction

$$\frac{\partial^2 u}{\partial x^2} = \frac{\partial u}{\partial y}$$

in which $a = 1$, $b = 0$ and $c = 0$ is parabolic.

The behaviour of the solutions of the three types of partial differential equations is quite different and the initial or boundary conditions that they have to obey are usually distinctive. In applications, they normally originate from attempts to model totally different phenomena. One may expect the problems in which the partial differential equation is elliptic in some parts of the (x, y)-plane and hyperbolic in others are particularly troublesome. We shall now undertake a more detailed study of the various types.

10.4 Elliptic partial differential equations

In the elliptic case we shall take the slope of $\phi = $ constant as

$$\frac{dy}{dx} = \frac{1}{a}\{b + (b^2 - ac)^{1/2}\} \tag{10.4.1}$$

and that of $\psi = $ constant as

$$\frac{dy}{dx} = \frac{1}{a}\{b - (b^2 - ac)^{1/2}\}. \tag{10.4.2}$$

Since the square root is purely imaginary, the two slopes are complex conjugates. It follows that ϕ and ψ are complex conjugates, i.e., $\psi = \phi^*$ and $\eta = \xi^*$, the asterisk denoting a complex conjugate.

In terms of ϕ and ψ, (10.4.1) and (10.4.2) can be expressed as

$$\frac{\partial \phi}{\partial x} = -\frac{1}{a}\{b + (b^2 - ac)^{1/2}\}\frac{\partial \phi}{\partial y}, \tag{10.4.3}$$

$$\frac{\partial \psi}{\partial x} = -\frac{1}{a}\{b + (b^2 - ac)^{1/2}\}\frac{\partial \psi}{\partial y}. \tag{10.4.4}$$

The characteristics have been chosen so that $a_1 = 0$ and $c_1 = 0$, but notice that neither a nor c can be zero because of the condition for ellipticity. For b_1 substitute from (10.4.3) and (10.4.4) into (10.3.5) to obtain

$$\begin{aligned} b_1 &= \frac{1}{a}\frac{\partial \phi}{\partial y}\frac{\partial \psi}{\partial y}\{b^2 - (b^2 - ac)\} - \frac{b}{a}\frac{\partial \phi}{\partial y}\frac{\partial \psi}{\partial y}2b + c\frac{\partial \phi}{\partial y}\frac{\partial \psi}{\partial y} \\ &= \frac{2}{a}\frac{\partial \phi}{\partial y}\frac{\partial \psi}{\partial y}(ac - b^2). \end{aligned} \tag{10.4.5}$$

Since ϕ and ψ are complex conjugates (10.4.5) may be rewritten as

$$b_1 = \frac{2}{a}\left|\frac{\partial \phi}{\partial y}\right|^2(ac - b^2)$$

showing that b_1 is real and has the same sign as a.

Furthermore, (10.3.7) and (10.3.8) indicate that $f_1 = e_1^*$ so that the partial differential equation has been transformed to

$$2b_1 \frac{\partial^2 u}{\partial \xi \partial \eta} + e_1 \frac{\partial u}{\partial \xi} + e_1^* \frac{\partial u}{\partial \eta} + g_1 u = 0. \tag{10.4.6}$$

In this form, sometimes called the **characteristic form** of an elliptic equation, the partial differential equation involves complex coordinates. To avoid these a further transformation is made. Let

$$\lambda = \tfrac{1}{2}(\xi + \eta), \qquad \mu = \tfrac{1}{2i}(\xi - \eta).$$

Then λ is the real part of ξ and μ is the imaginary part of ξ so that both λ and μ are real. Moreover

$$\frac{\partial u}{\partial \xi} = \frac{1}{2} \frac{\partial u}{\partial \lambda} + \frac{1}{2i} \frac{\partial u}{\partial \mu}, \qquad \frac{\partial u}{\partial \eta} = \frac{1}{2} \frac{\partial u}{\partial \lambda} - \frac{1}{2i} \frac{\partial u}{\partial \mu},$$

$$\frac{\partial^2 u}{\partial \xi \partial \eta} = \frac{1}{4} \frac{\partial^2 u}{\partial \lambda^2} + \frac{1}{4} \frac{\partial^2 u}{\partial \mu^2}$$

so that (10.4.6) becomes

$$\frac{1}{2} b_1 \left(\frac{\partial^2 u}{\partial \lambda^2} + \frac{\partial^2 u}{\partial \mu^2} \right) + \frac{1}{2} \frac{\partial u}{\partial \lambda} (e_1 + e_1^*) + \frac{1}{2i} \frac{\partial u}{\partial \mu} (e_1 - e_1^*) + g_1 u = 0.$$

By division by $\frac{1}{2} b_1$ we derive

$$\frac{\partial^2 u}{\partial \lambda^2} + \frac{\partial^2 u}{\partial \mu^2} + e_2 \frac{\partial u}{\partial \lambda} + f_2 \frac{\partial u}{\partial \mu} + g_2 u = 0 \tag{10.4.7}$$

where e_2, f_2 and g_2 are all real; in particular,

$$g_2 = 2g_1/b_1 = ga \left/ (ac - b^2) \left| \frac{\partial \phi}{\partial y} \right|^2 \right. \tag{10.4.8}$$

from (10.3.9) and (10.4.5).

Equation (10.4.7) is known as the **normal form** of an elliptic partial differential equation. The normal form appears in its simplest guise as

$$\frac{\partial^2 u}{\partial x^2} + \frac{\partial^2 u}{\partial y^2} + k^2 u = 0 \tag{10.4.9}$$

where k^2 is a real constant. If $k^2 = 0$ it reduces to Laplace's equation and occurs in problems in potential theory. If k^2 is positive it arises in the study of two-dimensional harmonic waves such as those produced by a vibrating membrane. The case when k^2 is negative typically results from trying to solve the two-dimensional wave equation by certain methods.

Usually, solutions of elliptic partial differential equations are sought that satisfy prescribed boundary conditions. There are four of common occurrence.

(1) *Dirichlet problem.* (a) In the *interior problem* u is given at every point of a closed curve C and has to be found inside C. (b) For the *exterior problem* u is given on a closed curve C and has to be determined outside C. A supplementary condition specifying the behaviour at infinity is normally imposed. For example, there might be a requirement that $\partial u/\partial r \to 0$ as $r = (x^2 + y^2)^{1/2} \to \infty$ in potential theory.

(2) *Neumann problem.* (a) In the interior problem the derivative of u along the normal to a closed curve C, i.e., $\partial u/\partial n$, is prescribed and u is required inside C. Often $\partial u/\partial n$ cannot be specified arbitrarily on C; for instance, if $k^2 = 0$ in (10.4.9), $\int_C (\partial u/\partial n)ds = 0$ by the divergence theorem so that any values ascribed to $\partial u/\partial n$ on C must be compatible with this relation. (b) $\partial u/\partial n$ is given on the closed curve C and u is to be found outside, usually subject to a condition at infinity as in (1b).

(3) *Mixed problem.* Here the closed curve C is split into two portions C_1 and C_2. On C_1, u is given and on $C_2, \partial u/\partial n$ is specified. Again, both the interior and exterior cases may arise.

(4) *Impedance or Robin problem.* In this case $hu + \partial u/\partial n$ is prescribed on the closed curve C, h being a known function. It can be regarded as including the three preceding cases.

Before leaving the subject we give a theorem which guarantees that the solution to the boundary value problem is unique in suitable circumstances.

THEOREM 10.4.1
Let the partial differential equation (10.3.1) be elliptic ($ac > b^2$) in a simply connected domain D with $g \le 0$ and a, c positive. Then, if $u = 0$ on a simple closed curve C in $D, u = 0$ at all points inside C.

Remark that a and c must have the same sign in order to meet the condition for an elliptic equation.

PROOF Change the variables so that the equation goes over to its normal form (10.4.7). The transformation maps D into a domain D_1 and C into a simple closed curve C_1. Note that the mapping is one-to-one because the Jacobian

$$\frac{\partial \phi}{\partial x}\frac{\partial \psi}{\partial y} - \frac{\partial \psi}{\partial x}\frac{\partial \phi}{\partial y} = -\frac{2i}{a}(ac - b^2)^{1/2}\left|\frac{\partial \phi}{\partial y}\right|^2.$$

If $\partial \phi/\partial y$ vanishes so does $\partial \phi/\partial x$ on account of (10.4.3), contrary to a characteristic having a definite slope. Therefore the Jacobian is non-zero and the mapping is one-to-one.

Also, from (10.4.8) and the hypotheses of the theorem, $g_2 \le 0$.

Suppose that u is continuous and bounded inside C. It will have the same properties inside C_1. As a continuous function u attains both its upper bound M (which cannot be negative since $u = 0$ on C_1) and its lower bound m. Let M be achieved at $\lambda = \lambda_0, \mu = \mu_0$. It follows that

$$\left(\frac{\partial u}{\partial \lambda}\right)_0 = 0, \qquad \left(\frac{\partial u}{\partial \mu}\right)_0 = 0$$

where $(f)_0$ means the value of f at $\lambda = \lambda_0, \mu = \mu_0$. Putting $\lambda = \lambda_0, \mu = \mu_0$ in (10.4.7) we obtain

$$\left(\frac{\partial^2 u}{\partial \lambda^2}\right)_0 + \left(\frac{\partial^2 u}{\partial \mu^2}\right)_0 + g_2 M = 0. \tag{10.4.10}$$

But, for a maximum

$$\left(\frac{\partial^2 u}{\partial \lambda^2}\right)_0 \leq 0, \qquad \left(\frac{\partial^2 u}{\partial \mu^2}\right)_0 \leq 0, \tag{10.4.11}$$

$$\left(\frac{\partial^2 u}{\partial \lambda^2}\right)_0 \left(\frac{\partial^2 u}{\partial \mu^2}\right)_0 > \left(\frac{\partial^2 u}{\partial \lambda \partial \mu}\right)_0^2. \tag{10.4.12}$$

If $g = 0$ then $g_2 = 0$ and we could satisfy (10.4.10) and (10.4.11) but not (10.4.12); the same is true if $M = 0$. On the other hand, if $M > 0$ and $g < 0$, then $g_2 M < 0$ and to satisfy (10.4.10) one at least of $(\partial^2 u/\partial \lambda^2)_0, (\partial^2 u/\partial \mu^2)_0$ must be positive, contradicting (10.4.11). Thus the conditions for a maximum cannot be met inside C_1 and so u must not be greater than zero inside C_1.

A similar argument based on the lower bound m reveals that u cannot be less than zero inside C_1.

Hence u is zero inside C_1 and therefore zero inside C. The theorem is proved. ⬜

COROLLARY 10.4.1 (Uniqueness Property)
Under the conditions of Theorem 10.4.1 there is only one u which solves (10.3.1) and takes given values on C.

Another way of describing this corollary is to say that the solution of the interior Dirichlet problem is unique for an elliptic partial differential equation in which a and g have opposite signs.

PROOF Suppose there were two possible solutions u_1 and u_2. Then $u_1 - u_2$ is zero on C and also satisfies (10.3.1). Consequently, Theorem 10.4.1 implies that $u_1 - u_2$ vanishes inside C and the corollary has been demonstrated. ⬜

It is important to observe that when a and g have the same signs Theorem 10.4.1 fails, in general. A counter-example is furnished by

$$\frac{\partial^2 u}{\partial x^2} + \frac{\partial^2 u}{\partial y^2} + (m^2 + n^2)u = 0$$

where m and n are integers. The solution

$$u = A \sin mx \sin ny$$

vanishes on the perimeter of the square $0 \le x \le \pi, 0 \le y \le \pi$ but is clearly not zero at all interior points.

10.5 Parabolic partial differential equations

In the parabolic case $ac = b^2$, there is only one family of real characteristics whose slope, according to (10.3.13), is given by

$$\frac{dy}{dx} = \frac{b}{a}.$$

One variable is fixed by $\xi = \phi(x, y)$ where

$$-\frac{\partial \phi}{\partial x} = \frac{b}{a} \frac{\partial \phi}{\partial y} \tag{10.5.1}$$

and the other variable $\eta = \psi(x, y)$ is selected in the same way as for first-order equations where only one set of characteristics is available. Thus $a_1 = 0$ and, assuming $a \neq 0, c = b^2/a$ so that,

$$b_1 = \frac{\partial \psi}{\partial x}\left(a\frac{\partial \phi}{\partial x} + b\frac{\partial \phi}{\partial y}\right) + \frac{\partial \psi}{\partial y}\left(b\frac{\partial \phi}{\partial x} + \frac{b^2}{a}\frac{\partial \phi}{\partial y}\right) = 0$$

from (10.5.1). Hence (10.3.3) reduces to

$$c_1 \frac{\partial^2 u}{\partial \eta^2} + e_1 \frac{\partial u}{\partial \xi} + f_1 \frac{\partial u}{\partial \eta} + g_1 u = 0. \tag{10.5.2}$$

If $a \equiv 0$ then $b \equiv 0$ and the original equation (10.3.1) is already in the form (10.5.2) without introducing new variables. Thus (10.5.2) is the characteristic or normal form for a parabolic partial differential equation; here, in contrast to the elliptic case, there is no distinction between the characteristic and normal forms.

It may happen that ψ can be chosen so that $f_1 = 0$ and then (10.5.2) has the simpler structure of

$$\frac{\partial^2 u}{\partial \eta^2} + e_2 \frac{\partial u}{\partial \xi} + g_2 u = 0. \tag{10.5.3}$$

10.6 Hyperbolic partial differential equations

In the hyperbolic case there are two families of real characteristics, since $b^2 > ac$. We may take ϕ and ψ to be defined as in (10.4.1) and (10.4.2) when $a \neq 0$ but they are now real. Thus $a_1 = 0, c_1 = 0$ and b_1 is given by (10.4.5),

i.e.,

$$b_1 = 2\frac{\partial \phi}{\partial y}\frac{\partial \psi}{\partial y}\frac{ac - b^2}{a}.$$

Consequently, b_1 is non-zero. If $a \equiv 0$, put $\phi = x$ and specify the slope of $\psi = $ constant by $dy/dx = c/2b$. Then $b_1 = b\partial\psi/\partial y$ which is non-zero by virtue of b not being permitted to be zero in order to retain the hyperbolic character. Hence the characteristic or normal form of a hyperbolic partial differential equation is

$$\frac{\partial^2 u}{\partial \xi \partial \eta} + e_2\frac{\partial u}{\partial \xi} + f_2\frac{\partial u}{\partial \eta} + g_2 u = 0. \tag{10.6.1}$$

It is now convenient to collect together the normal forms of the three types:

Elliptic:

$$\frac{\partial^2 u}{\partial \lambda^2} + \frac{\partial^2 u}{\partial \mu^2} + e_2\frac{\partial u}{\partial \lambda} + f_2\frac{\partial u}{\partial \mu} + g_2 u = 0.$$

Parabolic:

$$\frac{\partial^2 u}{\partial \eta^2} + e_2\frac{\partial i}{\partial \xi} + f_2\frac{\partial u}{\partial \eta} + g_2 u = 0.$$

Hyperbolic:

$$\frac{\partial^2 u}{\partial \xi \partial \eta} + e_2\frac{\partial u}{\partial \xi} + f_2\frac{\partial u}{\partial \eta} + g_2 u = 0.$$

Note that, in all cases, the variables are real and the type fixed by the structure of the partial derivatives of the second order solely; the remaining terms in the partial differential equation play no role in determining the type, although their presence or absence will influence the solution.

10.7 The wave equation

The one-dimensional wave equation

$$\frac{\partial^2 u}{\partial x^2} - \frac{1}{a_0^2}\frac{\partial^2 u}{\partial t^2} = 0 \tag{10.7.1}$$

in which a_0 is a positive constant is hyperbolic. For the characteristics we have

$$\left(\frac{dt}{dx}\right)^2 = \frac{1}{a_0^2}$$

so that

$$\frac{dt}{dx} = \pm\frac{1}{a_0}$$

with solutions $x \pm a_0 t = $ constant. Hence the characteristic variables are $\xi = x - a_0 t, \eta = x + a_0 t$. In terms of them (10.7.1) becomes

$$\frac{\partial^2 u}{\partial \xi \partial \eta} = 0. \tag{10.7.2}$$

This tells us that $\partial u / \partial \eta$ is a function of η only and hence that

$$u = f(\xi) + g(\eta) \tag{10.7.3}$$

where f and g are arbitrary functions of ξ and η, respectively. The formula (10.7.3) constitutes the general solution of (10.7.2). Correspondingly,

$$u = f(x - a_0 t) + g(x + a_0 t) \tag{10.7.4}$$

provides the general solution of (10.7.1).

An initial value problem for the wave equation can be solved by means of the general solution. Suppose the initial conditions are

$$u(x, 0) = h(x), \qquad \left[\frac{\partial}{\partial t} u(x, t)\right]_{t=0} = k(x)$$

for $A \le x \le B$. Then, from (10.7.4),

$$f(x) + g(x) = h(x), \tag{10.7.5}$$
$$a_0\{-f'(x) + g'(x)\} = k(x), \tag{10.7.6}$$

for $A \le x \le B$. Integration of (10.7.6) gives

$$-f(x) + g(x) = \frac{1}{a_0} \int_A^x k(v)dv + C \tag{10.7.7}$$

where C is a constant. Combining (10.7.5) and (10.7.7) we obtain

$$g(x) = \frac{1}{2}\left(h(x) + \frac{1}{a_0}\int_A^x k(v)dv + C\right),$$

$$f(x) = \frac{1}{2}\left(h(x) - \frac{1}{a_0}\int_A^x k(v)dv - C\right)$$

for $A \le x \le B$. Hence, if $A \le x - a_0 t \le B$ and $A \le x + a_0 t \le B$,

$$u = \frac{1}{2}\left(h(x - a_0 t) + \frac{1}{a_0}\int_A^{x-a_0 t} k(v)dv + C\right)$$
$$+ \frac{1}{2}\left(h(x + a_0 t) - \frac{1}{a_0}\int_A^{x+a_0 t} k(v)dv - C\right)$$

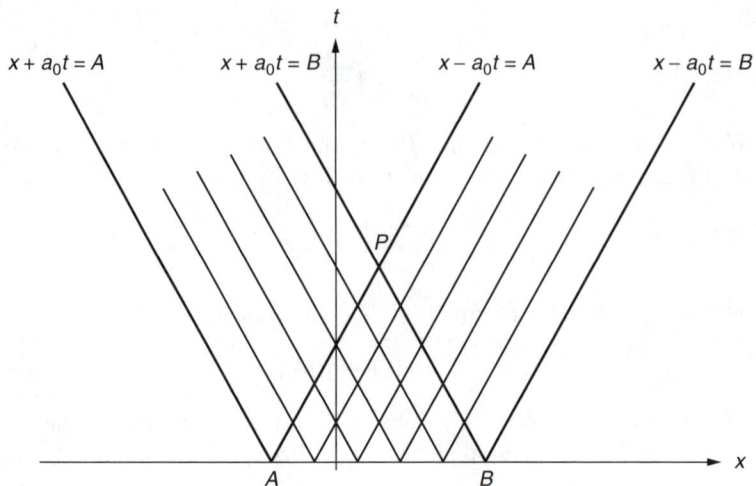

FIGURE 10.7.1: Region where solution of the initial value problem is available.

from which we deduce that

$$u = \tfrac{1}{2}\{h(x - a_0 t) + h(x + a_0 t)\} + \frac{1}{2a_0} \int_{x-a_0 t}^{x+a_0 t} k(v)\,dv. \qquad (10.7.8)$$

This constitutes the solution to the initial value problem subject to the restrictions imposed on $x - a_0 t$ and $x + a_0 t$. It is known as **d'Alembert's solution**.

The restrictions on $x \pm a_0 t$ confine the solution to the triangular region APB of Figure 10.7.1, which is the part of the $(x, y) =$ plane in $t > 0$ common to the characteristics which intersect AB. Nothing can be said about the solution outside this triangle without additional information. It may occur that these initial data are given on the whole x-axis; in these circumstances A is $-\infty$ and B is $+\infty$ so that d'Alembert's solution is valid in the entire half-plane $t > 0$.

Quite often, the initial value problem will be mixed up with a boundary value problem. Suppose that the initial data are prescribed as before but in the interval $(0,1)$ of the x-axis and that there are also boundary conditions $u(0, t) = 0, u(1, t) = 0$ for all t. To be consistent with these conditions it is necessary that $h(0) = 0, h(1) = 0$. The form (10.7.4) may still be used and, from the initial conditions,

$$f(x) = \frac{1}{2}\left(h(x) - \frac{1}{a_0}\int_0^x k(v)\,dv - C\right),$$

$$g(x) = \frac{1}{2}\left(h(x) + \frac{1}{a_0}\int_0^x k(v)\,dv + C\right)$$

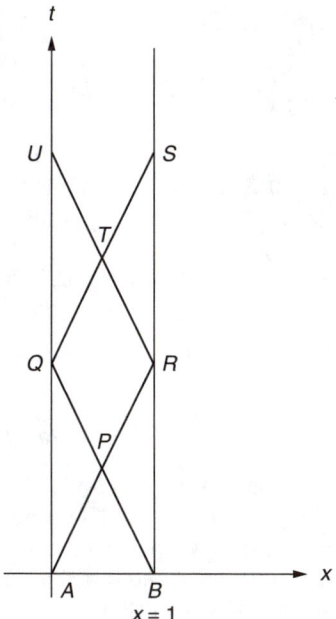

FIGURE 10.7.2: Reflection of characteristics by a boundary.

for x in $(0, 1)$. Thus (10.7.8) continues to be the solution in the triangle APB of Figure 10.7.2. We know f in PRB but not g because $x + a_0t - 1$ is positive there. Similarly, f is needed in APQ. The boundary conditions are used to supply the missing links; they imply that

$$f(-a_0t) + g(a_0t) = 0, \qquad f(1 - a_0t) + g(1 + a_0t) = 0$$

for $t \geq 0$ and therefore that

$$f(-y) + g(y) = 0, \tag{10.7.9}$$
$$f(1 - y) + g(1 + y) = 0 \tag{10.7.10}$$

for $y \geq 0$. From (10.7.9), it follows that

$$f(y) = -g(-y) = -\frac{1}{2}\left(h(-y) + \frac{1}{a_0}\int_0^{-y} k(v)dv + C\right) \tag{10.7.11}$$

when $-1 \leq y \leq 0$. This gives $f(x - a_0t)$ in the parallelogram $ARSQ$ where $-1 \leq x - a_0t \leq 0$. In particular, for APQ where $-1 \leq x - a_0t \leq 0$ and $0 \leq x + a_0t \leq 1$,

$$u = \tfrac{1}{2}\{h(x + a_0t) - h(a_0t - x)\} + \frac{1}{2a_0}\int_{a_0t-x}^{a_0t+x} k(v)dv.$$

In a similar way, (10.7.10) supplies

$$g(y) = -f(2-y) = -\frac{1}{2}\left(h(2-y) - \frac{1}{a_0}\int_0^{2-y} k(v)dv - C\right) \qquad (10.7.12)$$

for $1 \le y \le 2$; $g(y)$ for $2 \le y \le 3$ may also be obtained by means of the formula for $f(y)$ in (10.7.11). Via (10.7.12) we have $g(x + a_0t)$ in the parallelogram $BQUR$. Thus, in BPR where $0 \le x - a_0t \le 1$ and $1 \le x + a_0t \le 2$,

$$u = \tfrac{1}{2}\{h(x - a_0t) - h(2 - x - a_0t)\} + \frac{1}{2a_0}\int_{x-a_0t}^{2-x-a_0t} k(v)dv$$

and in $PQTR$ where both (10.7.11) and (10.7.12) are applicable:

$$u = -\tfrac{1}{2}\{h(a_0t - x) + h(2 - a_0t - x)\} + \frac{1}{2a_0}\int_{a_0t-x}^{2-x-a_0t} k(v)dv.$$

To find u in QTU it is necessary to employ (10.7.9) again but it is clear that we can determine $u(x,t)$ for $0 \le x \le 1$ and any t. The diagram makes it evident that the characteristics are being reflected at the boundaries and that, at each reflection, a new form of the solution has to be constructed.

A more general form of the wave equation is

$$\frac{\partial^2 u}{\partial x^2} = \frac{1}{\{a(x)\}^2}\frac{\partial^2 u}{\partial t^2} \qquad (10.7.13)$$

where a is a function of x only. The characteristic coordinates are now given by

$$\xi = t - \int^x \frac{dv}{a(v)}, \qquad \eta = t + \int^x \frac{dv}{a(v)}.$$

Hence

$$\frac{\partial u}{\partial x} = -\frac{1}{a}\frac{\partial u}{\partial \xi} + \frac{1}{a}\frac{\partial u}{\partial \eta}, \qquad \frac{\partial u}{\partial t} = \frac{\partial u}{\partial \xi} + \frac{\partial u}{\partial \eta},$$

$$\frac{\partial^2 u}{\partial x^2} = \left(\frac{\partial^2 u}{\partial \xi^2} - 2\frac{\partial^2 u}{\partial \xi \partial \eta} + \frac{\partial^2 u}{\partial \eta^2}\right)\frac{1}{a^2} - \left(\frac{\partial u}{\partial \eta} - \frac{\partial u}{\partial \xi}\right)\frac{a'}{a^2}.$$

$$\frac{\partial^2 u}{\partial t^2} = \frac{\partial^2 u}{\partial \xi^2} + 2\frac{\partial^2 u}{\partial \xi \partial \eta} + \frac{\partial^2 u}{\partial \eta^2}.$$

Accordingly (10.7.13) transforms to

$$4\frac{\partial^2 u}{\partial \xi \partial \eta} + a'\left(\frac{\partial u}{\partial \eta} - \frac{\partial u}{\partial \xi}\right) = 0. \qquad (10.7.14)$$

To express a' in terms of ξ and η, notice that

$$\eta - \xi = 2\int^x \frac{dv}{a(v)}$$

so that x is a function of $\eta - \xi$. Thus a' will also be some function of $\eta - \xi$.

In the particular case when $a = Ax, a' = A$ and the substitution $\xi = X/A, \eta = Y/A$ converts (10.7.14) to

$$4\frac{\partial^2 u}{\partial X \partial Y} + \frac{\partial u}{\partial Y} - \frac{\partial u}{\partial X} = 0.$$

More generally, if $a = Ax^\nu, a' = \nu Ax^{\nu-1}$ and, when $\nu \neq 1$,

$$\eta - \xi = -2/A(\nu - 1)x^{\nu-1}$$

with the consequence that

$$a' = \frac{-2\nu}{\nu - 1}\frac{1}{\eta - \xi}.$$

In this case (10.7.14) becomes

$$\frac{\partial^2 u}{\partial \xi \partial \eta} = \frac{\nu}{2(\nu - 1)}\frac{1}{\xi - \eta}\left(\frac{\partial u}{\partial \xi} - \frac{\partial u}{\partial \eta}\right). \tag{10.7.15}$$

This is related to the Euler-Darboux equation to be discussed in Section 10.9, where explicit solutions are found for certain values of ν.

10.8   Typical problems for the hyperbolic equation

In order to see what problems could be solved by tracing characteristics it is plausible to assume, in view of the importance of the second derivative in the classification, that, when the conditions are such that

$$\frac{\partial^2 u}{\partial \xi \partial \eta} = 0 \tag{10.8.1}$$

can be resolved, a similar problem will be reasonably posed for (10.6.1) and thereby for the general hyperbolic partial differential equation. Some examples will now be given where resolution is possible, together with the corresponding interpretation for the general equation.

(a)   The two-characteristic problem

Suppose that u is given on the portion of the characteristic $0 \leq \xi \leq a, \eta = 0$, say $u = h(\xi)$, and on the piece of characteristic $\xi = 0, 0 \leq \eta \leq b$ by $u = k(\eta)$ (Figure 10.8.1). From (10.7.3)

$$u = f(\xi) + g(\eta).$$

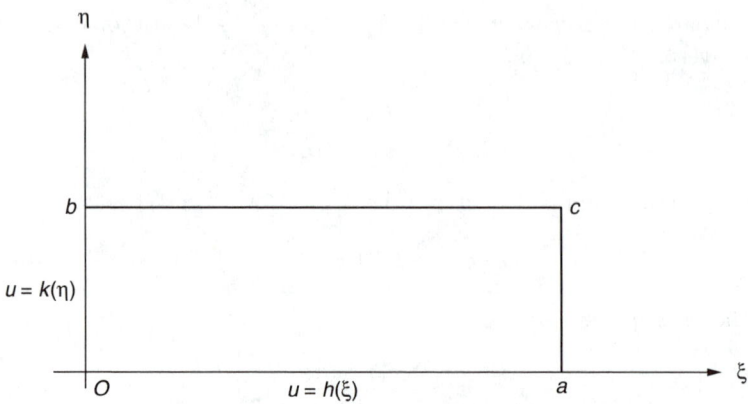

FIGURE 10.8.1: The two-characteristic problem.

By putting $\eta = 0$ we obtain

$$f(\xi) + g(0) = h(\xi) \tag{10.8.2}$$

for $0 \leq \xi \leq a$ and by placing $\xi = 0$

$$f(0) + g(\eta) = k(\eta) \tag{10.8.3}$$

for $0 \leq \eta \leq b$. Hence

$$\begin{aligned} u &= h(\xi) + k(\eta) - f(0) - g(0) \\ &= h(\xi) + k(\eta) - h(0) \end{aligned}$$

on making $\xi = 0$ in (10.8.2). The same result is obtained from (10.8.3) since $h(0) = k(0)$ for the data on u to be consistent at the origin.

This solution will hold everywhere in the rectangle $Oacb$, which is the region common to the characteristics that intersect the data lines.

Translating this back to the (x, y)-plane and general partial differential equation, we see that giving u on the two characteristics $O'A, O'B$ (Figure 10.8.2) determines u in the shaded region $O'ACB$ where AC and BC are the other characteristics through A and B, respectively.

(b)   The mixed problem

In the mixed problem, $u = h(\xi)$ on $0 \leq \xi \leq a, \eta = 0$ as before but either u or a linear combination of its first partial derivatives is given on an arc Oc in the first quadrant, the arc being met at most once by any characteristic.

Suppose that the equation of Oc is $\eta = \theta(\xi)$. Take a new variable $\chi = \theta(\xi)$. Then Oc becomes the straight line $\eta = \chi$ and the form of (10.8.1) is unaltered so that the characteristics remain parallel to the coordinate axes. There is

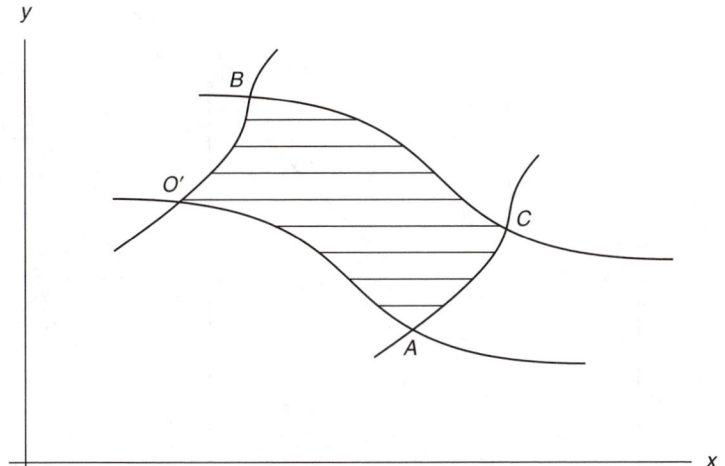

FIGURE 10.8.2: The two-characteristic problem in the general case.

therefore no loss of generality in selecting Oc as the straight line $\eta = \xi$ in the first place.

Suppose that $u = k(\xi)$ on Oc with $k(0) = h(0)$. Then we have

$$f(\xi) + g(0) = h(\xi), \qquad (10.8.4)$$
$$f(\xi) + g(\xi) = k(\xi) \qquad (10.8.5)$$

for $0 \le \xi \le a$. Therefore

$$u = h(\xi) + k(\eta) - h(\eta)$$

for $0 \le \xi \le a, 0 \le \eta \le a$. Consequently, u has been determined everywhere within the square with side Oa (Figure 10.8.3).

Instead of u being specified on Oc, the linear combination

$$\alpha(\xi)\frac{\partial u}{\partial \xi} + \beta(\xi)\frac{\partial u}{\partial \eta} = k(\xi)$$

might be given on Oc. In that case (10.8.5) will be replaced by

$$\alpha(\xi)f'(\xi) + \beta(\xi)g'(\xi) = k(\xi).$$

Since f is known from (10.8.4), g can be calculated by integration and again u has been discovered in the square.

Going back to the (x, y)-plane we can say that when u is given on the characteristic arc $O'A$ (Figure 10.8.4) and either u or a linear combination of its first partial derivatives is specified on the arc $O'C$, which lies between the characteristics $O'A, O'B$ and is met at most once by any characteristic, then u is determined in the shaded region $O'ACB$ where AC and BC are characteristics.

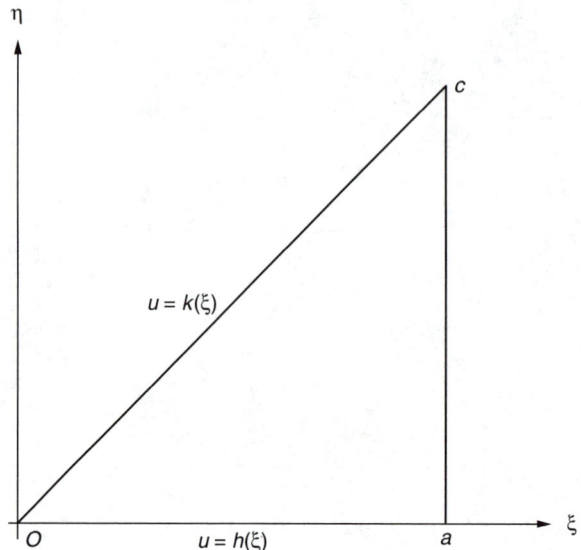

FIGURE 10.8.3: The mixed problem.

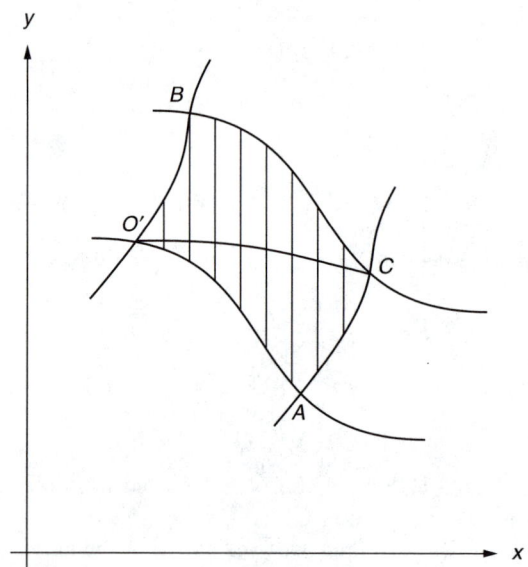

FIGURE 10.8.4: The general mixed problem.

(c) Cauchy's problem

Cauchy's problem is to determine u when u, $\partial u/\partial \xi$ and $\partial u/\partial \eta$ are given on Oc, where Oc has the same properties as in the previous section, i.e., it is not a characteristic and is met at most once by a characteristic. As before, there is no loss of generality in taking Oc to be $\eta = \xi$.

Suppose that on Oc

$$u = h(\xi), \qquad \frac{\partial u}{\partial \xi} = k(\xi), \qquad \frac{\partial u}{\partial \eta} = l(\xi).$$

There is a connection between h, k and l because

$$h'(\xi) = \frac{du}{d\xi} = \frac{\partial u}{\partial \xi}\frac{d\xi}{d\xi} + \frac{\partial u}{\partial \eta}\frac{d\eta}{d\xi} = k(\xi) + l(\xi) \qquad (10.8.6)$$

since the equation of Oc is $\eta = \xi$.

The imposition of the given conditions supplies

$$f(\xi) + g(\xi) = h(\xi),$$

$$f'(\xi) = k(\xi), \qquad g'(\xi) = l(\xi).$$

The three are consistent on account of (10.8.6). With f found from the second equation and g from the first, u is obtained in the same square as in the mixed problem.

For the general equation, data on u and its first partial derivative on $O'C$ in Figure 10.8.4 will fix u in the characteristic domain $O'ACB$.

A return to the reflection problem for the wave equation in Section 10.7 is pertinent here, to examine it from the point of view of this section. AB of Figure 10.7.2 is not a characteristic and u, $\partial u/\partial t$ are given on it. Therefore there is a Cauchy problem and, from the above, a solution is available in APB. From this solution u is known on AP, which is a characteristic arc, and $u = 0$ on AQ, which is not a characteristic but lies between the two characteristics through A. Hence there is a mixed problem and u can be found in APQ. Similarly, a mixed problem gives u in BPR. Now u is known on PQ and PR, which are both characteristic arcs, and so by solving a two-characteristic problem we determine u in $PRTQ$. In this way, a solution to a reflection problem for a general hyperbolic equation can be built up by means of characteristics.

It is possible to find u when u is given on two arcs Oc, Od in the first quadrant when they are not characteristics. If Oc is in the first quadrant and Od is in the fourth there is insufficient information for a unique solution. For if we give u any values on the ξ-axis we have a mixed problem in the first quadrant and another one in the fourth, so that a solution can be found.

When elliptic partial differential equations were discussed it was pointed out that Dirichlet's problem was an appropriate one. It will now be shown why it is not suitable for hyperbolic equations in general.

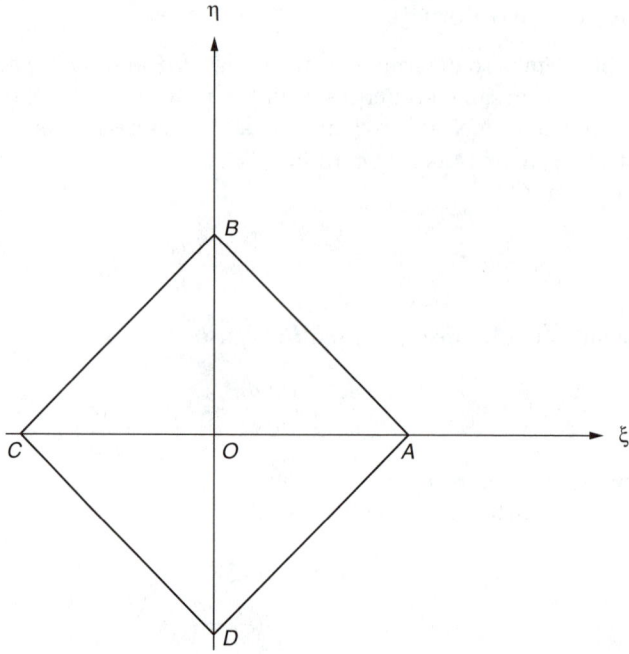

FIGURE 10.8.5:   The Dirichlet problem for the hyperbolic equation.

Let u be prescribed on the boundary $ABCD$ of Figure 10.8.5. The aim is to find u inside; so assume some values for u on OB. Then the form of u in OAB is known by solving a mixed problem and similarly in OBC. From the discovered values of u on OC, u is determined in OCD via a mixed problem. Similarly, values in OAD can be found. In order that the solution be correct, the values of u on OD from OCD and OAD must agree. This places a condition on the assumed values on OB. There is no guarantee that this condition can be fulfilled. If it cannot be met there is no solution to the Dirichlet problem. If the condition on OB can be satisfied there will be at least one solution and there may be several since there may be more than one way of complying with the condition.

We conclude that the Dirichlet problem for the hyperbolic differential equation must be treated with great caution.

10.9   The Euler-Darboux equation

This section is devoted to the hyperbolic partial differential equation

$$\frac{\partial^2 u}{\partial x \partial y} = \frac{m}{x - y}\left(\frac{\partial u}{\partial x} - \frac{\partial u}{\partial y}\right) \tag{10.9.1}$$

where m is a positive integer. This form has already occurred in Section 10.7 and it is called the **Euler-Darboux equation**.

Now

$$\frac{\partial^2}{\partial x \partial y}\{(x-y)v\} = \frac{\partial}{\partial x}\left((x-y)\frac{\partial v}{\partial y} - v\right) = (x-y)\frac{\partial^2 v}{\partial x \partial y} + \frac{\partial v}{\partial y} - \frac{\partial v}{\partial x}.$$

$$(10.9.2)$$

Consequently, when $m=1$, (10.9.1) after multiplication by $x-y$ may be written, on account of (10.9.2), as

$$\frac{\partial^2}{\partial x \partial y}\{(x-y)u\} = 0.$$

It follows that

$$u = \frac{f(x) + g(y)}{x - y},$$

with f and g arbitrary, provides the general solution of the Euler-Darboux equation when $m = 1$.

Multiply (10.9.1) by $x - y$ and apply the derivative $\partial^2/\partial x \partial y$. Then, by virtue of (10.9.2),

$$(x-y)\frac{\partial^2}{\partial x \partial y}\frac{\partial^2 u}{\partial x \partial y} + \left(\frac{\partial}{\partial y} - \frac{\partial}{\partial x}\right)\frac{\partial^2 u}{\partial x \partial y} = m\left(\frac{\partial}{\partial x} - \frac{\partial}{\partial y}\right)\frac{\partial^2 u}{\partial x \partial y}.$$

Thus, if u satisfies (10.9.1), $\partial^2 u/\partial x \partial y$ is a solution of (10.9.1) with m replaced by $m+1$. The solution already derived for $m = 1$ now permits the statement that

$$u = \frac{\partial^{2m-2}}{\partial x^{m-1}\partial y^{m-1}}\left(\frac{f(x) + g(y)}{x - y}\right),$$

with f and g arbitrary, solves (10.9.1).

Exercises

10.1 If the characteristics of

$$a(x,y)\frac{\partial u}{\partial x} + b(x,y)\frac{\partial u}{\partial y} = 0$$

are $\phi(x,y) = $ constant, show that the general solution is $u = G\{\phi(x,y)\}$ where G is an arbitrary function.

10.2 Solve

$$\frac{\partial u}{\partial x} + \frac{\partial u}{\partial y} = 2$$

given that $u = 2x^2$ when $y = 0$.

10.3 Solve

$$x\frac{\partial u}{\partial x} + y\frac{\partial u}{\partial y} = u$$

given that $u = 3x$ on $x + y = 1$.

10.4 Make the substitution $x = \rho\cos\theta, y = \rho\sin\theta$ in

$$(x - y)\frac{\partial u}{\partial x} + (x + y)\frac{\partial u}{\partial y} = 4.$$

Hence find the solution in $0 \le \theta < 2\pi$ such that $u = 2r$ when $\theta = 0$.

10.5 If $u = 3y^2$ when $x = 0$ solve

$$\frac{\partial u}{\partial x} - (x - y - 1)\frac{\partial u}{\partial y} = 3x(y - x).$$

10.6 Given that $u = y^2$ on $x = 0$, find u in $y > x > 0$ such that

$$y^2\frac{\partial u}{\partial x} + xy\frac{\partial u}{\partial y} = x.$$

10.7 Find the solution in $x > 0, y > 0$ of

$$x\frac{\partial u}{\partial x} - 2y\frac{\partial u}{\partial y} = x^2 + y^2$$

such that $u = x^2$ on $y = 1$.

10.8 Find the solution in $x > 0$ of

$$x\frac{\partial u}{\partial x} + y\frac{\partial u}{\partial y} = u - 2xy$$

given that $u = 2y^2 + 2$ on $x = 2$.

10.9 If

$$x\frac{\partial u}{\partial x} + y\frac{\partial u}{\partial y} = xy$$

and $u = \frac{1}{2}x^2$ on $y = x$ determine u.

10.10 In what regions are the following partial differential equations elliptic, hyperbolic or parabolic?

(a) $\dfrac{\partial^2 u}{\partial x^2} + 4\dfrac{\partial^2 u}{\partial y^2} = 0,$

(b) $7\dfrac{\partial^2 u}{\partial x \partial y} - 3\dfrac{\partial u}{\partial y} = 0,$

(c) $x^2 \dfrac{\partial^2 u}{\partial x^2} + 4y\dfrac{\partial^2 u}{\partial x \partial y} + \dfrac{\partial^2 u}{\partial y^2} + 2\dfrac{\partial u}{\partial x} = 0,$

(d) $3y\dfrac{\partial^2 u}{\partial x^2} - x\dfrac{\partial^2 u}{\partial y^2} = 0,$

(e) $\dfrac{\partial^2 u}{\partial x^2} + 4\dfrac{\partial u}{\partial x} = 0,$

(f) $x^2 y^2 \dfrac{\partial^2 u}{\partial x^2} + 2xy\dfrac{\partial^2 u}{\partial x \partial y} + \dfrac{\partial^2 u}{\partial y^2} = 0.$

10.11 Make the change of variables $x = \rho \cos \theta, y = \rho \sin \theta$ in

$$(x^2 \cos^2 \alpha - y^2 \sin^2 \alpha)\frac{\partial^2 u}{\partial x^2} + 2xy\frac{\partial^2 u}{\partial x \partial y} + (y^2 \cos^2 \alpha - x^2 \sin^2 \alpha)\frac{\partial^2 u}{\partial y^2} = 0$$

and show that $\ln \rho \pm \theta \cot \alpha$ are characteristic coordinates. Express the equation in characteristic form.

10.12 Express the partial differential equation

$$\frac{\partial^2 u}{\partial x^2} - e^{-x}\frac{\partial^2 u}{\partial y^2} = 0$$

in characteristic form.

10.13 Find the two families of characteristics of

$$2x^2 \frac{\partial^2 u}{\partial x^2} - 5xy\frac{\partial^2 u}{\partial x \partial y} + 2y^2 \frac{\partial^2 u}{\partial y^2} + 2x\frac{\partial u}{\partial x} + 2y\frac{\partial u}{\partial y} = 0.$$

Convert the equation to characteristic form and hence find u given that $u = 2x^2 - 6$ and $\partial u/\partial y = 3 - x^2$ on $y = 1$.

10.14 The characteristic form of

$$\{a(x)\}^2 \frac{\partial^2 u}{\partial x^2} = \frac{\partial^2 u}{\partial y^2}$$

is

$$\frac{\partial^2 u}{\partial \xi \partial \eta} = \frac{1}{2(\xi - \eta)}\left(\frac{\partial u}{\partial \xi} - \frac{\partial u}{\partial \eta}\right).$$

Show that $a(x) = Ae^{Bx}$ where A and B are constants.
If the characteristic form is

$$\frac{\partial^2 u}{\partial \xi \partial \eta} = \frac{\nu}{\xi - \eta}\left(\frac{\partial u}{\partial \xi} - \frac{\partial u}{\partial \eta}\right)$$

and $\nu \neq \frac{1}{2}$, find a formula for a.

10.15 The solution of

$$\frac{\partial^2 u}{\partial x^2} + 3\frac{\partial^2 u}{\partial y^2} + 2x = 0$$

is required in the square $0 < x < a, 0 < y < a$. You are asked to advise which of the following sets of boundary conditions on the perimeter of the square are suitable:

(a) u given on all four sides;

(b) u given on two adjacent sides only;

(c) u given on three sides and its normal derivative given on the middle side of the three.

Decide for (a), (b) and (c) whether there is a solution and, if so, whether there is more than one.

10.16 Repeat Exercise 10.15 for

$$\frac{\partial^2 u}{\partial x^2} + 3\frac{\partial^2 u}{\partial y^2} + u = 0.$$

10.17 Repeat Exercise 10.15 for

$$\frac{\partial^2 u}{\partial x^2} - \frac{\partial^2 u}{\partial y^2} - 2u = 0.$$

10.18 Repeat Exercise 10.15 for

$$\frac{\partial^2 u}{\partial x \partial y} + 3u = 0.$$

Chapter 11

Evolutionary Equations

11.1 The heat equation

The equation

$$\frac{\partial u}{\partial t} = D\nabla^2 u, \qquad (11.1.1)$$

where D is a positive constant and ∇^2 is the Laplacian, arises in the study of problems of heat conduction and for this reason is referred to as the **heat equation**. However, (11.1.1) arises in many other areas in which diffusion processes occur (the subject of Chapter 12) and in this context is referred to as the **classic diffusion equation**.

In this section we shall study (11.1.1) in one space dimension and where $D = 1$, that is, we consider the equation

$$\frac{\partial u}{\partial t} = \frac{\partial^2 u}{\partial x^2}. \qquad (11.1.2)$$

In addition to the techniques described in Chapter 10, the method of *separation of variables* is a powerful tool with which to solve linear partial differential equations. To begin with, we seek solutions of (11.1.2) in the form

$$u(x,t) = X(x)T(t). \qquad (11.1.3)$$

Substitution of this into (11.1.2) leads to the identity

$$X\frac{dT}{dt} = T\frac{d^2X}{dx^2}$$

or

$$\frac{1}{T}\frac{dT}{dt} = \frac{1}{X}\frac{d^2X}{dx^2}. \qquad (11.1.4)$$

Now $\frac{1}{T}\frac{dT}{dt}$ is a function of t only, while $\frac{1}{X}\frac{d^2X}{dx^2}$ is a function of x only. Consequently, both sides of (11.1.4) must be equal to a constant, say λ. Thus X

and T must satisfy the ordinary differential equations

$$\frac{dT}{dt} - \lambda T = 0,$$

$$\frac{d^2 X}{dx^2} - \lambda X = 0. \tag{11.1.5}$$

These have the general solutions

$$X(x) = \exp \pm \sqrt{\lambda} x, \quad T(t) = \exp \lambda t.$$

Hence (11.1.2) has a particular solution of the form

$$u(x, t) = A \exp(\pm\sqrt{\lambda} x + \lambda t). \tag{11.1.6}$$

Now suppose we are given initial data of the form

$$u(x, 0) = A \exp i\alpha x; \tag{11.1.7}$$

then a solution of (11.1.2) that satisfies (11.1.7) is

$$u(x, t) = A \exp(i\alpha x - \alpha^2 t).$$

It is of interest to see how solutions of this form can be utilised in order to provide a solution to the more general initial value problem

$$\frac{\partial u}{\partial t} = \frac{\partial^2 u}{\partial x^2}, \quad -\infty < x < \infty,$$
$$u(x, 0) = f(x). \tag{11.1.8}$$

To carry out this programme we shall need the **Fourier Integral Theorem**.

THEOREM 1
Let $f(x)$ and $\frac{df}{dx}$ be continuous and $\int_{-\infty}^{\infty} |f(x)| dx < \infty$.
Then

$$f(x) = \frac{1}{2\pi} \int_{-\infty}^{\infty} \left(\int_{-\infty}^{\infty} f(\xi) e^{-i\alpha\xi} d\xi \right) e^{i\alpha x} d\alpha. \tag{11.1.9}$$

The quantity

$$\hat{f}(\alpha) = \frac{1}{\sqrt{2\pi}} \int_{-\infty}^{\infty} f(\xi) e^{-i\alpha\xi} d\xi \tag{11.1.10}$$

is called the **Fourier transform** of f and (11.1.9) provides the **reciprocal relation**

$$f(x) = \frac{1}{\sqrt{2\pi}} \int_{-\infty}^{\infty} \hat{f}(\alpha) e^{i\alpha x} d\alpha. \tag{11.1.11}$$

Fourier transforms are another of an important class of techniques available for the solution of partial differential equations. To see this in the present situation, consider a special solution of the heat equation in the form

$$u(x, t; \alpha) = A(\alpha)e^{i\alpha x - \alpha^2 t}, \tag{11.1.12}$$

where $A(\alpha)$ is an arbitrary function of α. Since (11.1.2) is a linear partial differential equation, we know that a linear combination of solutions of the form (11.2.12) is also a solution. This, of course, is the essence of the separation of variables method. With this in mind, we take the reasonable step that the integral with respect to α of (11.2.12) may also be a solution.

Consider the function

$$u(x, t) = \int_{-\infty}^{\infty} u(x, t; \alpha) d\alpha = \int_{-\infty}^{\infty} A(\alpha)e^{i\alpha x - \alpha^2 t} d\alpha. \tag{11.1.13}$$

What we must do now is first to find a specific form for $A(\alpha)$ so that $u(x, t)$ defined by (11.1.13) satisfies the initial condition in (11.1.8) and then to show that (11.1.13) is in fact the (unique) solution to the problem (11.1.8).

Setting $t = 0$ in (11.1.13) we see that we must have

$$f(x) = \int_{-\infty}^{\infty} A(\alpha)e^{i\alpha x} d\alpha$$

or

$$\frac{f(x)}{\sqrt{2\pi}} = \frac{1}{\sqrt{2\pi}} \int_{-\infty}^{\infty} A(\alpha)e^{i\alpha x} d\alpha.$$

In other words, $A(\alpha)$ must be taken as the Fourier transform of $f(x)/\sqrt{2\pi}$. That is,

$$A(\alpha) = \frac{1}{2\pi} \int_{-\infty}^{\infty} f(\xi)e^{-i\alpha\xi} d\xi$$

and so from (11.1.13) we have

$$u(x, t) = \frac{1}{2\pi} \int_{-\infty}^{\infty} f(\xi)d\xi \int_{-\infty}^{\infty} e^{i\alpha(x-\xi) - \alpha^2 t} d\alpha. \tag{11.1.14}$$

The result (11.1.14) can be simplified.

Consider

$$I = \int_{-\infty}^{\infty} e^{i\alpha(x-\xi) - \alpha^2 t} d\alpha$$

$$= \int_{-\infty}^{\infty} \exp\left[-t\left(\alpha - \frac{i(x-\xi)}{2t}\right)^2 - \frac{(x-\xi)^2}{4t}\right] d\alpha.$$

Now set

$$\alpha - \frac{i(x - \xi)}{2t} = \frac{\beta}{\sqrt{t}}$$

to obtain

$$I = \frac{1}{\sqrt{t}} \exp\left(-\frac{(x - \xi)^2}{4t}\right) \int_{-\infty}^{\infty} e^{-\beta^2} d\beta.$$

The integral $\int_{-\infty}^{\infty} e^{-\beta^2} d\beta$ is known as the *error integral* and takes the value $\sqrt{\pi}$. Thus

$$I = \sqrt{(\pi/t)} \exp\left(-\frac{(x - \xi)^2}{4t}\right)$$

and so on using this in (11.1.14) we obtain the expression

$$u(x, t) = \int_{-\infty}^{\infty} \frac{\exp[-(x - \xi)^2/4t]}{\sqrt{4\pi t}} f(\xi) d\xi. \qquad (11.1.15)$$

The function

$$K(x - \xi, t) = \frac{\exp[-(x - \xi)^2/4t]}{\sqrt{4\pi t}}$$

is of special significance in the study of diffusion problems and is called the **fundamental solution** of the heat equation. It has a number of interesting properties that are important if we wish to study the regularity of solutions to heat conduction problems. For example, it is easy to see that $K \to 0$ as $t \to 0$ except at the point $\xi = x$, where $K \to \infty$ like the function $1/\sqrt{t}$. Furthermore,

$$\int_{-\infty}^{\infty} K(x - \xi, t) d\xi = 1$$

for all $t \geq 0$

To show that (11.1.15) does indeed solve the problem (11.1.8) can be done by formal substitution. However, it is important to realise that such a procedure requires careful consideration of questions of convergence and the operation of differentiation under the integral. Furthermore, there must be some restriction on the initial data. That this is so is due to the fact that (11.1.15) has been arrived at through the use of the Fourier integral theorem wherein we have assumed that

$$\int_{-\infty}^{\infty} |u(x, 0)| dx = \int_{-\infty}^{\infty} |f(x)| dx < 0.$$

This restriction can be relaxed; in fact (11.1.15) holds if

$$|f(x)| \leq M e^{Nx^2}.$$

Rigorous proofs of the existence and uniqueness of the solution (11.1.15) to the initial value problem (11.1.8) are to be found in most books devoted to the study of partial differential equations of which John (1975) is to be recommended.

11.2 Separation of variables

The class of problems treated here are those in which we wish to solve the heat equation in the presence of boundary as well as initial conditions. Such problems may be attacked by the use of "finite" Fourier transforms or, what is essentially the same thing, by the separation of variables method. Thus, instead of seeking solutions in integral form, as was done in the previous section, we derive solutions expressed in the form of infinite Fourier series.

The following is typical of such problems.

Example 11.2a

$$\frac{\partial u}{\partial t} = \frac{\partial^2 u}{\partial x^2}, \quad 0 < x < l, \ t > 0,$$
$$u(x, 0) = f(x),$$
$$u(0, t) = u(l, t) = 0. \tag{11.2.1}$$

As before, we seek solutions in the form

$$u(x, t) = X(x)T(t)$$

in which the functions $X(x)$ and $T(t)$ must satisfy the ordinary differential equations

$$\frac{dT}{dt} + \lambda T = 0,$$
$$\frac{d^2 X}{dx^2} + \lambda X = 0. \tag{11.2.2}$$

Notice that these equations differ from those in (11.1.5) by the change from $+\lambda$ to $-\lambda$. This is done purely for convenience and, for the present problem, makes the analysis a little easier to handle.

We seek solutions of the second equation in (11.2.2) that satisfy the boundary conditions

$$u(0, t) = u(l, t) = 0. \tag{11.2.3}$$

In other words, we have a typical eigenvalue problem (see Chapter 3, Section 3.6) giving rise to eigenvalues $\lambda_n = n^2\pi^2/l^2, n = 1, 2, \ldots$, and corresponding eigenfunctions $X_n(x) = \sin(n\pi x/l)$. Thus for each n the functions

$$u_n(x,t) = \exp\left(-\frac{n^2\pi^2 t}{l^2}\right)\sin\left(\frac{n\pi x}{l}\right)$$

are solutions of the heat equation that satisfy the boundary conditions (11.2.3). Since the heat equation is linear, a superposition of solutions of the form $u_n(x,t)$ is also a solution. Thus we look for a solution of the initial boundary value problem in the form of the infinite series

$$u(x,t) = \sum_{n=1}^{\infty} A_n \exp\left(-\frac{n^2\pi^2 t}{l^2}\right)\sin\left(\frac{n\pi x}{l}\right). \qquad (11.2.4)$$

If we set $t = 0$ in (11.2.4), then the initial condition is satisfied if we can choose the unknown coefficients $A_n, n = 1, 2, \ldots$, so that

$$f(x) = \sum_{n=1}^{\infty} A_n \sin\left(\frac{n\pi x}{l}\right). \qquad (11.2.5)$$

In other words we seek a "Fourier sine" series expansion of $f(x)$.

From Chapter 3 we know that the set of functions $\sin(n\pi x/l), n = 1, 2, \ldots$, form an orthogonal set on the interval $0 \le x \le l$, that is,

$$\int_0^l \sin\left(\frac{m\pi x}{l}\right)\sin\left(\frac{n\pi x}{l}\right)dx = \begin{cases} 0, & m \ne n. \\ l/2, & m = n. \end{cases}$$

Thus if we formally multiply (11.2.5) by $\sin\left(\frac{m\pi x}{l}\right)$ for some m and integrate over $0 \le x \le l$ we find that

$$\int_0^l f(x)\sin\left(\frac{m\pi x}{l}\right)dx = \frac{1}{2}A_m$$

or

$$A_m = \frac{2}{l}\int_0^l f(x)\sin\left(\frac{m\pi x}{l}\right)dx.$$

This gives the result

$$u(x,t) = \frac{2}{l}\sum_{n=1}^{\infty}\exp\left(-\frac{n^2\pi^2 t}{l^2}\right)\sin\left(\frac{n\pi x}{l}\right)\int_0^l f(y)\sin\left(\frac{n\pi y}{l}\right)dy. \qquad (11.2.6)$$

By construction it is clear that the series (11.2.6) satisfies the boundary and initial conditions of our problem. That it also satisfies the heat equation follows from the fact that the series is convergent for all $x, 0 \leq x \leq l$, and $t \geq 0$ and can be differentiated term by term.

The method of separation of variables is applicable not only to the heat equation but also to problems involving Laplace's equation and the wave equation treated under a variety of initial and boundary conditions. To help to appreciate this we consider the following problem. $\Box$

Example 11.2b

$$\frac{\partial u}{\partial t} = \frac{\partial^2 u}{\partial x^2} - au, \quad 0 < x < \pi, \ t > 0, \tag{11.2.7}$$

where a is a positive constant and $u(x, t)$ is required to satisfy the conditions

$$u(0, t) = \frac{\partial u(\pi, t)}{\partial x} = 0,$$
$$u(x, 0) = x(\pi - x). \tag{11.2.8}$$

Proceeding as before we set

$$u(x, t) = X(x)T(t)$$

to find

$$\frac{1}{T}\frac{dT}{dt} = \frac{1}{X}\frac{d^2 X}{dx^2} - a$$

or

$$\frac{dT}{dt} + aT = -\lambda = \frac{d^2 X}{dx^2}.$$

In other words, the functions $X(x), T(t)$ must satisfy the ordinary differential equations

$$\frac{d^2 X}{dx^2} + \lambda X = 0, \tag{11.2.9}$$

$$\frac{dT}{dt} + (a + \lambda)T = 0. \tag{11.2.10}$$

We seek solutions of (11.2.9) that satisfy the boundary conditions

$$X(0) = \frac{dX(\pi)}{dx} = 0. \tag{11.2.11}$$

First of all suppose $\lambda = 0$; then (11.2.9) has the general solution $X(x) = A + Bx$. On using the boundary conditions (11.2.11), it follows that $A = B = 0$ and so $X \equiv 0$ in this case. When $\lambda \neq 0$ the general solution of (11.2.9) is

$$X(x) = A \sin \sqrt{\lambda} x + B \cos \sqrt{\lambda} x$$

and

$$\frac{dX}{dx} = A\sqrt{\lambda}\cos\sqrt{\lambda}x - B\sqrt{\lambda}\sin\sqrt{\lambda}x.$$

Consequently, the boundary conditions (11.2.11) demand that $A, B, \sqrt{\lambda}$ be chosen so that

$$B = 0,$$
$$A\lambda\cos\sqrt{\lambda}\pi - B\lambda\sin\sqrt{\lambda}\pi = 0.$$

Thus for $A \neq 0$ we must have $\cos\sqrt{\lambda}\pi = 0$, and so

$$\sqrt{\lambda}\pi = (2n-1)\pi/2, \quad n = 1, 2, \ldots.$$

That is the eigenvalues are

$$\lambda_n = \frac{1}{4}(2n-1)^2,$$

and the corresponding eigenfunctions are

$$X_n(x) = \sin\left(\frac{(2n-1)x}{2}\right).$$

Therefore we have a solution $u_n(x,t)$ of (11.2.7) satisfying the boundary conditions in (11.2.8) in the form

$$u_n(x,t) = \exp\{-[(2n-1)^2 + 4a]t/4\}\sin[(2n-1)x/2].$$

The complete solution to the problem is then sought in the form

$$u(x,t) = \sum_{n=1}^{\infty} A_n \exp\{-[(2n-1)^2 + 4a]t/4\}\sin[(2n-1)x/2]. \qquad (11.2.12)$$

By imposing the initial condition in (11.2.8) we determine the unknown coefficients A_n from the Fourier series

$$x(\pi - x) = \sum_{n=1}^{\infty} A_n \sin\left(\frac{(2n-1)x}{2}\right). \qquad (11.2.13)$$

From the previous example we see that

$$A_n = \frac{2}{\pi}\int_0^{\pi} x(\pi - x)\sin\left(\frac{(2n-1)x}{2}\right)dx.$$

The integral appearing on the right-hand side is best evaluated by "integration by parts." Consider

$$
\begin{aligned}
I_n &= \int_0^\pi x(\pi - x)\sin\left(\frac{(2n-1)x}{2}\right) dx \\
&= \left[-x(\pi - x)\frac{2}{2n-1}\cos\left(\frac{(2n-1)}{2}x\right)\right]_0^\pi \\
&\quad + \frac{2}{2n-1}\int_0^\pi (\pi - 2x)\cos\left(\frac{(2n-1)}{2}x\right) dx \\
&= \left[(\pi - 2x)\frac{4}{(2n-1)^2}\sin\left(\frac{(2n-1)}{2}x\right)\right]_0^\pi \\
&\quad + \frac{8}{(2n-1)^2}\int_0^\pi \sin\left(\frac{(2n-1)}{2}x\right) dx \\
&= \frac{4\pi}{(2n-1)^2}\cos(n\pi) - \frac{16}{(2n-1)^3}\left[\cos\left(\frac{(2n-1)}{2}x\right)\right]_0^\pi \\
&= \frac{4\pi}{(2n-1)^2}\cos(n\pi) + \frac{16}{(2n-1)^3}.
\end{aligned}
$$

Thus

$$
A_n = 8[(4/\pi)(2n-1)^{-3} + (-1)^n(2n-1)^{-2}],
$$

and we have the solution to the problem in the form

$$
\begin{aligned}
u(x,t) = 8\sum_{n=1}^\infty &[(4/\pi)(2n-1)^{-3} + (-1)^n(2n-1)^{-2}] \\
&\times \exp\{-[(2n-1)^2 + 4a]t/4\}\sin[(2n-1)x/2].
\end{aligned}
$$

Consider the problem in two dimensions. ⬜

Example 11.2c

$$
\frac{\partial u}{\partial t} = \frac{\partial^2 u}{\partial x^2} + \frac{\partial^2 u}{\partial y^2}, \quad 0 < x < a, \; 0 < y < b, \; t > 0,
$$

$u(x,y,t) = 0$ on the boundary of the rectangle defined by $0 \le x \le a, 0 \le y \le b$ and

$$
u(x,y,0) = f(x,y). \tag{11.2.14}
$$

The procedure here is precisely the same as in the previous examples except that we begin by seeking a solution in the form

$$
u(x,y,t) = T(t)S(x,y)
$$

to find that T and S must satisfy the equations

$$\frac{dT}{dt} + \lambda T = 0, \tag{11.2.15}$$

$$\frac{\partial^2 S}{\partial x^2} + \frac{\partial^2 S}{\partial y^2} + \lambda S = 0. \tag{11.2.16}$$

Now (11.2.16) is an elliptic equation of the type considered in Chapter 10 and here we seek a solution $S(x, y)$ which satisfies the boundary conditions

$$S(0, y) = S(a, y) = 0, \quad 0 \le y \le b,$$
$$S(x, 0) = S(x, b) = 0, \quad 0 \le x \le a. \tag{11.2.17}$$

This problem is solved by applying the separation of variables technique once more. That is, we set

$$S(x, y) = X(x)Y(y)$$

and find that $X(x), Y(y)$ must satisfy

$$\frac{1}{X}\frac{d^2 X}{dx^2} + \frac{1}{Y}\frac{d^2 Y}{dy^2} + \lambda = 0,$$

i.e.,

$$\frac{1}{X}\frac{d^2 X}{dx^2} = -\mu = -\left(\frac{1}{Y}\frac{d^2 Y}{dy^2} + \lambda\right),$$

or

$$\frac{d^2 X}{dx^2} + \mu X = 0, \tag{11.2.18}$$

$$\frac{d^2 Y}{dy^2} + (\lambda - \mu)Y = 0. \tag{11.2.19}$$

In addition we require

$$X(0) = X(a) = 0, \tag{11.2.20}$$
$$Y(0) = Y(b) = 0. \tag{11.2.21}$$

The problem (11.2.18), (11.2.20) is of the same form as the problem (11.2.2), (11.2.3) and so we have the eigenvalues

$$\mu_m = \frac{m^2 \pi^2}{a^2}, \quad m = 1, 2 \ldots$$

and corresponding eigenfunctions

$$X_m(x) = \sin\left(\frac{m\pi x}{a}\right).$$

For a typical eigenvalue μ_m, the problem (11.2.19), (11.2.21) becomes

$$\frac{d^2 Y}{dy^2} + (\lambda - \mu_m)Y = 0,$$

$$Y(0) = Y(b) = 0.$$

This is another typical eigenvalue problem of the type already considered and leads to the eigenvalues

$$\lambda_n - \mu_m = \frac{n^2 \pi^2}{b^2}, \quad n = 1, 2 \ldots$$

and corresponding eigenfunctions

$$Y_n(y) = \sin\left(\frac{n\pi y}{b}\right)$$

In other words, a solution of (11.2.16), (11.2.17) is of the form

$$S_{mn}(x, y) = \sin\left(\frac{m\pi x}{a}\right) \sin\left(\frac{n\pi y}{b}\right),$$

together with the eigenvalue

$$\lambda_{mn} = \left(\frac{m^2}{a^2} + \frac{n^2}{b^2}\right)\pi^2.$$

It is important to notice here that since S_{mn} depends on both m and n, the general solution of the problem (11.2.14) must be sought as a "double" series where the summation extends over both m and n. Thus we write

$$u(x, y, t) = \sum_{m=1}^{\infty} \sum_{n=1}^{\infty} A_{mn} \sin\left(\frac{m\pi x}{a}\right) \sin\left(\frac{n\pi y}{b}\right)$$

$$\times \exp\left[-\left(\frac{m^2}{a^2} + \frac{n^2}{b^2}\right)\pi^2 t\right]. \tag{11.2.22}$$

By imposing the initial condition we determine A_{mn} from the "double" Fourier series

$$f(x, y) = \sum_{m=1}^{\infty} \sum_{n=1}^{\infty} A_{mn} \sin\left(\frac{m\pi x}{a}\right) \sin\left(\frac{n\pi y}{b}\right). \tag{11.2.23}$$

In order to determine the coefficients A_{mn} we apply the "one-dimensional" arguments used above twice.

Thus if we consider y as fixed then

$$\frac{a}{2} \sum_{n=1}^{\infty} A_{mn} \sin\left(\frac{n\pi y}{b}\right) = \int_0^a f(x, y) \sin\left(\frac{m\pi x}{a}\right) dx. \tag{11.2.24}$$

Now the right-hand side of (11.2.24) is a function of y and so (11.2.24) can be considered as a Fourier series in the single variable y. Thus if we set

$$\int_0^a f(x, y) \sin\left(\frac{m\pi x}{a}\right) dx = g(y),$$

then the coefficients A_{mn} (m fixed) in (11.2.24) are given by

$$\frac{ab}{4} A_{mn} = \int_0^b g(y) \sin\left(\frac{n\pi y}{b}\right) dy$$

i.e.,

$$A_{mn} = \frac{4}{ab} \int_0^a \int_0^b f(x, y) \sin\left(\frac{m\pi x}{a}\right) \sin\left(\frac{n\pi y}{b}\right) dx dy. \tag{11.2.25}$$

In conclusion then we have the possible series solution to problem (11.2.14) in the form (11.2.22) where the coefficients A_{mn} are given by (11.2.25). Again we must verify that (11.2.22) does indeed solve the problem (11.2.14). By construction, the initial and boundary conditions are satisfied and the uniform convergence of the series (11.2.22) justifies differentiation of the series and this in turn allows for verification of the solution by direct substitution into the two-dimensional heat equation. $\square$

11.3 Simple evolutionary equations

An evolutionary equation is one of the form

$$\frac{\partial u}{\partial t} = D\nabla^2 u + f(\mathbf{x}, t, u), \tag{11.3.1}$$

where the function f may be quite a complicated function of its arguments and could even depend on the spatial derivatives of u. Equations of the form (11.3.1) have already arisen many times in the context of biological modelling (see Chapter 4) and will be taken up again in Chapter 12.

The expression "simple evolutionary equations" used for this section is somewhat illusory. It does not mean that the problems we shall consider are particularly simple or easy, but refers to equations in which the function f depends only on the dependent variable u and is of a simple form such as a polynomial. Thus the equations we shall consider will be of the type

$$\frac{\partial u}{\partial t} = D\nabla^2 u + f(u). \tag{11.3.2}$$

If $f(u)$ is linear, say

$$f(u) = Au + B,$$

where A and B are constants, then in many instances the equation (11.3.2) can be solved by the separation of variables method. However if, as in many of the applications considered in this book, $f(u)$ is non-linear, say

$$f(u) = u(1-u)(u-a), \quad 0 < u < 1, \tag{11.3.3}$$

then the problem is much more intractable. Indeed, it is not usually possible to obtain "general" analytic solutions and one must solve such problems numerically (see Mitchell and Griffiths, 1980). Despite this, however, many evolutionary equations have "special" or "particular" solutions, which are of fundamental importance to our understanding of biological phenomena modelled by evolutionary equations. In this section we shall explore some of these particular solutions and demonstrate their use in the section to follow.

Suppose that $f(u)$ is a polynomial in u and that $f(u)$ has real roots at $u = \alpha, \beta, \gamma, \ldots$. Then it is obvious that in this case $u = \alpha, u = \beta, u = \gamma$, etc., are all constant solutions of (11.3.2). Such solutions are of importance in the treatment of the pure initial value problem

$$\frac{\partial u}{\partial t} = D\frac{\partial^2 u}{\partial x^2} + f(u), \quad -\infty < x < \infty,$$
$$u(x,0) = \phi(x). \tag{11.3.4}$$

Here, for example, it often happens that the solution $u(x,t)$ evolves as $t \to \infty$ into one of the asymptotic states $u = \alpha, \beta, \gamma, \ldots$. Which state is reached depends crucially on the form of the initial data ϕ and that of f. Indeed some of these constant asymptotic states are stable to small perturbations of ϕ while others are not.

As an example of this, consider the problem (11.3.4) with $f(u)$ given by (11.3.3). Then it can be shown (see Section 11.4) that if

$$sup_{x \in (-\infty,\infty)}\phi(x) < a,$$

$u(x,t) < a$ for all $x \in (-\infty,\infty), t > 0$. From this it may be proved that $u(x,t) \to 0$ (exponentially) as $t \to \infty$. By a similar argument it can also be shown that if

$$inf_{x \in (-\infty,\infty)}\phi(x) > a,$$

$supu(x,t) > a$ for all $t > 0$ and that $u(x,t) \to 1$ as $t \to \infty$. In other words, the problem (11.3.4) has $u = 0$ or $u = 1$ as asymptotic states depending on the magnitude of the initial data. The number a is called a "threshold" parameter.

Another fundamentally important set of solutions to equation (11.3.4) is the class of "travelling wave" solutions. Here we look for solutions of the form

$$u(x,t) = V(x + ct),$$

where c is a constant, positive or negative, called the "wave speed." If we make the transformation $\xi = x + ct$ in (11.3.4) then we see that $V(\xi)$ satisfies the non-linear ordinary differential equation

$$V'' + cV' + f(V) = 0, \tag{11.3.5}$$

where the primes represent differentiation with respect to ξ.

We have encounted travelling waves several times already (Chapters 4, 7 and 8) and have analysed them using the phase plane methods of Chapter 5. They are also of fundamental importance in the study of asymptotic states of the problem (11.3.4). Indeed we may ask the question: When does the solution $u(x, t)$ "evolve" into the travelling wave $V(x + ct)$? Mathematically we can state the problem as: determine conditions on the initial data $\phi(x)$ so that

$$\lim_{t \to \infty} |u(x, t) - V(x + ct)| = 0.$$

In practice, there may be several such travelling waves, some stable and some unstable, and the determination of the appropriate stable travelling wave is often a challenging task.

Travelling waves can be classified as follows:

(a) wave trains — $V(\xi)$ periodic;

(b) wave fronts — $V(-\infty)$ and $V(\infty)$ exist and are unequal;

(c) pulses — $V(-\infty)$ and $V(\infty)$ exist and are equal and $V(\xi)$ is not constant.

Apart from the use of phase plane methods the problem of finding travelling wave solutions is usually solved by numerical methods. However, in some cases, if we are fortunate, travelling waves can be determined analytically.

Example 11.3a

$$f(u) = u(1 - u)(u - a).$$

Here

$$V(x + ct) = V(\xi) = \left(1 + e^{-\xi/\sqrt{2}}\right)^{-1}, \tag{11.3.6}$$

where

$$c = \sqrt{2}\left(\frac{1}{2} - a\right).$$

In this example $V(-\infty) = 0$, $V(\infty) = 1$ and so in this case $V(\xi)$ is a wave front travelling from right to left with speed $c = \sqrt{2}(\frac{1}{2} - a)$. $\quad\Box$

Example 11.3b

$$f(u) = \begin{cases} u, & 0 \le u < \frac{1}{2}, \\ 1 - u, & \frac{1}{2} \le u. \end{cases} \tag{11.3.7}$$

If we make the substitution $\xi = x + ct$ then we obtain equation (11.3.6) with f defined by (11.3.8). Again we look for a non-decreasing (i.e., monotone increasing) solution satisfying the conditions $0 \leq V(\xi) \leq 1, V(-\infty) = 0, V(\infty) = 1$.

As long as $V < \frac{1}{2}$ we have to solve

$$V'' - cV' + V = 0,$$

which has the solution

$$V = Ae^{\alpha_1 \xi} + Be^{\alpha_2 \xi}, \tag{11.3.8}$$

where A and B are constants to be determined and

$$\alpha_1 = \frac{c + \sqrt{(c^2 - 4)}}{2}, \quad \alpha_2 = \frac{c - \sqrt{(c^2 - 4)}}{2},$$

provided $c^2 \neq 4$. In order to satisfy the condition $V(-\infty) = 0$, it is clear that we must have $c > 0$. If $c < 2$ then (11.3.9) is oscillatory and so we would not have a monotone solution. Consequently we must have $c \geq 2$. Suppose $c > 2$. Then we have a solution monotone increasing with ξ. If without loss of generality $V = 1/2$ when $\xi = 0$ then we require

$$A + B = \frac{1}{2}. \tag{11.3.9}$$

For $V \geq \frac{1}{2}$ we must solve the equation

$$V'' - cV' + 1 - V = 0,$$

which has the general solution

$$V = De^{\beta_1 \xi} + Ee^{\beta_2 \xi} + 1, \tag{11.3.10}$$

where D, E are constants to be determined and

$$\beta_1 = \frac{c + \sqrt{(c^2 + 4)}}{2}, \quad \beta_2 = \frac{c - \sqrt{(c^2 + 4)}}{2}.$$

The requirement $V(\infty) = 1$ demands that $D = 0$. In addition, the condition $V(0) = \frac{1}{2}$ gives

$$E + 1 = \frac{1}{2}$$

i.e.,

$$E = -\frac{1}{2}.$$

Furthermore, if we require V to be continuous at $\xi = 0$ then a differentiation of (11.3.9) and (11.3.11) provides the extra condition

$$\alpha_1 A + \alpha_2 B = \beta_2 E = -\frac{1}{2}\beta_2. \tag{11.3.11}$$

Equations (11.3.9), (11.3.10) can now be solved to give

$$A = \frac{\sqrt{(c^2 + 4)} - 2c + \sqrt{(c^2 - 4)}}{4\sqrt{(c^2 - 4)}},$$

$$B = \frac{\sqrt{(c^2 - 4)} + 2c - \sqrt{(c^2 + 4)}}{4\sqrt{(c^2 - 4)}}.$$

Thus with A, B and E determined, (11.3.8), (11.3.10) give the complete solution to the problem for any $c > 2$. That is,

$$V(\xi) = \frac{\sqrt{(c^2 + 4)} - 2c + \sqrt{(c^2 - 4)}}{4\sqrt{(c^2 - 4)}} \exp\left[\left(\frac{c + \sqrt{(c^2 - 4)}}{2}\right)\xi\right]$$

$$+ \frac{\sqrt{(c^2 - 4)} + 2c - \sqrt{(c^2 + 4)}}{4\sqrt{(c^2 - 4)}} \exp\left[\left(\frac{c - \sqrt{(c^2 - 4)}}{2}\right)\xi\right], \quad -\infty < \xi < 0$$

$$= -\frac{1}{2}\exp\left[\left(\frac{c - \sqrt{(c^2 + 4)}}{2}\right)\xi\right] + 1, \quad 0 \le \xi < \infty.$$

Now suppose $c = 2$; then the general solution (11.3.8) must be replaced by one of the form

$$V = Ae^{\xi} + B\xi e^{\xi} \tag{11.3.12}$$

and (11.3.10) becomes

$$V = De^{(1+\sqrt{2})\xi} + Ee^{(1-\sqrt{2})\xi} + 1. \tag{11.3.13}$$

By imposing the boundary and continuity conditions we have

$$A = \frac{1}{2} = -E, \quad D = 0$$

and

$$A + B = -\frac{1}{2}(1 - \sqrt{2}),$$

i.e.,

$$B = -\frac{1}{2}(2 - \sqrt{2}).$$

Then, in this case

$$V = \begin{cases} \frac{1}{2}e^{\xi} - \frac{1}{2}(2 - \sqrt{2})\xi e^{\xi}, & -\infty < \xi \le 0, \\ 1 - \frac{1}{2}e^{(1-\sqrt{2})\xi}, & 0 \le \xi < \infty. \end{cases}$$

In conclusion then, (11.3.4) with f given by (11.3.7) has a one parameter family of wavefront solutions for any wave speed $c \ge 2$.

Another class of solutions to (11.3.4), which is often useful, are those independent of x. In this case we have to solve the ordinary differential equation

$$\frac{du}{dt} = f(u). \tag{11.3.14}$$

Suppose we have the initial condition

$$u(0) = u_0 = constant;$$

then (11.3.14) can be solved implicitly as

$$t = \int_{u_0}^{u} \frac{d\eta}{f(\eta)}, \tag{11.3.15}$$

provided u_0 and u are such that the integral exists. □

Example 11.3c

Suppose $f(u) = u(1 - u)$. Then (11.3.15) has the solution ($u_0 \ne 0$)

$$t = \int_{u_0}^{u} \frac{d\eta}{\eta(1 - \eta)} = \int_{u_0}^{u} \left(\frac{1}{\eta} + \frac{1}{1 - \eta}\right) d\eta,$$

$$= \log \left|\frac{u(1 - u_0)}{u_0(1 - u)}\right|.$$

We therefore obtain the explicit solution

$$u = \frac{u_0 e^t}{1 - u_0 + u_0 e^t}.$$

This solution shows that if $0 < u_0 < 1$ then $u(t) < 1$ and $u \to 1$ as $t \to \infty$. Similarly if $u_0 > 1$ then $u(t) > 1$ and again $u \to 1$ as $t \to \infty$. Note that if $u_0 = 0$ then $u(t) \equiv 0$ and if $u_0 = 1$ then $u(t) \equiv 1$.

Finally consider those solutions of (11.3.4) that are independent of t. In this case we have to solve the ordinary differential equation

$$\frac{d^2 u}{dx^2} + f(u) = 0. \tag{11.3.16}$$

In many situations it is possible to solve this equation in the following manner: if $\frac{du}{dx} \neq 0$ then (11.3.16) can be expressed as

$$\frac{1}{2}\frac{d}{dx}\left(\frac{du}{dx}\right)^2 + f(u)\frac{du}{dx} = 0,$$

i.e.,

$$\left(\frac{du}{dx}\right)^2 + 2\int^u f(\eta)d\eta = A,$$

for some arbitrary constant A. Thus if we set

$$\int^u f(\eta)d\eta = F(u)$$

then

$$\frac{du}{dx} = \pm\sqrt{[A - 2F(u)]}, \qquad (11.3.17)$$

and this first-order differential equation can be solved, at least in principle, in the same way as equation (11.3.14). ▯

Example 11.3d

Consider the positive solutions to the boundary value problem

$$\frac{d^2u}{dx^2} + u(1 - u) = 0, \qquad (11.3.18)$$

$$u(-l/2) = u(l/2) = 0. \qquad (11.3.19)$$

If u is a solution to this problem with $u > 0$ for $-l/2 < x < l/2$ and since $u = 0$ at the end points it follows that u must take on its maximum value m at some intermediate point $x = a$ where $-l/2 < a < l/2$. In other words, $0 < u(x) \leq u(a) = m$ for $-l/2 < x < l/2$ and $u'(a) = 0, u''(a) \leq 0$. From this observation, equation (11.3.18) shows that

$$u''(a) = -m(1 - m) \leq 0$$

and so $0 < m \leq 1$. In fact $0 < m < 1$ for if $m = 1$ then $u(x) = 1$ is the unique solution to (11.3.18) with $u(a) = 1, u'(a) = 0$. However, such a solution does not satisfy the boundary conditions (11.3.19). With $0 < m < 1$ we see that $u''(x) < 0$ for all x in the interval $-l/2 < x < l/2$ and so $u(x) < m$ except at $x = a$.

Let us now attempt to construct the solution to the boundary value problem via the method outlined above. Here we can write

$$F(u) = \int_0^u V(1 - V)dV = \frac{u^2}{2} - \frac{u^3}{3}$$

and so (11.3.17) can be written as

$$\left(\frac{du}{dx}\right)^2 + 2F(u) = A.$$

When $x = a$, $\frac{du}{dx} = 0$ and so $A = 2F(m)$. Thus

$$\left(\frac{du}{dx}\right)^2 + 2F(u) = 2F(m). \tag{11.3.20}$$

Notice also that F is a strictly increasing function of u for $0 < u < 1$. Thus if $x \neq a$ we have from (11.3.20)

$$\frac{du}{dx} = \begin{cases} \sqrt{2}\sqrt{[F(m) - F(u)]}, & -\frac{1}{2}l \leq x < a, \\ -\sqrt{2}\sqrt{[F(m) - F(u)]}, & a < x \leq \frac{1}{2}l. \end{cases}$$

In the first case

$$\int_u^m \frac{d\eta}{\sqrt{[F(m) - F(\eta)]}} = \sqrt{2}(a - x), \quad -\frac{1}{2}l \leq x < a, \tag{11.3.21}$$

and in the second

$$\int_u^m \frac{d\eta}{\sqrt{[F(m) - F(\eta)]}} = \sqrt{2}(x - a), \quad a < x \leq \frac{1}{2}l. \tag{11.3.22}$$

The implicit solution defined by (11.3.21), (11.3.22) is not yet in a desirable form since it involves the two unknown constants a and m. On using the boundary conditions we have

$$\int_0^m \frac{d\eta}{\sqrt{[F(m) - F(\eta)]}} = \sqrt{2}(a + l/2),$$

$$\int_0^m \frac{d\eta}{\sqrt{[F(m) - F(\eta)]}} = \sqrt{2}(l/2 - a),$$

and these identities are only compatible if $a = 0$. That is, $u(x)$ takes its maximum value at the midpoint of the interval $-l/2 \leq x \leq l/2$. With $a = 0$ we see that l and m are related through

$$l = \sqrt{2} \int_0^m \frac{d\eta}{\sqrt{[F(m) - F(\eta)]}}. \tag{11.3.23}$$

In summary then, the solution to the boundary value problem is given implicitly by the formula

$$\int_0^m \frac{d\eta}{\sqrt{[F(m) - F(\eta)]}} = \sqrt{2}|x|, \tag{11.3.24}$$

where m, as a function of l, is given by (11.3.23).

Further important information can be obtained about the problem if we analyse the relationship between m and l defined by (11.3.23). In particular it can be shown that

(a) l is an increasing function of m for $0 \le m < 1$;

(b) $l \to \infty$ as $m \to 1$ from below;

(c) $l \to \pi$ as $m \to 0$ from above.

These facts imply that there is no positive value of m satisfying (11.3.23) if $l < \pi$ while for $l \ge \pi, m$ increases from 0 to 1 as l increases from π to ∞. Thus, for $l \le \pi$, (11.3.18) and (11.3.19) have only the trivial solution $u = 0$, but for $l > \pi$ the solution (11.3.24) appears. This type of behaviour is fundamental to the subject of **bifurcation theory** and $l = \pi$ is called the **bifurcation point**. ⬚

11.4 Comparison theorems

Consider the ordinary differential equation

$$\frac{d^2 u}{dx^2} = F(x), \tag{11.4.1}$$

defined on the interval $-\infty \le a \le x \le b \le \infty$, where $F(x) \ge 0$ and u is a continuous function of x. If $F(x) > 0$ then we conclude that $\frac{d^2 u}{dx^2} > 0$ for all x in the interval $a \le x \le b$. This means that $u(x)$ is concave and therefore takes its maximum value M at either $x = a$ or $x = b$ or both. Now suppose that $F(x) \ge 0$ and that $u \le N$ at the end points $x = a, b$. Then for any $\epsilon > 0$ the function $V = u + \epsilon x^2$ satisfies the equation

$$\frac{d^2 V}{dx^2} = F(x) + 2\epsilon > 0$$

and so, from the above observation, V must attain its maximum at least at one of the endpoints $x = a, b$. Thus

$$V(x) \le max[u + \epsilon x^2] \le N + \epsilon max(|a|^2, |b|^2).$$

Since $u \le V$ we have $u \le N + \epsilon max(|a|^2, |b|^2)$ for all x in the interval $a \le x \le b$ and for any $\epsilon > 0$. Letting $\epsilon \to 0$, we conclude that

$$u(x) \le N, \quad a \le x \le b.$$

That is, if u is a solution to (11.4.1) with $F \ge 0$, the value of u in the interval $a \le x \le b$ cannot exceed its maximum value at the end points $x = a, b$.

This relatively simple observation is an example of the **maximum principle**. We shall see that a result of this type also holds for partial differential equations, but, more than this, maximum principles have far-reaching consequences when applied to the study of qualitative properties of the solutions to evolutionary equations. In this context, we prefer to use the term **comparison principle** or **theorem** for reasons that will become apparent. Before describing these results it is useful to note that the conclusions reached above do not violate the arguments used in Example 11.3d where the maximum u was attained at the midpoint of the interval $-\frac{1}{2}l \leq x \leq \frac{1}{2}l$. In that example it was shown that $0 < u < 1$ and so $F(x) = -u(x)[1 - u(x)] < 0$.

Let us now turn to the elliptic partial differential equation

$$\frac{\partial^2 u}{\partial x^2} + \frac{\partial^2 u}{\partial y^2} = F(x, y) \qquad (11.4.2)$$

defined on a bounded domain Ω with boundary $\partial\Omega$. The arguments we employ as well as the results obtained extend without too much difficulty to higher dimensions, as well as to more general elliptic partial differential operators.

Suppose that $F > 0$ in Ω and that $u(x, y)$ is continuous on $\Omega \cup \partial\Omega$. Then u attains its maximum value M somewhere on $\Omega \cup \partial\Omega$. If $u = M$ at some point (x_0, y_0) of Ω then we know that

$$\frac{\partial u}{\partial x} = \frac{\partial u}{\partial y} = 0$$

and

$$\frac{\partial^2 u}{\partial x^2} \leq 0, \quad \frac{\partial^2 u}{\partial y^2} \leq 0$$

at (x_0, y_0). That is,

$$\frac{\partial^2 u}{\partial x^2} + \frac{\partial^2 u}{\partial y^2} \leq 0$$

at (x_0, y_0), which contradicts the assumption that $F > 0$. Thus the maximum of u must occur on the boundary $\partial\Omega$. If $F \geq 0$ and $u \leq N$ on $\partial\Omega$ then we can apply our previous argument to the function

$$V = u + \epsilon(x^2 + y^2)$$

where $\epsilon > 0$ is arbitrary. Clearly, V satisfies the partial differential equation

$$\frac{\partial^2 V}{\partial x^2} + \frac{\partial^2 V}{\partial y^2} = F + 4\epsilon > 0$$

in Ω.

Our above conclusions apply and lead to the fact that

$$V \leq max_{\partial\Omega}[u + \epsilon(x^2 + y^2)]$$
$$\leq N + \epsilon R^2$$

where R is the radius of a circle enclosing Ω. Furthermore, since $u \leq V$ we have

$$u \leq N + \epsilon R^2$$

and on letting $\epsilon \to 0$ we obtain the result $u(x, y) \leq N$ in Ω. Thus we have the following **maximum principle**.

THEOREM 2

Let u be a continuous solution of (11.4.2) with $F \geq 0$. Then the value of u in Ω cannot exceed its maximum on $\partial\Omega$.

As an application of the maximum principle, consider the following uniqueness problem for the boundary value problem

$$\frac{\partial^2 u}{\partial x^2} + \frac{\partial^2 u}{\partial y^2} = F(x, y)$$

in Ω and $u = f$ on $\partial\Omega$.

Suppose the solution to this problem is not unique and that there exist distinct solutions u_1 and u_2 where $w = u_1 - u_2 \neq 0$. Since the partial differential equation is linear, it follows that w is a solution to the problem

$$\frac{\partial^2 w}{\partial x^2} + \frac{\partial^2 w}{\partial y^2} = 0$$

in Ω and $w = 0$ on $\partial\Omega$. It follows immediately from the maximum principle that $w \equiv 0$ and so $u_1 = u_2$.

Let us now turn to the heat equation (11.1.2) and consider the function $u(x, t)$ satisfying the inequality

$$L[u] \equiv \frac{\partial^2 u}{\partial x^2} - \frac{\partial u}{\partial t} \geq 0$$

defined on the rectangular $R = \{0 < x < a, 0 < t \leq T\}$. Suppose that M is the maximum of the values of u that occur on the sides $S = \{x = 0, 0 \leq t \leq T\} \cup \{0 \leq x \leq a, t = 0\} \cup \{x = a, 0 \leq t \leq T\}$. In addition, suppose that u takes its overall maximum value $M_1 > M$ at an interior point (x_0, t_0) of the rectangle. We shall show that this is impossible.

Define the auxiliary function

$$w(x) = \frac{M_1 - M}{2a^2}(x - x_0)^2.$$

Now since $u \leq M$ on the three sides S of the rectangle listed above we have

$$V(x, t) \equiv u(x, t) + w(x) \leq M + \frac{M_1 - M}{2} < M, \tag{11.4.3}$$

on S. Furthermore

$$V(x_0, t_0) = u(x_0, t_0) = M_1 \qquad (11.4.4)$$

and

$$L[V] = L[u] + L[w] = L[u] + \frac{M_1 - M}{a^2} > 0 \qquad (11.4.5)$$

throughout R. Conditions (11.4.3) and (11.4.4) show that V must assume its maximum either at an interior point of R or along the open interval

$$\{0 < x < a, t = T\}. \qquad (11.4.6)$$

If the maximum occurs at an interior point then

$$\frac{\partial^2 V}{\partial x^2} \leq 0, \quad \frac{\partial V}{\partial t} = 0,$$

and so at such a point $L[V] \leq 0$, which contradicts (11.4.5). If the maximum occurs at some point of the interval (11.4.6), then $\frac{\partial^2 V}{\partial x^2} \leq 0$ and (11.4.5) requires $\frac{\partial V}{\partial t}$ to be strictly negative. In other words, V must be larger at an earlier time so that the maximum cannot occur on the interval (11.4.6). Thus the assumption that $u(x_0, t_0) = M_1 > M$ cannot hold and so the maximum value of u must be attained somewhere on the sides S.

This conclusion is an example of the maximum principle for the heat equation. There is a corresponding minimum principle associated with solutions of $L[u] = 0$; we simply replace u by $-u$ in the above discussion and repeat the arguments.

A weakness of this type of maximum principle is that it permits the maximum of u to occur at interior points of a region as well as on the boundary. Thus it is desirable to establish a "strong" form of the maximum principle. In fact, we shall establish a strong maximum principle for a more general differential inequality, which will prove to be of basic importance to the comparison principle, which is our goal.

THEOREM 3 (Strong Maximum Principle)
Let $u(x, t)$ satisfy the differential inequality

$$(L + h)[u] \geq 0 \qquad (11.4.7)$$

in the rectangular region R and where the given function $h(x, t) \leq 0$ in R. If the maximum M of u is attained at an interior point (x_1, t_1) and if $M \geq 0$, then $u \equiv M$ on all line segments $t = $ constant, which lie directly below the horizontal segment of R containing (x_1, t_1).

The proof of this result follows from the following lemmas:

LEMMA 4

Let $u(x,t)$ satisfy the differential inequality (11.4.7) in the rectangular region R. Let D be a circular disc such that it and its boundary ∂D are contained in R. Suppose the maximum of u in R is $M \geq 0$, that $u < M$ in the interior of D and that $u = M$ at some point $P(x_1, t_1)$ on the boundary ∂D. Then the tangent to D at P is parallel to the x-axis. That means P is either at the top or the bottom of the disc D.

PROOF   Let the disc have centre $(\bar{x}, \bar{t})$ and let A be the radius of D (see Figure 11.4.1). The idea is to assume that P is not at the top or the bottom of the disc and reach a contraction.   ⬚

We can safely assume that P is the only boundary point where $u = M$. For, if otherwise, we can choose a smaller disc interior to D and tangent to D at P. This smaller disc has exactly one point on its boundary where $u = M$.
Construct a disc D' with centre (x_1, t_1) and of radius A' where

$$A' < |x_1 - \bar{x}| \tag{11.4.8}$$

and D' also lies inside R. Note that since P is not at the top or the bottom of D, $x_1 \neq \bar{x}$. Now the boundary of D' consists of two arcs $\partial D'$ (which includes its points of intersection with ∂D) and $\partial D''$ (as shown in Figure 11.4.1). Since $u < M$ on $\partial D'$ we can find a positive number η so that

$$u \leq M - \eta$$

on $\partial D'$.

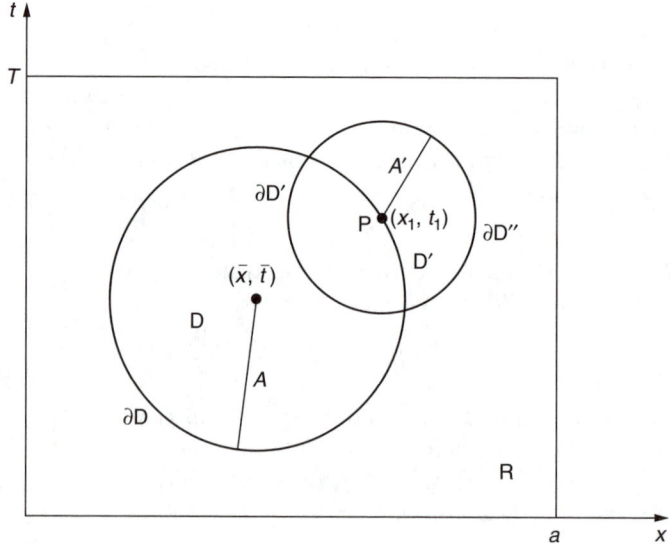

FIGURE 11.4.1:   The domain of lemma 4.

Furthermore, since $u \leq M$ throughout the rectangle R,

$$u \leq M$$

on $\partial D''$. Define the function

$$V(x,t) = \exp\{-\alpha[(x - \bar{x})^2 + (t - \bar{t})^2]\} - e^{-\alpha A^2}$$

where α is a positive number to be suitably chosen. $V(x,t)$ has the following properties: (i) $V(x,t) = 0$ on ∂D, (ii) $V(x,t) > 0$ inside D and (iii) $V(x,t) < 0$ outside D. If we substitute $V(x,t)$ into the differential expression (11.4.7) then

$$(L + h)[V] = 2\alpha \exp\{-[(x - \bar{x})^2 + (t - \bar{t})^2]\}[2\alpha(x - \bar{x})^2 - 1 + (t - \bar{t})]$$
$$+ h(x,t)V(x,t).$$

In the disc D' and on its boundary we have

$$|x - \bar{x}| > |x_1 - \bar{x}| - A' > 0$$

and so we can choose α large enough so that

$$(L + h)[V] > 0$$

for all points (x,t) in $D' \cup \partial D'$. Next form the function

$$W(x,t) = u(x,t) + \epsilon V(x,t),$$

where ϵ is a positive number to be suitably chosen. Then

$$(L + h)[W] = (L + h)[u] + \epsilon(L + h)[V] > 0 \qquad (11.4.9)$$

inside D'. Since $u \leq M - \eta$ on $\partial D'$ we can select ϵ so small so that

$$W = u + \epsilon V < M$$

on $\partial D'$, and since $V < 0$ on $\partial D''$ and $u \leq M$ we have

$$W = u + \epsilon V < M$$

on $\partial D''$. Thus $W < M$ on the entire boundary of D'.

In addition, we observe that since $V = 0$ on ∂D we have

$$W(x_1, t_1) = u(x_1, t_1) + \epsilon V(x_1, t_1)$$
$$= u(x_1, t_1) = M.$$

Hence the maximum of W on D' must occur at an interior point. This means that at such a point

$$\frac{\partial W}{\partial t} = 0, \quad \frac{\partial^2 W}{\partial x^2} \leq 0$$

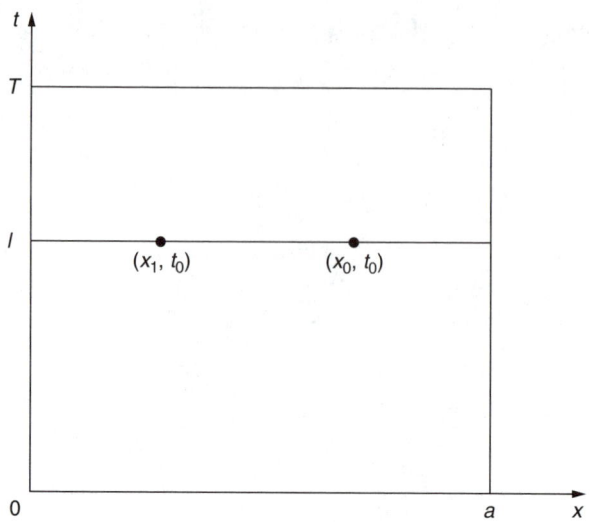

FIGURE 11.4.2: The domain of lemma 5.

and $W = M \geq 0$, and so since $h \leq 0$ we have

$$(L + h)[W] \leq 0$$

at the maximum and this contradicts (11.4.9).

Thus the assumption that P is not at the top or the bottom of D cannot hold. That is, we have reached a contradiction and the lemma is proved. Note that for P at the top or the bottom of D our arguments break down because $x_1 = \bar{x}$ and so we cannot construct the disc D' satisfying the condition (11.4.8).

LEMMA 5

Suppose that $u(x,t)$ satisfies (11.4.7) in the rectangular region R, $u < M$ at some interior point (x_0, t_0) of R and that $u \leq M$ throughout R. If l is the horizontal line through (x_0, t_0) (see Figure 11.4.2) then $u < M$ on l.

PROOF Suppose that $u = M$ at some interior point (x_1, t_0) on l and that $u < M$ at (x_0, t_0). As in the proof of lemma 5, we endeavour to reach a contradiction. For convenience we assume that $x_1 < x_0$ and move x_1 to the right if necessary so that $u < M$ for $x_1 < x \leq x_0$. Let d_0 be either the length $x_0 - x_1$ or the minimum of the distances from any point of the segment $x_1 \leq x \leq x_0, t = t_0$ to the boundary of the rectangle R, whichever is the smaller.

For $x_1 < x < x_0 + d_0$, we define $d(x)$ to be the distance from (x, t_0) to the nearest point in R where $u = M$. Since $u(x_1, t_0) = M, d(x) \leq x - x_1$.

According to lemma 4 this nearest point is directly above or below (x, t_0). That is, either $u(x, t_0 + d(x)) = M$ or $u(x, t_0 - d(x)) = M$.

Since the distance from a point $(x + \delta, t_0)$ to $(x, t_0 \pm d(x))$ is $\sqrt{[d(x)^2 + \delta^2]}$ we see that

$$d(x + \delta) \le \sqrt{[d(x)^2 + \delta^2]} < d(x) + \frac{\delta^2}{2d(x)}. \tag{11.4.10}$$

Similarly by replacing x by $x + \delta$ and δ by $-\delta$ we have

$$d(x + \delta) > \sqrt{[d(x)^2 - \delta^2]}. \tag{11.4.11}$$

Suppose then that $d(x) > 0$ and choose $0 < \delta < d(x)$. Now subdivide the interval $(x, x + \delta)$ into n equal parts and apply the inequalities (11.4.11) to get

$$d\left(x + \frac{j+1}{n}\delta\right) - d\left(x + \frac{j}{n}\delta\right) \le \frac{\delta^2}{2n^2 d(x + j\delta/n)}$$
$$\le \frac{\delta^2}{2n^2 \sqrt{[d(x)^2 - \delta^2]}},$$

$j = 0, 1, \ldots, n - 1$.

Summing from $j = 0$ to $n - 1$ gives

$$d(x + \delta) - d(x) \le \frac{\delta^2}{2n^2 \sqrt{[d(x)^2 - \delta^2]}}$$

for any integer n. Now if we let $n \to \infty$ then it follows that

$$d(x + \delta) \le d(x), \quad \delta > 0.$$

In other words $d(x)$ is a non-increasing function of x. Since $d(x) \le x - x_1$, which is arbitrarily small for x sufficiently close to x_1, we see that $d(x) \equiv 0$ for $x_1 < x < x_1 + d_0$. In other words $u(x, t_0) \equiv M$ on this interval, which contradicts our assumption that $u < M$ for $x_1 < x \le x_0$. This establishes the lemma. ▯

LEMMA 6
Suppose that $u(x, t)$ satisfies (11.4.7) in the rectangular region R and that $u < M$ where $M \ge 0$ in the horizontal strip contained in R and defined by $\{0 \le x \le l, t_0 < t < t_1\}$ for some numbers t_0 and t_1. Then $u < M$ on the line $t = t_1$ contained in R.

PROOF Again the proof is by contradiction. Suppose $P(x_1, t_1)$ is a point on the line $t = t_1$ where $u = M$. Construct a disc D with centre P and radius so small that the lower half of D is entirely in the portion of R where $t > t_0$. (See Figure 11.4.3.)

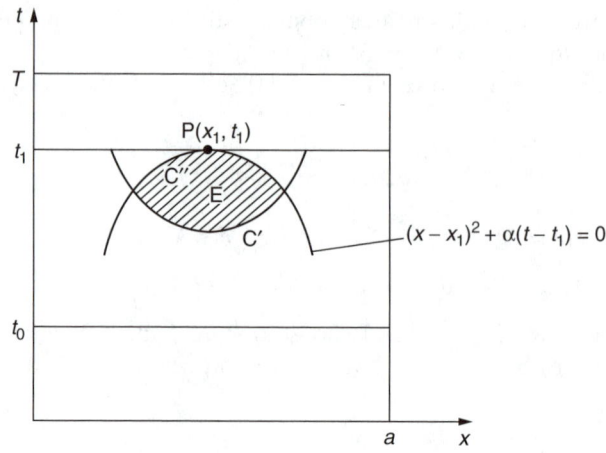

FIGURE 11.4.3:   The domain of lemma 6.

Introduce the auxiliary function

$$V(x,t) = \exp\{-[(x - x_1)^2 + \alpha(t - t_1)]\} - 1.$$

Substituting this into the differential expression in (11.4.7), we have

$$(L + h)[V] = \exp\{-[(x - x_1)^2 + \alpha(t - t_1)]\}[4(x - x_1)^2 - 2 + \alpha] + h(x,t)V(x,t).$$

Now choose α positive and sufficiently large so that

$$(L + h)[V] > 0$$

in D for $t \leq t_1$.

The parabola

$$(x - x_1)^2 + \alpha(t - t_1) = 0$$

is tangent to the line $t = t_1$ at P. We denote by C' the portion of ∂D that is below the parabola (including the endpoints) and C'' the portion of the parabola lying inside the disc D.

By hypothesis, $u < M$ on C' and so we can find a number $\eta > 0$ such that

$$u \leq M - \eta$$

on C'.

Now construct the function

$$W(x,t) = u(x,t) + \epsilon V(x,t)$$

where ϵ is a positive number yet to be chosen. Observe that $V = 0$ on C'' and so if ϵ is chosen sufficiently small we see that W enjoys the following properties:

(a) $(L+h)[W] = (L+h)[u] + \epsilon(L+h)[V] > 0$ in the shaded region E shown in Figure 11.4.3.

(b) $W = u + \epsilon V < M$ on C'.

(c) $W = u + \epsilon V \leq M$ on C''.

Condition (a) shows that W cannot attain its maximum in E, and so the maximum of W is M and it occurs at P. We conclude that

$$\frac{\partial W}{\partial t} \geq 0 \tag{11.4.12}$$

at P.

Also it is easy to verify that

$$\frac{\partial V}{\partial t} = -\alpha < 0 \tag{11.4.13}$$

at P.

Thus

$$\frac{\partial u}{\partial t} = \frac{\partial W}{\partial t} - \epsilon \frac{\partial V}{\partial t} > 0 \tag{11.4.14}$$

at P. However since the maximum of u on $t = t_1$ occurs at P we have

$$\frac{\partial u}{\partial x} = 0, \quad \frac{\partial^2 u}{\partial x^2} \leq 0,$$

and $u = M \geq 0$, and these inequalities violate the differential inequality (11.4.7). Thus we have again arrived at a contradiction and the lemma is proved. □

Armed with lemmas 4 and 6 we can now give the proof of Theorem 3.

PROOF OF THEOREM 3   Suppose there is a value of t, say $t_0 < t_1$, such that $u(x_1, t_1) < M$. Let τ be the least upper bound of values of $t < t_1$ for which $u(x_1, t) < M$. By continuity $u(x_1, \tau) = M$ while $u(x_1, t) < M$ for some interval $\tau_1 < t < \tau$. Then from lemma 4 it follows that $u < M$ in the strip $\{0 \leq x \leq l, \tau_1 < t < \tau\}$, and lemma 6 shows that $u(x_1, t) < M$ on $t = \tau$, which is a contradiction and so $u(x_1, t_0) \equiv M$. This proves the theorem. □

Having established the strong maximum principle, we now use it to prove a comparison theorem, which is an invaluable tool with which to study qualitative properties of evolutionary equations of the type (11.3.2).

THEOREM 7 (The Comparison Principle)
Let $u(x,t), v(x,t), A \leq u(x,t), v(x,t) \leq B$ be bounded and continuous functions satisfying the inequalities

$$u_t - u_{xx} - f(u) \geq v_t - v_{xx} - f(v), \quad \text{in } (a,b) \times [0,T],$$
$$A \leq v(x,0) \leq u(x,0) \leq B, \quad \text{in } (a,b)$$

where $-\infty \leq a < b \leq \infty$ *and* $0 < T \leq \infty$ *and* $f(u)$ *is a continuously differential function of* u.

Furthermore if $a > -\infty$ assume that

$$A \leq v(a,t) \leq u(a,t) \leq B \quad \text{on } [0,T]$$

and if $b < \infty$ *assume that*

$$A \leq v(a,t) \leq u(a,t) \leq B \quad \text{on } [0,T].$$

Then either $u \equiv v$ *in* $[a,b] \times [0,T]$ *or* $u > v$ *in* $(a,b) \times [0,T]$.

PROOF From the mean value theorem we can write

$$(u-v)_t - (u-v)_{xx} \geq f(u) - f(v) = f'(v + \theta(u-v))(u-v)$$

for some θ where $0 < \theta < 1$. Now take

$$\alpha = \max_{[A,B]} f'(u)$$

and define the function

$$W(x,t) = (u-v)e^{-\alpha t}.$$

It then follows that $W(x,t)$ satisfies the inequality

$$(W)_t - (W)_{xx} \geq \{f'(v + \theta(u-v)) - \alpha\}W.$$

Since the coefficient of W is non-positive, the conclusions of the theorem follow from an application of the strong maximum principle theorem. ▯

Example 11.4a

Let $u(x,t)$ satisfy the following problem:

$$u_t = u_{xx} + u(1-u),$$
$$u = 0 \quad for \ |x| = \frac{1}{2}l,$$
$$u(x,0) = u_0(x) > 0.$$

For this problem, several applications of the comparison theorem with $f(u) = u(1-u)$ can be used to show that $u(x,t) \geq 0$ and that if $l > \pi$ then

$$\lim_{t \to \infty} u(x,t) = V(x) \tag{11.4.15}$$

where $V(x)$ is the unique positive solution to Example 11.3d. The proof of these results proceeds in several steps, some of which will be given here and the remaining given as problems at the end of this chapter.

To show that $u(x,t) \geq 0$ take $v(x,t) = 0$ in the comparison theorem. In order to establish the result (11.4.15) we proceed to construct appropriate **upper** and **lower** functions $\bar{u}(x,t)$ and $\underline{u}(x,t)$, respectively, so that

$$\underline{u}(x,t) \leq u(x,t) \leq \bar{u}(x,t)$$

and such that

$$\lim_{t\to\infty} \underline{u}(x,t) = \lim_{t\to\infty} \bar{u}(x,t) = V(x). \qquad (11.4.16)$$

In order to construct $\bar{u}(x,t)$ we proceed as follows: let $M = max(1, M^*)$ where $M^* = max\ u_0(x)$ and $\bar{u}(x,t)$ denotes the solution to

$$\bar{u}_t = \bar{u}_{xx} + \bar{u}(1-\bar{u}), \quad |x| < l/2, \ t > 0,$$
$$\bar{u}(x.0) = M, \quad if\ |x| < l/2,$$
$$\bar{u} = 0, \quad for\ |x| = l/2, \ t > 0.$$

On setting $u = \bar{u}$ and $v = u(x,t)$ in the comparison theorem it is easy to see that

$$u(x,t) \leq \bar{u} \quad for\ |x| \leq l/2, \ t > 0.$$

It is easy to see that $\bar{u} \geq 0$. Next we show that $\bar{u}(x,t) \leq M$. This is done by taking $u = M$ and $v = \bar{u}$ in the comparison theorem. If, for arbitrary $h > 0$, we set $u = \bar{u}$ and $v = \bar{u}(x, t+h)$, then the comparison theorem shows that $\bar{u}(x,t) \geq \bar{u}(x, t+h)$. In other words, $\bar{u}(x,t)$ is a non-increasing function of t for each x. Furthermore, $\bar{u}(x,t)$ is bounded below and so

$$\lim_{t\to\infty} \bar{u}(x,t) = W(x)$$

exists and $0 \leq W(x) \leq M$ for $|x| \leq l/2$.

It can now be shown that $W(x)$ is a non-negative solution of Example 11.3d (see Exercise 11.16).

Now set $u = \bar{u}(x,t)$ and $v = V(x)$. Then the comparison theorem shows that $W(x) \geq V(x)$. However, $V(x)$ is the unique positive solution to Example 11.3d and so it must be the case that $W(x) = V(x)$. These arguments establish one part of the limit relation (11.4.16). The other is proved in a similar manner (see Exercises 11.15 and 17). ∎

Example 11.4b
Let $u(x,t)$ satisfy the problem

$$u_t = u_{xx} + u(1-u)(u-a), \quad -\infty < x < \infty, \ t > 0,$$
$$u(x,0) = \phi(x).$$

Suppose

$$\sup_{-\infty < x < \infty} \phi(x) < a$$

and let $v(x,t) = a$. Then the comparison principle applied to the functions $u(x,t), v(x,t)$ shows that $u(x,t) < a$ for all x and t.

Similarly, if

$$\inf_{-\infty < x < \infty} \phi(x) > a$$

then again the comparison principle says that $u(x,t) > a$ for all x and t. $\quad\square$

11.5 Notes

For further reading on partial differential equations see F. John, *Partial Differential Equations,* Springer-Verlag, Berlin, 1975; and H.F. Weinberger, *Partial Differential Equations,* Blaisdell, New York, 1965.

A comprehensive treatment of maximum and comparison principles is to be found in M.H. Protter and H.F. Weinberger, *Maximum Principles in Differential Equations,* Springer-Verlag, Berlin, 1984 or A. Friedman, *Partial Differential Equations of Parabolic Type,* Prentice-Hall, Englewood Cliffs, NJ, 1964.

For an introduction to the numerical solution of partial differential equations see A.R. Mitchell and D.F. Griffiths, *The Finite Difference Method in Partial Differential Equations,* John Wiley & Sons, New York, 1980.

Exercises

11.1 Show that the bounded solution of $u_t = u_{xx}$ with initial data

$$u(x,0) = \begin{cases} 1, & x > 0, \\ 0, & x < 0 \end{cases}$$

is given by

$$u(x,t) = \frac{1}{2}[1 + \phi(x/2\sqrt{t})]$$

where $\phi(x)$ is the error function defined by

$$\phi(x) = \frac{2}{\sqrt{\pi}} \int_0^x e^{-t^2}\, dt.$$

11.2 Find the solution $u(x,t)$ of the heat equation

$$u_t = u_{xx}, \quad x > 0, \ t > 0$$

with boundary, initial data

$$u(x,0) = 0 \quad for \ x > 0,$$
$$u(0,t) = 1 \quad for \ t > 0.$$

11.3 Find the solution of the initial value problem

$$u_t = u_{xx}, \quad 0 < x < 1, \ t > 0,$$
$$u(x,0) = f(x), \quad 0 < x < 1,$$
$$u(0,t) = u(1,t) = 0, \quad t > 0.$$

11.4 Solve

$$u_t = u_{xx}, \quad 0 < x < \pi, \ t > 0,$$
$$u(x,0) = x(\pi - x), \quad 0 < x < \pi,$$
$$u(0,t) = u(\pi,t) = 0, \quad t > 0.$$

11.5 Solve

$$u_t = u_{xx}, \quad 0 < x < \pi, \ t > 0,$$
$$u(x,0) = \sin x, \quad 0 < x < \pi,$$
$$\frac{\partial u}{\partial x}(0,t) = \frac{\partial u}{\partial x}(\pi,t) = 0, \quad t > 0.$$

11.6 Solve

$$u_t = u_{xx} + u_{yy}, \quad x,y \in (0,1) \times (0,1), \ t > 0,$$
$$u(x,0,t) = u(x,1,t) = 0,$$
$$u(0,y,t) = u(1,y,t) = 0,$$
$$u(x,y,0) = xy.$$

11.7 For the n-dimensional evolutionary equation

$$u_t = \nabla^2 u + f(u),$$

show that the wavefront $u(\mathbf{x} \cdot \nu - ct)$, where

$$\mathbf{x} \cdot \nu = \sum_{i=1}^{n} x_i \nu_i$$

and ν is an arbitrary unit vector, is a solution provided $u(\xi)$ satisfies the ordinary differential equation

$$u'' + cu' + f(u) = 0.$$

11.8 By following the method illustrated in Example 11.3b, find a wavefront solution to the equation

$$u'' - cu' + f(u) = 0,$$

satisfying the conditions $c > 0$, $0 \leq u \leq 1$, $u(-\infty) = 0$, $u(\infty) = 1$ where

$$f(u) = \begin{cases} 0, & 0 \leq u < 1/2, \\ u - 1/2, & 1/2 \leq u \leq 3/4, \\ 1 - u, & u > 3/4. \end{cases}$$

11.9 Establish the results (a), (b), (c) following (11.3.24).

11.10 Show that if

$$\frac{d^2 u}{dx^2} + A(x)\frac{du}{dx} = 0, \quad 0 < x < 1,$$

where A is continuous, then u attains its maximum in the interval $0 \leq x \leq 1$ either at $x = 0$ or at $x = 1$.

11.11 Show that if u is a solution of

$$\nabla^2 u + D(x, y)\frac{\partial u}{\partial x} + E(x, y)\frac{\partial u}{\partial x} = -F(x, y)$$

in some bounded two-dimensional region Ω with $F < 0$, then u cannot attain its maximum at any point of Ω.

11.12 Show that if u is continuous in a bounded two-dimensional region Ω with boundary $\partial\Omega$ and

$$\frac{\partial^2 u}{\partial x^2} + e^x \frac{\partial^2 u}{\partial y^2} - e^{x+y} u = 0 \quad in \ \Omega,$$

$$u \leq 0 \quad on \ \partial\Omega,$$

then $u \leq 0$ in Ω.

11.13 Show that if

$$\frac{\partial u}{\partial t} = \frac{\partial^2 u}{\partial x^2}, \quad 0 < x < 1,$$

$$\frac{\partial u}{\partial x}(0, t) = 0,$$

the maximum of u for $0 \leq x \leq 1$, $0 \leq t \leq T$ must occur at $t = 0$ or at $x = 1$.

11.14 The function $u(x,t) = -(x^2 + 2xt)$ is a solution of the equation

$$x\frac{\partial^2 u}{\partial x^2} - \frac{\partial u}{\partial t} = 0.$$

Is the maximum principle value in the region $-1 < x < 1$, $0 \le t \le 1/2$?

11.15 In Example 11.4a, let $v_\lambda(x)$ denote the positive solution of the problem

$$v'' + v(1-v) = 0, \quad |x| < \lambda/2, \ \pi < \lambda < l,$$
$$v = 0, \quad |x| = \lambda/2,$$

and let $\underline{u}(x,t)$ denote the solution of the problem

$$\frac{\partial u}{\partial t} = \frac{\partial^2 u}{\partial x^2} + \underline{u}(1 - \underline{u}), \quad |x| < l/2$$

$$\underline{u} = \begin{cases} v_\lambda(x), & |x| \le \lambda/2, \\ 0, & |x| > \lambda/2, \end{cases}$$

$$\underline{u}(x,t) = 0 \quad |x| = l/2, \ t \ge 0.$$

Use the comparison theorem to show that

$$\underline{u} \le u(x,t),$$

for $|x| \le l/2$, $t \ge 0$.

11.16 Use the comparison theorem to show that the "upper" solution $\bar{u}(x,t)$ constructed in Example 11.4a satisfies

$$\bar{u}(x,t) \ge v(x),$$

where $v(x)$ is the unique positive solution to Example 11.3d. Hence prove that $v(x) = w(x)$, where $w(x)$ is the limiting function in Example 11.4a.

11.17 By using the comparison theorem, show that $\underline{u}(x,t)$ and $v_\lambda(x)$ defined in Exercise 11.15 satisfy the inequality

$$\underline{u}(x,t) \ge v_\lambda(x) = \underline{u}(x,0), \quad |x| \le \lambda/2.$$

Hence show that $\underline{u}(x,t)$ increases to $v(x)$ as $t \to \infty$.

Chapter 12

Problems of Diffusion

12.1 Diffusion through membranes

The equations we shall study in this chapter are all particular cases of the general system

$$
\begin{bmatrix} \dfrac{\partial u_1}{\partial t} \\[1ex] \dfrac{\partial u_2}{\partial t} \\[1ex] \vdots \\[1ex] \dfrac{\partial u_m}{\partial t} \end{bmatrix} = \begin{bmatrix} D_1 & 0 & 0 & \cdots & 0 & 0 \\ 0 & D_2 & 0 & \cdots & 0 & 0 \\ \vdots & \vdots & \vdots & \cdots & 0 & 0 \\ 0 & 0 & 0 & \cdots & D_{m-1} & 0 \\ 0 & 0 & 0 & \cdots & 0 & D_m \end{bmatrix} \begin{bmatrix} \nabla^2 u_1 \\ \nabla^2 u_2 \\ \vdots \\ \nabla^2 u_m \end{bmatrix} + \begin{bmatrix} f_1(u_1, u_2, \ldots, u_m) \\ f_2(u_1, u_2, \ldots, u_m) \\ \vdots \\ f_m(u_1, u_2, \ldots, u_m) \end{bmatrix}.
$$

This system can be conveniently expressed in the matrix form

$$
\frac{\partial \mathbf{u}}{\partial t} = \mathbf{D}\nabla^2 \mathbf{u} + \mathbf{f}(\mathbf{u}), \tag{12.1.1}
$$

where $\mathbf{u} = \mathbf{u}(t, x_1, \ldots, x_n)$ is a vector-valued function of time t and the n variables $x_i, i = 1, \ldots, n$, and has m components $u_i(t, x_1, \ldots, x_n), i = 1, \ldots, m$. $\mathbf{D}$ is an $m \times m$ diagonal matrix, ∇^2 is the n-dimensional Laplacian (see Chapter 4) and f is an $m \times 1$ column vector. Such equations as (12.1.1) are called **non-linear diffusion equations** and the matrix $\mathbf{D}$ is called the **diffusion matrix**. Our treatment will be confined to $n = 1$ (one space dimension) and $m = 1$ or 2.

Most of the biological models we have described so far are formulated as non-linear diffusion equations and the reader may care to verify this by referring back to Chapter 4. Although we have used the term "diffusion" before, we have done so without motivation. Thus, before describing some of the salient features of diffusion equations, we shall consider the precise meaning of diffusion and the diffusion process.

Consider a fluid called the **solvent** in which some matter has been dissolved. This matter is called a **solute** and the combined fluid is called a **solution**. The solution may be conveniently characterised by its mass concentration, which we denote by C. Thus C is the **mass of dissolved matter per unit volume** of liquid.

One of the basic methods by which the solute molecules disperse and are transported through the solvent is by **diffusion**. This mechanism is usually a consequence of thermal motion of the individual solute molecules.

Let us now look at a solution in which simple molecular diffusion is taking place, the fluid being otherwise at rest. The chief mechanism of transport of the solute is governed by concentration differences. Thus we ask, what is the flux of solute molecules going through a unit area in a unit of time? The material flux per unit area j is according to **Fick's law** related to the rate of change $\frac{\partial C}{\partial x}$ of C by

$$j = -D\frac{\partial C}{\partial x}, \tag{12.1.2}$$

where we assume that C depends only on t and the space variable x. The relation (12.1.2) is referred to as **Fick's first law of diffusion** and D is called the **diffusion coefficient** and is a characteristic of the solute in the fluid. The minus sign in (12.1.2) is important in that it reflects the fact that the molecular flow is from a high concentration to a low concentration region.

Suppose now we consider the solution to occupy some region of three-dimensional space and so the concentration $C = C(t, x, y, z)$ will be a function of the time t and the spatial coordinates x, y and z. Clearly the flux will be a vector $\mathbf{j}$ with components j_x, j_y, j_z and, according to Fick's first law of diffusion,

$$j_x = -D\frac{\partial C}{\partial x}, \quad j_y = -D\frac{\partial C}{\partial y}, \quad j_z = -D\frac{\partial C}{\partial z}. \tag{12.1.3}$$

Here of course we are assuming the solvent to be *homogeneous and isotropic* so that D is independent of position and is the same in all directions. In other words D is assumed constant.

Suppose we have a fixed but arbitrary volume of fluid V with surface S (see Figure 12.1.1) and we consider the flow of solute through this volume. By the law of conservation of mass we know that the rate at which solute accumulates (or dissappears) within V must be balanced by the net influx (or efflux) across the bounding surface S. The total amount of solute at time t in V is $\int_V C(t, x, y, z)dV$, where dV is a typical volume element, and the total flux across the boundary S is $\int_S \mathbf{j} \cdot \mathbf{n}\, dS$. Consequently, by the law of conservation of mass,

$$\frac{\partial}{\partial t}\int_V C(t, x, y, z)dV = -\int_S \mathbf{j} \cdot \mathbf{n}\, dS,$$

i.e.,

$$\int_V \frac{\partial}{\partial t}C(t, x, y, z)dV + \int_S \mathbf{j} \cdot \mathbf{n}\, dS = 0.$$

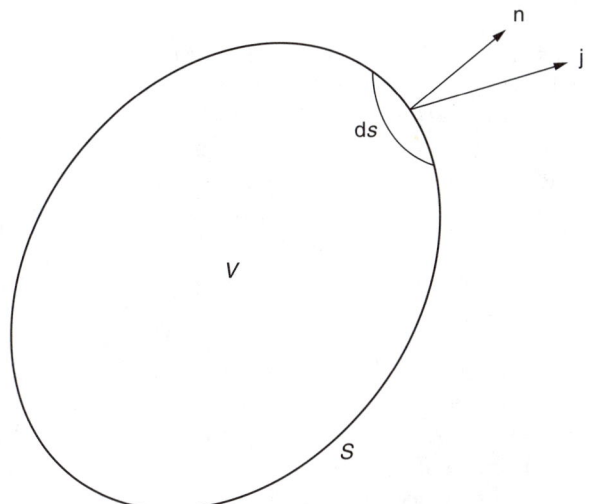

FIGURE 12.1.1: The conservation of mass.

By making use of Gauss' divergence theorem we can write this last result as

$$\int_V \left(\frac{\partial}{\partial t} C(t, x, y, z) + div\mathbf{j} \right) dV = 0,$$

where $div\mathbf{j}$ is the divergence of the vector $\mathbf{j}$. Written in terms of Cartesian coordinates it is defined as

$$div\mathbf{j} = \frac{\partial j_x}{\partial x} + \frac{\partial j_y}{\partial y} + \frac{\partial j_z}{\partial z}.$$

But V is arbitrary and so this equation must hold for all volumes, and this leads to

$$\frac{\partial}{\partial t} C + div\mathbf{j} = 0. \tag{12.1.4}$$

Finally, substituting for $\mathbf{j}$ from (12.1.3) we have

$$\frac{\partial}{\partial t} C - D\nabla^2 C = 0. \tag{12.1.5}$$

This is the classic diffusion equation. In our context it is an expression of **Fick's second law** and in physics it is the basic equation of heat conduction (cf. Chapter 11). In this case C is interpreted as temperature and D heat conductivity.

Let us now consider the application of Fick's laws to the problem of diffusion of a solute into a cell. The diffusion here depends critically on the transport of solution through the membrane surrounding the protoplasm of the cell, namely the **plasma membrane**. This membrane consists of a double layer of lipid molecules called the **bilipid layer**. This layer contains a structure of globular proteins on both its surfaces, some of which penetrate across the entire width of the bilipid layer. However, the process by which these proteins control the diffusion of the solute is not fully understood.

The simplest theory of transport into or out of a cell by diffusion is derived on the assumption that the intracellular diffusion is so rapid that the solute concentration there is uniform in space. In most experiments of diffusion into cells, the volume of extracellular space is made so large that the solute concentration there is not affected to any significant extent by any loss in cells and may be assumed constant. Consider then an idealised cell membrane model to consist of a homogeneous lipid layer separating two aqueous phases, the cell interior and the cell exterior (see Figure 12.1.2a). At each of the two lipid interfaces, a discontinuity in solute concentration exists at equilibrium. This discontinuity is a consequence of the molecular barrier that exists for a solute molecule entering the lipid phase from the aqueous phase. Suppose C_o represents the concentration on the outside of the cell and $\bar{C}_o$ the concentration inside the lipid layer and adjacent to the outer interface. The discontinuity between C_o and $\bar{C}_o$ is expressed in terms of the **partition coefficient** Γ, i.e.,

$$\bar{C}_o = \Gamma C_o,$$

where $\Gamma < 1$ unless the solute is more readily soluble in lipid than in water, in which case $\Gamma > 1$.

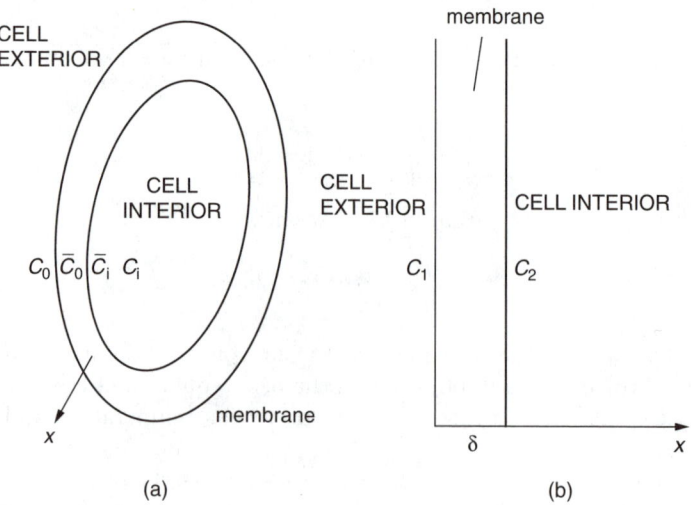

(a)   (b)

FIGURE 12.1.2: Idealised cell membrane model.

In a similar manner we have

$$\bar{C}_i = \Gamma C_i$$

where C_i represents the concentration in the cell interior and $\bar{C}_i$ the concentration inside the lipid layer and adjacent to the inner surface.

Assume the cell membrane thickness δ to be small so that to first order the concentration gradient in the lipid layer can be expressed as

$$\frac{dC}{dx} = \frac{(\bar{C}_o - \bar{C}_i)}{\delta}.$$

On using Fick's first law of diffusion we obtain

$$j = \frac{D}{\delta}(\bar{C}_i - \bar{C}_o), \tag{12.1.6}$$

where D is the diffusion constant of the solute in the lipid layer. In experiments, measurements are made of C_i and C_o and $\bar{C}_i, \bar{C}_o$ are in general inaccessible to observation and so we prefer to write

$$j = \frac{\Gamma D}{\delta}(C_i - C_o). \tag{12.1.7}$$

This is a good approximation to the flux j provided that C_i, C_o vary slowly.

Let m be the mass of solute contained in the cell; then at time t the law of conservation of mass gives

$$\frac{dm}{dt} = -jA, \tag{12.1.8}$$

the minus sign indicating that when j is positive, solute flowing outwards, m decreases. In (12.1.8) A denotes the surface area of the cell. Since C_i is assumed uniform and varies only with time, the mass of solute contained in the cell at time t is

$$m = C_i V \tag{12.1.9}$$

where V is the volume of the cell. Using this result in (12.1.8) we find

$$V\frac{dC_i}{dt} = -jA,$$

i.e.,

$$\frac{dC_i}{dt} = k(C_o - C_i) \tag{12.1.10}$$

where

$$k = \frac{\Gamma D A}{\delta V}.$$

Suppose at time $t = 0, C_i = C_{io}$; then (12.1.10) is readily solved to give

$$C_i(t) = C_o + (C_{io} - C_o)e^{-kt}.$$

Thus as $t \to \infty, C_i(t) \to C_o$ and the cell contains solute of concentration C_o entirely.

This simple result is derived on the assumption that the thickness δ of the lipid layer is small. If this is not the case then a more precise study must be made. To this end we consider diffusion through a lipid layer in the form of a slab as shown in Figure 12.1.2(b). For simplicity we shall assume that there is no solute in the membrane initially. Thus we have to solve the diffusion equation (12.1.5) in one spatial dimension x subject to the boundary conditions

$$C(0, t) = C_1,$$
$$C(\delta, t) = C_2, \qquad (12.1.11)$$

and the initial condition

$$C(x, 0) = 0, \quad 0 < x < \delta. \qquad (12.1.12)$$

The concentrations C_1, C_2 at the two outer walls of the membrane are assumed to be constant.

The problem defined by (12.1.5), (12.1.11) and (12.1.12) is typical of the heat conduction problems that we have considered in Chapter 11. Thus, if we apply the method of separation of variables, the most general solution of (12.1.5) satisfying the boundary conditions (12.1.11) can be written in the form

$$C(x, t) = C_1 + (C_2 - C_1)\frac{x}{\delta} + \sum_{n=1}^{\infty} A_n \sin\left(\frac{n\pi x}{\delta}\right) \exp\left(-\frac{n^2\pi^2 Dt}{\delta}\right), \quad (12.1.13)$$

where the constants $A_n, n = 1, 2, \ldots$, are to be determined from the initial condition (12.1.12).

On setting $t = 0$ in (12.1.12) a simple Fourier analysis (see Chapter 11) shows that

$$A_n = \frac{2}{n\pi}(C_2 \cos n\pi - C_1), \quad n = 1, 2, \ldots,$$

and so the complete solution to the problem is given by

$$C(x, t) = C_1 + (C_2 - C_1)\frac{x}{\delta}$$
$$+ \frac{2}{\pi} \sum_{n=1}^{\infty} \frac{(C_2 \cos n\pi - C_1)}{n} \sin\left(\frac{n\pi x}{\delta}\right) \exp\left(-\frac{n^2\pi^2 Dt}{\delta}\right). \quad (12.1.14)$$

From this expression we can calculate the rate j at which the diffusing substance emerges at the interface $x = 0$ per unit area per unit time. That is,

$$j = -D \left(\frac{\partial C}{\partial x} \right)_{x=0},$$

i.e.,

$$j = -(C_2 - C_1)\frac{D}{\delta} - \frac{2}{\delta} \sum_{n=1}^{\infty} (C_2 \cos n\pi - C_1) \exp\left(-\frac{n^2 \pi^2 Dt}{\delta} \right).$$

The first term on the right represents the steady-state flux, while the remaining terms represent a transient flux that is significant only for short times.

12.2 Energy and energy estimates

In most processes that are governed by diffusion equations of the type introduced in equation (12.1.1) one is interested in how the concentration $\mathbf{u}$ behaves for large times. It is well known that for the linear case wherein $\mathbf{f}(\mathbf{u}) \equiv 0$ any initial concentration $\mathbf{u}_0$ is smoothed out and, in fact, $\mathbf{u}(x, t)$ often tends exponentially to zero as t tends to infinity. In our case $\mathbf{f}(\mathbf{u}) \neq 0$ and the behaviour of $\mathbf{u}(\mathbf{x}, t)$ depends crucially on the form of $\mathbf{f}(\mathbf{u})$. Furthermore, in many cases, rather specially designed methods have to be developed to examine asymptotic behaviour.

A method that is widely applicable in many cases is the so-called "energy method." Here we shall illustrate its use in a few fairly simple examples, and examine the method in more detail in the next few sections when we take up the subjects of nerve impulse transmissions and chemical reactions once more.

Consider the diffusion problem defined by

$$\frac{\partial u}{\partial t} = D\frac{\partial^2 u}{\partial x^2} + f(u), \tag{12.2.1}$$

$$u(x, 0) = u_0(x),$$

$$\frac{\partial}{\partial x}u(0, t) = \frac{\partial}{\partial x}u(1, t) = 0, \tag{12.2.2}$$

and $f(u)$ is assumed to be such that this problem has solutions that are bounded for all time $t \geq 0$.

Define an "energy function" $E(t)$ by

$$E(t) = \frac{1}{2} \int_0^1 \left(\frac{\partial u}{\partial x} \right)^2 dx. \tag{12.2.3}$$

Differentiating this with respect to t gives

$$\frac{dE}{dt} = \int_0^1 \frac{\partial u}{\partial x} \frac{\partial^2 u}{\partial x \partial t} dx, \tag{12.2.4}$$

which, on using the differential equation, can be written in the form

$$\frac{dE}{dt} = \int_0^1 \frac{\partial u}{\partial x} \frac{\partial}{\partial x} \left(D \frac{\partial^2 u}{\partial x^2} + f(u) \right) dx,$$

$$= \int_0^1 D \frac{\partial u}{\partial x} \frac{\partial^3 u}{\partial x^3} dx + \int_0^1 \left(\frac{\partial u}{\partial x} \right)^2 \frac{\partial f}{\partial u} dx. \tag{12.2.5}$$

Integrating the first integral appearing in (12.2.5) by parts we find

$$\frac{dE}{dt} = \left[D \frac{\partial u}{\partial x} \frac{\partial^2 u}{\partial x^2} \right]_0^1 - D \int_0^1 \left(\frac{\partial^2 u}{\partial x^2} \right)^2 dx + \int_0^1 \left(\frac{\partial u}{\partial x} \right)^2 \frac{\partial f}{\partial u} dx$$

$$= -D \int_0^1 \left(\frac{\partial^2 u}{\partial x^2} \right)^2 dx + \int_0^1 \left(\frac{\partial u}{\partial x} \right)^2 \frac{\partial f}{\partial u} dx. \tag{12.2.6}$$

Let $m = max|\partial f / \partial u|$ over all solution values (assumed bounded) of u; then (12.2.6) can be replaced by the inequality

$$\frac{dE}{dt} \leq -D \int_0^1 \left(\frac{\partial^2 u}{\partial x^2} \right)^2 dx + m \int_0^1 \left(\frac{\partial u}{\partial x} \right)^2 dx,$$

i.e.,

$$\frac{dE}{dt} \leq -D \int_0^1 \left(\frac{\partial^2 u}{\partial x^2} \right)^2 dx + 2mE(t). \tag{12.2.7}$$

In order to proceed further, we would like to express the first integral appearing in (12.2.7) in a more recognisable form. For example, suppose we could prove that for some constant λ,

$$\int_0^1 \left(\frac{\partial^2 u}{\partial x^2} \right)^2 dx \geq \lambda \int_0^1 \left(\frac{\partial u}{\partial x} \right)^2 dx. \tag{12.2.8}$$

Then we could replace (12.2.7) by the inequality

$$\frac{dE}{dt} \leq 2(m - \lambda D)E(t), \tag{12.2.9}$$

from which we conclude that, since $E(t) \geq 0$,

$$\frac{d}{dt} \left(e^{-2(m-\lambda D)t} E(t) \right) \leq 0.$$

An integration then reveals that

$$E(t) \leq E(0)e^{2(m-\lambda D)t},$$

where $E(0) \geq 0$. From this we conclude that if the diffusion coefficient D is sufficiently large, i.e., $D > m/\lambda$, then $E(t)$ decays exponentially to zero as $t \to \infty$ and hence $\partial u/\partial x$ tends to zero for all $x \in [0,1]$ as $t \to \infty$. This implies that u becomes spatially homogeneous regardless of the initial concentration $u_0(x)$.

To arrive at the above result we have relied on the assumption that an inequality of the form (12.2.8) holds. Such an inequality is called the **Poincaré inequality**; it occurs widely in problems where estimation depends on establishing certain inequalities and is proved as follows.

Consider the ordinary differential equation

$$\frac{d^2y}{dx^2} + \lambda y = 0 \tag{12.2.10}$$

subject to the boundary conditions

$$\frac{dy(0)}{dx} = \frac{dy(1)}{dx} = 0. \tag{12.2.11}$$

This constitutes a typical "eigenvalue problem" and has eigensolutions

$$\phi_n(x) = \cos n\pi x, \quad n = 0, 1, 2, \ldots,$$

provided $\lambda = \lambda_n = n^2\pi^2$. Such eigenfunctions and eigenvalues have already been encountered in Chapter 11 and in Section 12.1 above. Indeed we know that for any function $w(x)$ that satisfies the conditions (12.2.11) and is twice differentiable, $\frac{d^2w}{dx^2}$ can be expanded as

$$\frac{d^2w}{dx^2} = \sum_{n=0}^{\infty} a_n \cos n\pi x \tag{12.2.12}$$

where

$$a_0 = \int_0^1 \frac{d^2w}{dx^2} dx = \left[\frac{dw}{dx}\right]_0^1 = 0$$

and

$$a_n = 2\int_0^1 \frac{d^2w}{dx^2} \cos n\pi x \, dx, \quad n \geq 1.$$

If we integrate (12.2.12) twice and use the boundary conditions (12.2.11) we find

$$w(x) = \sum_{n=0}^{\infty} -\frac{a_n}{n^2\pi^2} \cos n\pi x + C. \tag{12.2.13}$$

Thus

$$\int_0^1 \left(\frac{dw}{dx}\right)^2 dx = \left[w\frac{dw}{dx}\right]_0^1 - \int_0^1 w\frac{d^2w}{dx^2}dx$$

$$= \int_0^1 \left(\sum_{n=0}^{\infty} \frac{a_n}{n^2\pi^2}\cos n\pi x\right)\left(\sum_{n=0}^{\infty} a_n \cos n\pi x\right) dx$$

$$-C\int_0^1 \left(\sum_{n=0}^{\infty} a_n \cos n\pi x\right) dx$$

$$= \frac{1}{2}\sum_{n=0}^{\infty} \frac{a_n^2}{n^2\pi^2}$$

$$\leq \frac{1}{2\pi^2}\sum_{n=0}^{\infty} a_n^2. \tag{12.2.14}$$

Now if we multiply (12.2.12) by d^2w/dx^2 and integrate we find

$$\int_0^1 \left(\frac{d^2w}{dx^2}\right)^2 dx = \sum_{n=0}^{\infty} a_n \int_0^1 \frac{d^2w}{dx^2}\cos n\pi x\, dx = \frac{1}{2}\sum_{n=0}^{\infty} a_n^2. \tag{12.2.15}$$

This is another useful result in the application of Fourier series and is called the **Parseval equality**. Using (12.2.15) in the estimate (12.2.14) leads us to the final result,

$$\int_0^1 \left(\frac{dw}{dx}\right)^2 dx \leq \frac{1}{\pi^2}\int_0^1 \left(\frac{d^2w}{dx^2}\right)^2 dx. \tag{12.2.16}$$

Comparing this result with the estimate (12.2.8) we see that $\lambda = \pi^2$ and so if the diffusion coefficient $D > m/\pi^2$ then $E(t)$ decays at least as fast as $exp\, 2(m - \pi^2 D)t$.

One should remark that this decay estimate depends on how good the bound m on $|\partial f/\partial u|$ is; indeed m may be only a crude estimate and so D may have a lower bound smaller than m/π^2. Throughout we have only considered a problem defined in one space dimension; this is no real restriction because the Poincaré inequality (12.2.16) holds in any space dimension with, of course, a different constant λ.

Suppose now that instead of the boundary conditions (12.2.2) we impose the conditions

$$u(0,t) = u(1,t) = 0; \tag{12.2.17}$$

then the energy function (12.2.3) is not really appropriate and is replaced by

$$E(t) = \frac{1}{2}\int_0^1 u^2 dx. \tag{12.2.18}$$

Proceeding in the same way as before we have

$$\frac{dE}{dt} = \int_0^1 u \frac{\partial u}{\partial t} dx$$

$$= \int_0^1 u \left(D \frac{\partial^2 u}{\partial x^2} + f(u) \right) dx$$

$$= D \int_0^1 u \frac{\partial^2 u}{\partial x^2} dx + \int_0^1 u f(u) dx$$

$$= D \left[u \frac{\partial u}{\partial x} \right]_0^1 - D \int_0^1 \left(\frac{\partial u}{\partial x} \right)^2 dx + \int_0^1 u f(u) dx$$

$$= -D \int_0^1 \left(\frac{\partial u}{\partial x} \right)^2 dx + \int_0^1 u f(u) dx. \tag{12.2.19}$$

Unlike the problem treated above, the inequality (12.2.8) does not help us to estimate usefully the first integral appearing in (12.2.9). However, if we look at (12.2.16) and observe that the substitution $V(x) = dw/dx$ satisfies the conditions

$$V(0) = V(1) = 0,$$

then we have an inequality for V of the form

$$\int_0^1 \left(\frac{\partial V}{\partial x} \right)^2 dx \geq \pi^2 \int_0^1 V^2 dx. \tag{12.2.20}$$

Using this in (12.2.19) gives

$$\frac{dE}{dt} \leq -2D\pi^2 E(t) + \int_0^1 u f(u) dx.$$

Now suppose $|uf(u)| < mu^2$ for all solution values of u. Then we have

$$\frac{dE}{dt} \leq -2D\pi^2 E(t) + m \int_0^1 u^2 dx$$

$$= 2(m - D\pi^2) E(t),$$

and once again we see that if $D > m/\pi^2$, $E(t) \to 0$ as $t \to \infty$, which implies $u(x,t) \to 0$ as $t \to \infty$, and this means that the concentration u decays exponentially to zero regardless of the form of the initial conditions.

In a similar way we may deduce a corresponding behaviour if $uf(u) \leq -du^2$ for all $x \in [0,1]$ and $t \geq 0$. Here we do not need the inequality (12.2.20) but simply ignore the first term on the right-hand side of (12.2.19) to arrive at the estimate

$$\frac{dE}{dt} \leq \int_0^1 u f(u) dx$$

$$\leq -d \int_0^1 u^2 dx = -2dE(t),$$

and so

$$E(t) \leq E(0)e^{-2dt}.$$

The conclusions above hold also in this case. It is useful to note that unlike the problem in which there are "zero flux" conditions at $x = 0$, $x = 1$, we have obtained our asymptotic results by imposing certain constraints on the form of the reaction term $f(u)$ other than requiring it to be bounded for all solution values.

We shall look at this again in the following section.

12.3 Global behaviour of nerve impulse transmissions

In the study of the various models proposed to govern nerve impulse transmissions, an important feature that requires investigation is the long-term behaviour of the axon potential. This is a formidable task if we consider the Hodgkin-Huxley model. However, the FitzHugh-Nagumo system is a little more tractable and allows for the application of the energy arguments that we used in the previous section.

The problem to discuss is defined as

$$\frac{\partial^2 u}{\partial x^2} = \frac{\partial u}{\partial t} - u(1-u)(u-a) + w,$$

$$\frac{\partial w}{\partial t} = bu - \gamma w, \quad x \geq 0, \ t \geq 0, \tag{12.3.1}$$

$$u(0,t) = P(t), \quad u(x,0) = 0,$$

$$w(0,t) = b \int_0^t P(s)e^{-\gamma s}ds, \quad w(x,0) = 0. \tag{12.3.2}$$

Recall that in the conditions (12.3.2), $P(t)$ represents the stimulus emanating at $x = 0$. In practice, this stimulus acts over a finite interval of time and so we make the additional assumption that

$$P(t) = 0 \quad \text{for } t \geq T$$

for some time T. Furthermore we assume that $u(x,t)$, $w(x,t)$ tend sufficiently rapidly to zero as x tends to infinity so that the integrals $\int_0^\infty u^2(x,t)dx$, $\int_0^\infty w^2(x,t)dx$ exist and are finite for all t.

If we multiply the first of equations (12.3.1) by bu and the second by w and subtract the resulting equations, we obtain

$$bu\frac{\partial^2 u}{\partial x^2} - w\frac{\partial w}{\partial t} = bu\frac{\partial u}{\partial t} - bu^2(1-u)(u-a) + \gamma w^2,$$

i.e.,

$$bu\frac{\partial u}{\partial t} + w\frac{\partial w}{\partial t} = bu\frac{\partial^2 u}{\partial x^2} + bu^2(1-u)(u-a) - \gamma w^2$$

or

$$\frac{1}{2}\frac{\partial}{\partial t}(bu^2 + w^2) = b\frac{\partial}{\partial x}\left(u\frac{\partial u}{\partial x}\right) - b\left(\frac{\partial u}{\partial x}\right)^2 + bu^2(1-u)(u-a) - \gamma w^2.$$

$$(12.3.3)$$

Integrating this expression with respect to x and using the initial conditions (12.3.2) we find, for $t \geq T$, that

$$\frac{1}{2}\frac{d}{dt}\int_0^\infty (bu^2 + w^2)dx = \int_0^\infty b\frac{\partial}{\partial x}\left(u\frac{\partial u}{\partial x}\right)dx - b\int_0^\infty \left(\frac{\partial u}{\partial x}\right)^2 dx$$

$$+ \int_0^\infty [bu^2(1-u)(u-a) - \gamma w^2]dx$$

$$= -b\int_0^\infty \left(\frac{\partial u}{\partial x}\right)^2 dx + \int_0^\infty [bu^2(1-u)(u-a) - \gamma w^2]dx$$

$$\leq \int_0^\infty [bu^2(1-u)(u-a) - \gamma w^2]dx. \qquad (12.3.4)$$

Suppose now that $u(x,t) < a$ for all x, $t > 0$; then we can find a constant δ which depends on a so that

$$u^2(1-u)(u-a) \leq -\delta u^2.$$

This estimate used in (12.3.4) gives the inequality

$$\frac{1}{2}\frac{d}{dt}\int_0^\infty (bu^2 + w^2)dx \leq - \int_0^\infty (b\delta u^2 + \gamma w^2)dx.$$

If $C = min\{\delta, \gamma\}$ then

$$\frac{d}{dt}\int_0^\infty (bu^2 + w^2)dx \leq -2C\int_0^\infty (bu^2 + w^2)dx.$$

From this we conclude that

$$\int_0^\infty (bu^2 + w^2)dx \leq Ke^{-2Ct}. \qquad (12.3.5)$$

In summarising this result we can say that if $u(x,t)$, $w(x,t)$ are sufficiently smooth and if the action potential always remains below the potential value a, then the mean square of $u(x,t)$ decays exponentially in time. In other words, a represents a threshold effect. Recall that we have already identified a as a threshold parameter in Chapter 7 where we considered the dynamics of the space-clamped FitzHugh-Nagumo model. Thus it is reasonable to conclude that, unless the potential $u(x,t)$ exceeds a, a pulse will not propagate. Whether the axon reaches threshold or not depends on the stimulus $P(t)$.

From numerical and biological evidence there is support for the conjecture that a strong stimulus of short duration or a weak stimulus of long duration is sub-threshold. Mathematically, this conjecture can be shown to be correct as follows.

Suppose $a > b/\gamma$ and that the stimulus $P(t)$ is bounded, continuous and satisfies the conditions

$$\text{(i)} \ \ P(t) = P(0) = 0 \quad \text{for all } t \geq T > 0,$$

$$\text{(ii)} \ \ \sup_{t \geq 0} |P(t)| \leq M.$$

Then

$$\sup_{x \geq 0} |u(x,t)| \leq k \int_0^T |P(t)| dt$$

for all $t \geq T$. Furthermore there are constants C, k and λ such that if $\int_0^T |P| dt < \lambda$ then

$$\sup_{x \geq 0} |u(x,t)| \leq k \exp(-Ct), \quad t \geq 0.$$

This result, the proof of which is beyond the scope of this book, says that if a is not too small but the mean absolute value of the stimulus P is sufficiently small then the action potential u will decay exponentially. One can prove a further result, namely that if M in (ii) is sufficiently small then

$$\sup_{x \geq 0} |u(x,t)| \leq k \exp(-Ct), \quad t \geq 0.$$

Throughout our discussion of nerve axons we have always taken the axon to be infinite in extent. A real nerve is of finite extent and the correct problem to be studied is a mixed initial-boundary problem.

Suppose the axon is of length l. At $x = 0$ we retain the conditions (12.3.2) while at $x = l$ we impose the condition

$$u(l,t) = 0, \quad t \geq 0,$$

or

$$\frac{\partial u(l,t)}{\partial x} - \alpha u(l,t) = 0, \quad t \geq 0, \ \alpha \leq 0. \tag{12.3.6}$$

What the precise conditions prevail in practice at $x = l$ is not known.

In order to analyse the problem governed by (12.3.1), (12.3.2) and (12.3.6) we again suppose that the stimulus $P(t)$ is non-zero over a finite time interval, i.e., $P(t) = 0$, $t \geq T$.

Multiply the first of equations (12.3.1) by u and the second by $b^{-1}w$ and subtract the results to get

$$\frac{1}{2}\frac{\partial}{\partial t}\left(u^2 + \frac{w^2}{b}\right) = u\frac{\partial^2 u}{\partial x^2} + u^2(1-u)(u-a) - \frac{\gamma}{b}w^2.$$

Integrating this over the interval $[0, l]$ we find

$$\frac{1}{2}\int_0^l \frac{\partial}{\partial t}\left(u^2 + \frac{w^2}{b}\right)dx = \int_0^l \left(u\frac{\partial^2 u}{\partial x^2} + u^2(1-u)(u-a) - \frac{\gamma}{b}w^2\right)dx.$$

$$(12.3.7)$$

Now

$$\int_0^l u\frac{\partial^2 u}{\partial x^2}dx = \left[u\frac{\partial u}{\partial x}\right]_0^l - \int_0^l \left(\frac{\partial u}{\partial x}\right)^2 dx.$$

If $u(0, t) = 0 = u(l, t)$, $t \geq T$, then

$$\int_0^l u\frac{\partial^2 u}{\partial x^2}dx = -\int_0^l \left(\frac{\partial u}{\partial x}\right)^2 dx.$$

Similarly if

$$u(0, t) = 0 = \frac{\partial}{\partial x}u(l, t) - \alpha u(l, t), \quad t \geq T,$$

then

$$\int_0^l u\frac{\partial^2 u}{\partial x^2}dx = \alpha u^2(l, t) - \int_0^l \left(\frac{\partial u}{\partial x}\right)^2 dx.$$

Thus, in either case, (12.3.7) gives rise to the inequality

$$\frac{1}{2}\frac{\partial}{\partial t}\int_0^l \left(u^2 + \frac{w^2}{b}\right)dx \leq \int_0^l \left[u^2(1-u)(u-a) - \frac{\gamma}{b}w^2 dx - \left(\frac{\partial u}{\partial x}\right)^2\right]dx.$$

$$(12.3.8)$$

In order to analyse (12.3.8) we first observe the quadrature term $(1-u)(u-a)$ has a maximum at $u = (1+a)/2$ given by $(1-a)^2/4$ and so we can write

$$\frac{1}{2}\frac{\partial}{\partial t}\int_0^l \left(u^2 + \frac{w^2}{b}\right)dx \leq \int_0^l \left[\frac{(1-a)^2}{4}u^2 - \frac{\gamma}{b}w^2 - \left(\frac{\partial u}{\partial x}\right)^2\right]dx. \quad (12.3.9)$$

Next, we estimate the last term in (12.3.9) using the Poincaré inequality discussed in the previous section. For example, if we use the boundary conditions

$$u(0,t) = 0 = u(l,t), \quad t \geq T,$$

then

$$\int_0^l \left(\frac{\partial u}{\partial x}\right)^2 dx \geq \left(\frac{\pi}{l}\right)^2 \int_0^l u^2 dx. \qquad (12.3.10)$$

A similar inequality is obtained if we use the boundary condition

$$\frac{\partial}{\partial x} u(l,t) - \alpha u(l,t) = 0.$$

Thus from (12.3.10) we have

$$\frac{1}{2}\frac{\partial}{\partial t}\int_0^l \left(u^2 + \frac{w^2}{b}\right) dx \leq \int_0^l \left[\left(\frac{(1-a)^2}{4} - \frac{\pi^2}{l^2}\right) u^2 - \frac{\gamma}{b} w^2\right] dx.$$

Now if $(1-a)^2/4 - \pi^2/l^2 < 0$, i.e., $l < 2\pi/(1-a)$, then with

$$C = \max\left\{\frac{\pi^2}{l^2} - \frac{(1-a)^2}{4}, \gamma\right\}$$

we have

$$\frac{\partial}{\partial t}\int_0^l \left(u^2 + \frac{w^2}{b}\right) dx \leq -2C \int_0^l \left(u^2 + \frac{w^2}{b}\right) dx.$$

From this we can argue, as we did above, to conclude that

$$\int_0^l \left(u^2 + \frac{w^2}{b}\right) dx$$

decays exponentially in time for $t \geq T$. In other words, unless the axon is long enough, all stimuli decay exponentially with time. The reader should bear in mind, from our previous discussion, that the converse is not necessarily true. That is, even an axon of infinite length cannot propagate a stimulus unless that stimulus satisfies certain criteria. The precise nature of $P(t)$ necessary for the propagation of an action potential is not fully understood.

12.4 Global behaviour in chemical reactions

In this section we take up once more a study of the simplified model of the Belousov-Zhabotinskii reaction, which we dealt with in Chapter 8. The objective here is to enlarge on the discussion of travelling wavefronts.

The model to be considered is defined through the equations

$$\frac{\partial u}{\partial t} = u(1 - u - rv) + \frac{\partial^2 u}{\partial x^2},$$

$$\frac{\partial v}{\partial t} = -buv + \frac{\partial^2 v}{\partial x^2}, \quad -\infty < x < \infty, \ t > 0, \qquad (12.4.1)$$

where u represents the non-dimensionalised concentration of bromous acid ($HBrO_2$) and v represents the non-dimensionalised concentration of bromide ion (Br^-). The constants r and b are positive and are of the order 10 and 2.5×10, respectively. Furthermore, the concentrations u and v satisfy the inequalities

$$0 \le u \le 1, \quad 0 \le v \le 1.$$

The type of wavefront solutions we seek are those which satisfy the boundary conditions

$$u(-\infty, t) = 0, \quad u(\infty, t) = 1,$$
$$v(-\infty, t) = 1, \quad v(\infty, t) = 0. \qquad (12.4.2)$$

There is one special case that can be treated fully; this is the situation in which

$$v = 1 - u. \qquad (12.4.3)$$

For now we need only consider the equations

$$\frac{\partial u}{\partial t} = \frac{\partial^2 u}{\partial x^2} + (1 - r)u(1 - u),$$

$$\frac{\partial u}{\partial t} = \frac{\partial^2 u}{\partial x^2} + bu(1 - u), \qquad (12.4.4)$$

which are the same if $b = 1 - r$ with $r < 1$.

We have shown in Chapter 8 that the equation

$$\frac{\partial \phi}{\partial t} = \frac{\partial^2 \phi}{\partial x^2} + k\phi(1 - \phi)$$

has travelling wave solutions $\phi(x + ct)$ satisfying the conditions $\phi(-\infty) = 0$, $\phi(\infty) = 1$, provided $c \ge 2\sqrt{k}$. Consequently the system (12.4.4) has travelling wave solutions of speed

$$c \ge 2\sqrt{(1 - r)} = 2\sqrt{b}, \quad \text{if } r \le 1.$$

If, however, we impose initial data of the form

$$u(x,0) = h(x), \qquad (12.4.5)$$

where $h(x)$ is zero for $x < 0$ and equals one for $x > 0$, then Kolmogorov, Petrovskii and Piskounov, as long ago as 1937, proved that $u(x,t)$ satisfying (12.4.4), (12.4.5) evolves into a travelling wave $u(x + ct)$ as $t \to \infty$, where $c = 2\sqrt{(1-r)} = 2\sqrt{b}$ if $r \leq 1$.

This result can be established by following comparison type arguments similar to those used in Chapter 11, and depends very much on the non-linear term $u(1 - u)$. We shall not go further into the matter here.

We now demonstrate that any wave solution, in which $u \geq 0$, $v \geq 0$, must have a wave speed $c(r, b) \leq 2$ for all $b \geq 0$ and all $r \geq 0$. To do this, we shall introduce the function $w(x,t)$ defined as the solution to the problem

$$\frac{\partial w}{\partial t} = \frac{\partial^2 w}{\partial x^2} + w(1 - w),$$

$$w(x,0) = h(x),$$

$$w(-\infty, t) = 0, \quad w(\infty, t) = 1, \qquad (12.4.6)$$

where $h(x)$ is of the same form as described above.

We now know that $w(x,t)$ evolves into a travelling wave speed 2. We write

$$\bar{w}(x,t) = u(x,t) - w(x,t), \qquad (12.4.7)$$

where $u(x,t)$ is the unique solution of (12.4.1) which has the same initial and boundary conditions as $w(x,t)$. Subtracting the equation in (12.4.6) from the first of equations (12.4.1), we find

$$\frac{\partial^2 \bar{w}}{\partial x^2} - \frac{\partial \bar{w}}{\partial t} + [1 - (w + u)]\bar{w} = ruv \qquad (12.4.8)$$

with the restriction $w \geq 0$, $u \geq 0$ and with the class of initial conditions we have $u \leq 1$, $w \leq 1$. This last statement is a consequence of the comparison theorem in Chapter 11. Thus $1 - (w + u) \leq 1$. What we would like to do now is to employ the maximum principle of Chapter 11 in order to estimate $\bar{w}$. However, since the coefficient of $\bar{w}$ (12.4.8) is less than one, some modification must first be carried out. If we set

$$W = \bar{w}e^{-Kt}$$

for some positive constant K then (12.4.8) becomes

$$\frac{\partial^2 W}{\partial x^2} - \frac{\partial W}{\partial t} + [1 - (w + u) - K]W = ruve^{-Kt} \geq 0.$$

Choose $K > 1$ so that $1 - (w + u) - K < 0$. On noting that W satisfies the same boundary and initial conditions as $\bar{w}$, the maximum principle can now be invoked to conclude that W has its maximum at $t = 0$ or at $|x| = \infty$.

Thus $\bar{w}$ has its maximum at either of these possible points. However, the maximum value of $\bar{w}$ occurs at the maximum value of $u(x, t) - w(x, t)$, which is zero at $t = 0$ and $x = \pm\infty$. From this we conclude that

$$u(x, t) \leq w(x, t) \tag{12.4.9}$$

for all x and $t > 0$. But this implies that the solution $u(x, t)$ is at all points less than or equal to $w(x, t)$; in particular this is true for the travelling wave solution $w(x, t) = f(x + ct)$ of (12.4.6). Thus, if $u(x, t)$ does evolve into a travelling wave solution with speed c then $c \leq 2$; for if the wave speed were greater than two the inequality (12.4.9) would be violated.

To complete our discussion we consider the behaviour of the wave speed $c(r, b)$ for limiting values of r and b. If $b = 0$ then the second equation of (12.4.1) reduces to the heat equation

$$\frac{\partial v}{\partial t} = \frac{\partial^2 v}{\partial x^2},$$

which does not have travelling wave solutions and so neither can the equation for $u(x, t)$. Thus we conclude that $c(r, b) \to 0$ as $b \to 0$ with $r > 0$.

If $b \to \infty$ then (12.4.1) implies $v = 0$ (the trivial solution $u = 0$ is excluded). In this case $c(b, r) \to 2$ as $b \to \infty$ with $r \geq 0$. Next, if $r = 0$ then (12.4.1) decouples and, in particular, we have to discuss the equation

$$\frac{\partial u}{\partial t} = \frac{\partial^2 u}{\partial x^2} + u(1 - u).$$

This equation is simply the equation (12.4.6) for which travelling wave solutions exist with speed $c = 2$. Thus with $r = 0$ both u and v evolve into travelling waves with speed $c = 2$, i.e., $c(r, b) \to 2$ as $r \to 0, b > 0$. If $r \to \infty$ then $u \equiv 0$ or $v \equiv 0$ and in either case there can be no wave solution and so $c(r, b) \to 0$ as $r \to \infty$. These results are summarised in Figure 12.4.1

There are a number of important problems that remain in the discussion of the Belousov-Zhabotinskii reaction and diffusion problems in general. For example, there is the question of existence of periodic solutions, as well as a thorough study of those classes of initial data $h(x)$ for which u and v evolve into travelling waves. That is to say, how do solutions behave under small disturbances of initial and/or boundary data? Such problems are largely beyond the scope of this book, but the interested reader is encouraged to consult the literature cited in the notes at the end of this chapter.

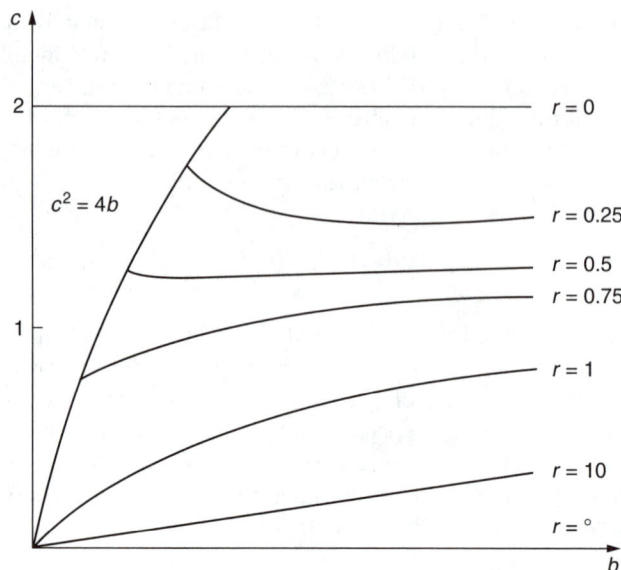

FIGURE 12.4.1:   Wave speed c as a function of b for various values of r.

12.5   Turing diffusion driven instability and pattern formation

Here we discuss the role of pattern in developmental biology and the mechanisms proposed to analyse it from a mathematical point of view. To begin with we introduce the following biological concepts:

Embryology: embryology is that part of biology that is concerned with the formation and development of the embryo from fertilization until birth. Embryological development is a sequential process and follows a ground plan layed down early in gestation.

Morphogenesis: morphogenesis is that part of embryology that is concerned with the development of pattern and form. How the development of the ground plan is established is not really understood.

Pattern and form: Whatever pattern is observed in the animal world it is almost certain that the process that produced it is unknown. The mechanism must be genetically controlled but the genes themselves cannot create the pattern.

This promotes the question: how is genetic information translated into patterns?

There have been proposed two major concepts that provide mechanisms for pattern formation: (i) positional information due to Lewis Wolpert (1969) and (ii) diffusion driven instability due to Alan Turing (1952).

Wolpert's idea of positional information suggests that cells are pre-programmed to read a chemical (or morphogen) concentration and differentiate accordingly into different kinds of cells leading to the development of, for example, cartilage, bone, tissue, hair, etc. This point of view separates the developmental process into the following steps.

(i) The creation of a morphogen concentration of spatial pattern (a morphogen is the chemical associated with creating morphogenesis).

(ii) The establishment of positional information that depends on a chemical pre-specification so that the cell can read out its position in the coordinates of chemical concentration and differentiate, undergo appropriate cell shape change or migrate.

(iii) Once the pre-pattern is established, morphogenesis is a slave process.

To date it is very much an open problem as to how this idea is to be implemented in mathematical modelling.

In 1952 Alan Turing suggested that under certain conditions chemicals can react and diffuse in such a way as to produce steady state heterogeneous spatial patterns of chemical concentration. It is this counterintuitive idea that we pursue here.

To begin with let us return to the discussion leading to the derivation of Fick's second law. This time, however, instead of considering the flow of solute through an arbitrary volume of fluid, we first suppose the volume V to contain a mixture of chemicals of concentration $c_i, i = 1, \ldots, n$ and further that there is a chemical source $\mathbf{f}$ that is vector valued and depends only on the chemical concentrations c_i. Using the law of conservation of mass we argue as we did before. That is, the rate of change of a chemical concentration in V must equal the rate of change of flow of chemical across the surface S together with the chemical concentration created in V. Consequently for each chemical c_i we have

$$\frac{\partial}{\partial t} \int_V c_i(\mathbf{x}, t)dV = -\int_S \mathbf{J}_i \cdot \mathbf{n}\, dS + \int_V f_i\, dV, \qquad (12.5.1)$$

where $\mathbf{J}_i = -D_i \nabla c_i$ with D_i being the diffusion constant for c_i. Again, by noting that V is arbitrary we arrive at the evolutionary equation

$$\frac{\partial c_i}{\partial t} = D_i \Delta c_i + f_i(c_1, \ldots, c_n), \quad i = 1, 2, \ldots, n. \qquad (12.5.2)$$

In matrix notation

$$\frac{\partial \mathbf{c}}{\partial t} = D\Delta \mathbf{c} + \mathbf{f}(\mathbf{c}), \qquad (12.5.3)$$

where D is a diagonal diffusivity matrix with entries $D_i, i = 1, \ldots, n, \mathbf{c} = (c_1, \ldots, c_n), \mathbf{f}(\mathbf{c}) = (f_1(c_1, \ldots, c_n), \ldots, f_n(c_1, \ldots, c_n))$.

Throughout we shall consider the case of chemicals c_1, c_2, $n = 2$. That is, we consider the system

$$\frac{\partial c_1}{\partial t} = D_1 \Delta c_1 + f_1(c_1, c_2),$$

$$\frac{\partial c_2}{\partial t} = D_2 \Delta c_1 + f_2(c_1, c_2). \tag{12.5.4}$$

Suppose in the first instance that the spatial domain is one-dimensional and infinite in extent. The situation in which the domain is finite will be considered later. In this case we have the system

$$\frac{\partial c_1}{\partial t} = D_1 \frac{\partial^2 c_1}{\partial x^2} + f_1(c_1, c_2),$$

$$\frac{\partial c_2}{\partial t} = D_2 \frac{\partial^2 c_1}{\partial x^2} + f_2(c_1, c_2). \tag{12.5.5}$$

Suppose that there exists a steady state to this system in which the chemicals are well mixed and uniform. Turing's profound idea is simple. He suggested that if in the absence of diffusion (i.e., $D_1 = D_2 = 0$ in (12.5.5)), c_1 and c_2 approach a linearly stable uniform steady state as $t \to \infty$; then under certain conditions spatially inhomogeneous patterns can evolve by "diffusion driven instability" for $D_1 \neq D_2 \neq 0$.

It is this fundamental idea that we shall explore.

Suppose there are solutions

$$c_1 = c_{1,0}, \quad c_2 = c_{2,0} \tag{12.5.6}$$

of (12.5.5) which are independent of time t and space x. Notice that if $c_{1,0}$ and $c_{2,0}$ are steady states then, from (12.5.5), they must satisfy

$$f_1(c_1, c_2) = f_2(c_1, c_2) = 0. \tag{12.5.7}$$

Consequently there may be several steady states, each determined as a solution of (12.5.7).

We now ask a fundamental question:

Can spatially inhomogeneous patterns spontaneously develop from arbitrarily small perturbations from a uniform **linearly stable** steady state?

To investigate this question let

$$c_1(x, t) = c_{1,0} + d_1,$$

$$c_2(x, t) = c_{2,0} + d_2, \tag{12.5.8}$$

where $|d_i| \ll 1, i = 1, 2$.

We now "linearise" (12.5.5) by expanding $f_i(c_1, c_2)$ as a Taylor series about $(c_{1,0}, c_{2,0})$. That is,

$$f_i(c_1, c_2) = f_i(c_{1,0}, c_{2,0}) + d_1 \frac{\partial f_i}{\partial c_1}(c_{1,0}, c_{2,0}) + d_2 \frac{\partial f_i}{\partial c_2}(c_{1,0}, c_{2,0})$$
$$+ \text{higher order terms in } d_1, d_2, \quad i = 1, 2.$$

To a first order approximation in which we neglect the higher order terms and noting that $f_i(c_{1,0}, c_{2,0}) = 0$ we arrive at the linear system

$$\frac{\partial d_i}{\partial t} = D_i \frac{\partial^2 d_i}{\partial x^2} + d_1 \frac{\partial f_i}{\partial c_1}(c_{1,0}, c_{2,0}) + d_2 \frac{\partial f_i}{\partial c_2}(c_{1,0}, c_{2,0}), \quad i = 1, 2. \quad (12.5.9)$$

For notational convenience and ease of presentation we define the quantities

$$a_{ij} \equiv \frac{\partial f_i}{\partial c_i}(c_{1,0}, c_{2,0}), \quad i, j = 1, 2. \quad (12.5.10)$$

Then (12.5.9) can be written as

$$\frac{\partial d_1}{\partial t} = D_i \frac{\partial^2 d_1}{\partial x^2} + a_{11}d_1 + a_{12}d_2,$$

$$\frac{\partial d_2}{\partial t} = D_i \frac{\partial^2 d_2}{\partial x^2} + a_{21}d_1 + a_{22}d_2, \quad (12.5.11)$$

or, in matrix form,

$$\frac{\partial \mathbf{d}}{\partial t} = D \frac{\partial^2 \mathbf{d}}{\partial x^2} + A\mathbf{d}, \quad (12.5.12)$$

where $\mathbf{d} = (d_1, d_2)^T$, $D = diag(D_1, D_2)$ and A is the 2×2 matrix with entries $a_{i,j}$.

The system (12.5.12) can be solved by the method of separation of variables as described in Chapter 11.

We shall look for solutions of the form

$$\begin{bmatrix} d_1 \\ d_2 \end{bmatrix} = \begin{bmatrix} \alpha_1 \\ \alpha_2 \end{bmatrix} e^{\sigma t} \cos kx, \quad (12.5.13)$$

in which $\alpha_i(i = 1, 2)$, σ and k are constants. This form of solution is not the most general solution to (12.5.12). Indeed we could choose a solution of the form

$$\begin{bmatrix} d_1 \\ d_2 \end{bmatrix} = \begin{bmatrix} \beta_1 \\ \beta_2 \end{bmatrix} e^{\sigma t} \sin kx, \quad (12.5.14)$$

or even a linear combination of such solutions. Whichever choice is made leads to the same analysis and so there is no loss in generality. The importance of the form (12.5.13) will become apparent when we consider diffusion driven instability for systems defined on finite spatial domains in which boundary conditions have to be taken into account.

On substituting (12.5.13) into (12.5.12) and dividing the resulting equations throughout by $e^{\sigma t} \cos kx$ we arrive at the pair of algebraic equations

$$\alpha_1 \sigma = a_{11}\alpha_1 + a_{12}\alpha_2 - D_1 k^2 \alpha_1,$$

$$\alpha_2 \sigma = a_{21}\alpha_1 + a_{22}\alpha_2 - D_2 k^2 \alpha_2,$$

i.e.,

$$\alpha_1(\sigma - a_{11} + D_1 k^2) - \alpha_2 a_{12} = 0,$$
$$-\alpha_1 a_{21} + \alpha_2(\sigma - a_{22} + D_2 k^2) = 0. \tag{12.5.15}$$

If we think of this pair of equations as simultaneous equations for the determination of (α_1, α_2) then for a non-trivial solution to exist we must have

$$\begin{vmatrix} \sigma - a_{11} + D_1 k^2 & -a_{12} \\ -a_{21} & \sigma - a_{22} + D_2 k^2 \end{vmatrix} = 0,$$

i.e.,

$$\sigma^2 + \sigma((D_1 + D_2)k^2 - (a_{11} + a_{22})) + ((a_{11} - D_1 k^2)(a_{22} - D_2 k^2) - a_{12} a_{21}) = 0. \tag{12.5.16}$$

This is the characteristic equation for the determination of σ. Of fundamental importance is the fact that σ determines the growth or decay in time of the solution (12.5.13). That is, if $\Re\sigma > 0$ then the solution (12.5.13) grows exponentially in time and so is **unstable**. This means that the homogeneous steady state $(c_{1,0}, c_{2,0})$ is unstable to small spatially inhomogeneous perturbations. Likewise if $\Re\sigma < 0$ then the steady state $(c_{1,0}, c_{2,0})$ is stable to small spatially inhomogeneous perturbations.

Recall that in the absence of diffusion we require the uniform steady state $(c_{1,0}, c_{2,0})$ to be linearly stable. So on setting $D_1 = D_2 = 0$ in (12.5.16) we require the roots σ_1, σ_2, say, of

$$\sigma^2 - \sigma(a_{11} + a_{22}) + (a_{11} a_{22} - a_{12} a_{21}) = 0, \tag{12.5.17}$$

to satisfy $\Re\sigma_i < 0, i = 1, 2$. Now

$$\sigma_1 + \sigma_2 = a_{11} + a_{22},$$
$$\sigma_1 \sigma_2 = (a_{11} a_{22} - a_{12} a_{21}),$$

from which it is clear that whether σ_1, σ_2 are real or complex, $\Re\sigma_i < 0$ if and only if

$$a_{11} + a_{22} < 0,$$
$$a_{11} a_{22} - a_{12} a_{21} > 0. \tag{12.5.18}$$

That these inequalities hold is crucial as to whether spatially inhomogeneous patterns can evolve by diffusion driven instability.

When $D_1, D_2 \neq 0$ we require the roots σ_1', σ_2' of (12.5.16) to be such that $\Re\sigma_1'$ and/or $\Re\sigma_2'$ to be non-negative.

Now

$$\sigma_1' + \sigma_2' = (a_{11} + a_{22}) - (D_1 + D_2)k^2 \tag{12.5.19}$$

and

$$\sigma_1'\sigma_2' \equiv H(k^2) = (a_{11} - D_1 k^2)(a_{22} - D_2 k^2) - a_{12} a_{21}. \tag{12.5.20}$$

Since D_1, D_2 are positive and noting (12.5.18) it is clear that

$$\sigma_1' + \sigma_2' < 0. \tag{12.5.21}$$

If σ_1', σ_2' are complex then (12.5.21) implies that $\Re\sigma_i' < 0, (i = 1, 2)$ and $\sigma_1'\sigma_2' = |\sigma_1'|^2 > 0$ and so the perturbation (12.5.12) decays exponentially and patterns do not form. So the only possibility is that σ_1', σ_2' are real and $\sigma_1'\sigma_2' < 0$, i.e., $H(k^2) < 0$.

Let us examine $H(k^2)$ more closely. As a quadratic equation in k^2 we have

$$H(k^2) = D_1 D_2 k^4 - (D_1 a_{22} + D_2 a_{11})k^2 + a_{11} a_{22} - a_{12} a_{21}. \tag{12.5.22}$$

When $k = 0, H(k^2) > 0$ and if $H(k^2) < 0$ for some values of k^2 then (12.5.22) must have real positive roots and so

$$D_1 a_{22} + D_2 a_{11} > 0 \tag{12.5.23}$$

and

$$(D_1 a_{22} + D_2 a_{11})^2 \geq 4D_1 D_2 (a_{11} a_{22} - a_{12} a_{21}). \tag{12.5.24}$$

In general the graph of $H(k^2)$ against k^2 will take one of the forms illustrated in Figures 12.5.1a,b,c.

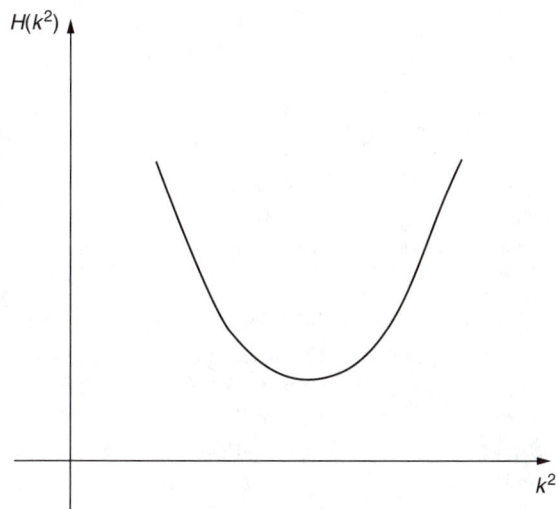

FIGURE 12.5.1a: Steady state stable; no pattern formation.

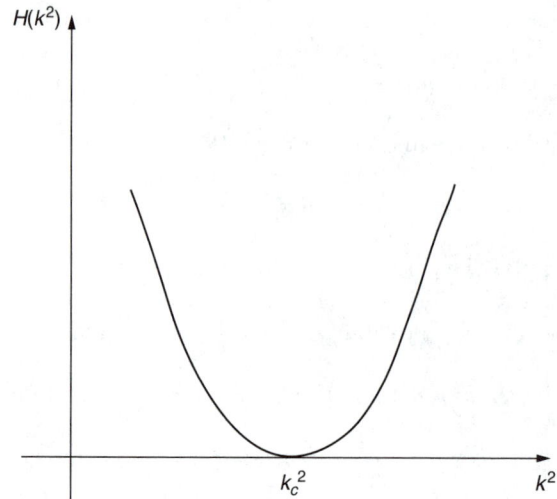

FIGURE 12.5.1b: Critical case; onset of instability.

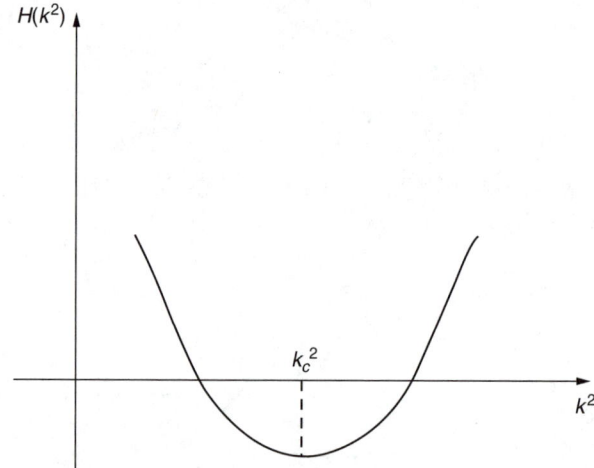

FIGURE 12.5.1c: Steady state unstable; patterns form.

Figure 12.5.1.a) $H(k^2) > 0$ and the steady state is stable to small perturbations; b) $H(k^2)$ has a single critical root k_c^2 indicating the onset of instability of the steady state to small perturbations; c) $H(k^2) < 0$ for a range of k^2 for which the steady state is unstable to small perturbations.

From (12.5.23), (12.5.24) and Figures 12.5.1a,b,c we arrive at some very important conclusions about the type of reaction diffusion systems that lead to diffusion driven instability. From (12.5.18) and (12.5.23) we immediately conclude that $D_1 \neq D_2$. In other words one morphogen must diffuse faster than the other. Again from these inequalities we see that a_{11} and a_{22} must be of opposite sign.

Suppose $a_{11} > 0$ and $a_{22} < 0$. Then from (12.5.10) we have

$$a_{11} = \frac{\partial f_1}{\partial c_1}(c_{1,0}, c_{2,0}) > 0, \qquad a_{22} = \frac{\partial f_2}{\partial c_2}(c_{1,0}, c_{2,0}) > 0,$$

which means that c_1 is **activated** by its own rate of production whereas c_2 is **inhibited** by its rate of production. In this case we naturally refer to c_1 as the **activator** and c_2 as the **inhibitor**.

Futhermore we also see that

$$\frac{D_2}{D_1} > -\frac{a_{22}}{a_{11}} > 1,$$

or

$$D_2 > D_1. \tag{12.5.25}$$

In other words the inhibitor c_2 diffuses faster than the activator c_1.

With the choice $a_{11} > 0, a_{22} < 0$, the fact that (see (12.5.18)) $a_{11}a_{22} - a_{12}a_{21} > 0$ means that $a_{12}a_{21} < 0$. If $a_{12} < 0, a_{21} > 0$ then we have an **activator-inhibitor** system whereas if $a_{12} > 0, a_{21} < 0$ we have a **positive feedback** system.

Turning now to Figures 12.5.1a,b,c, we see that in a) there are no values of k^2 for which $H(k^2) < 0$ and so small perturbations are stable and no patterns form. Figure 12.5.1b shows the critical case in which there is precisely one value k_c^2 of k^2 occurring when $H(k^2) = 0$. This value is easily calculated to be

$$k_c^2 = \frac{1}{2}\left(\frac{a_{11}}{D_1} + \frac{a_{22}}{D_2}\right), \tag{12.5.26}$$

where D_1 and D_2 are critical diffusion constants that satisfy

$$(D_1 a_{22} + D_2 a_{11})^2 = 4D_1 D_2(a_{11}a_{22} - a_{12}a_{21}).$$

In other words the critical wave number and critical diffusion constants are related by

$$k_c^2 = \left[\frac{(a_{11}a_{22} - a_{12}a_{21})}{D_1 D_2}\right]^{1/2}. \tag{12.5.27}$$

In Figure 12.5.1c, $H(k^2) < 0$ and so there are a continuum of spatial patterns that can evolve by diffusion driven instability for a continuum of k^2 satisfying

$$k_1^2 < k^2 < k_2^2, \tag{12.5.28}$$

where $H(k_i^2) = 0$, $i = 1, 2$.

To illustrate the general analysis we consider the specific example:

$$\frac{\partial c_1}{\partial t} = \frac{c_1^2}{c_2} - bc_1 + D_1\frac{\partial^2 c_1}{\partial x^2},$$

$$\frac{\partial c_2}{\partial t} = c_1^2 - c_2 + D_2\frac{\partial^2 c_2}{\partial x^2}, \quad (x,t) \in (-\infty,\infty) \times (0,\infty), \quad (12.5.29)$$

where b is a positive constant.

The steady states of the system (12.5.29) are determined by

$$\frac{c_1^2}{c_2} = bc_1, \qquad c_1^2 = c_2. \tag{12.5.30}$$

The non-trivial steady state is therefore

$$c_{1,0} = \frac{1}{b}, \qquad c_{2,0} = \frac{1}{b^2}. \tag{12.5.31}$$

As above we linearise the system about this steady state by writing

$$c_1 = \frac{1}{b} + d_1, \qquad c_2 = \frac{1}{b^2} + d_2. \tag{12.5.32}$$

Furthermore, from (12.5.10), we can write down the coefficients $a_{ij}, i,j = 1,2$ of the linearised system (12.5.11) as

$$a_{11} = b, \qquad a_{12} = -b^2,$$
$$a_{21} = \frac{2}{b}, \qquad a_{22} = -1. \tag{12.5.33}$$

Note that a_{11} and a_{22} are of opposite sign and furthermore the inequalities (12.5.18) demand that

$$0 < b < 1. \tag{12.5.34}$$

For the evolution of pattern we require $H(k^2)$ given by (12.5.22) to be non-positive, i.e.,

$$D_1 D_2 k^4 - (bD_2 - D_1)k^2 + b \le 0, \tag{12.5.35}$$

together with the inequalities (see (12.5.23) and (12.5.24))

$$bD_2 - D_1 > 0 \tag{12.5.36}$$

and

$$(bD_2 - D_1)^2 \ge 4D_1 D_2 b. \tag{12.5.37}$$

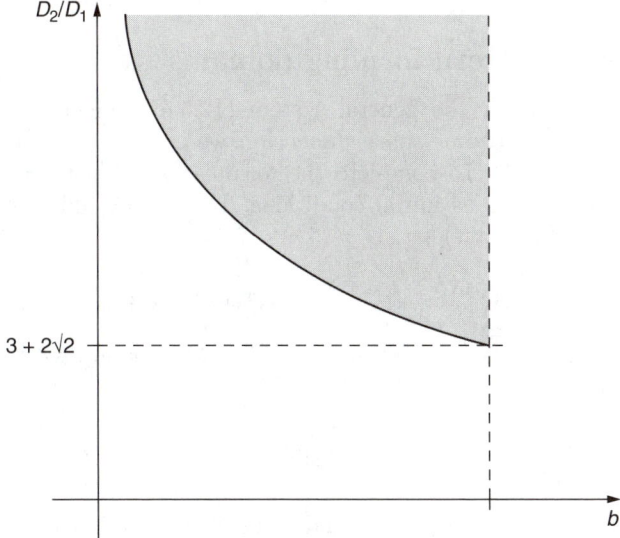

FIGURE 12.5.2: Turing parameter space for the system (12.5.29).

Consequently

$$\frac{D_2}{D_1} > \frac{1}{b} > 1,$$

showing that the inhibitor c_2 diffuses faster than the activator c_1.
    Also from (12.5.37) we deduce that

$$\frac{D_2}{D_1}b \geq 3 + 2\sqrt{2}. \qquad (12.5.38)$$

This inequality defines the "Turing parameter space," shown in Figure 12.5.2, in which diffusion driven instability occurs.
    Next, the critical wave number k_c is given by

$$k_c^2 = \frac{1}{2}\left(\frac{b}{D_1} - \frac{1}{D_2}\right) \qquad (12.5.39)$$

and a continuum of spatial patterns can evolve for all values of k^2 satisfying

$$k_1^2 < k^2 < k_2^2$$

where $H(k_i^2) = 0$, $i = 1, 2$. In other words

$$2D_1 D_2 k_1^2 = (bD_2 - D_1) - \sqrt{(bD_2 - D_1)^2 - 4D_1 D_2 b}$$

and

$$2D_1 D_2 k_1^2 = (bD_2 - D_1) + \sqrt{(bD_2 - D_1)^2 - 4D_1 D_2 b}.$$

12.6 Finite pattern forming domains

Let us now consider the general system (12.5.4) defined on a bounded domain. In the case of one-space dimension we consider the domain to be the interval $x \in [0, l]$. To complete the formulation of the problem we require boundary as well as initial conditions. That is, in addition to (12.5.4) we assume $c_1(x, t)$, $c_2(x, t)$ satisfy

$$c_1(x, 0) = f_1(x), \qquad c_2(x, 0) = f_2(x), \qquad (12.6.1)$$

where $f_1(x)$ and $f_2(x)$ are given concentrations. Also we impose the "**no flux**" boundary conditions

$$\frac{\partial c_i}{\partial x}(x, t) = 0, \quad x = 0, l, \ i = 1, 2. \qquad (12.6.2)$$

The reason for choosing zero flux boundary conditions is that we are primarily interested in self organisation of patterns. No flux boundary conditions mean that there is to be no external input of morphogens.

To develop an analysis of Turing driven instabilities in this case we proceed in precisely the same way as we did above for an infinite domain. The crucial point is that the perturbations d_1 and d_2, which solve the system (12.5.11), must now satisfy the conditions

$$\frac{\partial d_i}{\partial x}(x, t) = 0, \quad x = 0, l, \ i = 1, 2. \qquad (12.6.3)$$

A little thought shows that d_1, d_2 must be of the form (12.5.13) where

$$\cos kx = 0 \qquad (12.6.4)$$

for $x = 0, l$. Consequently, rather than being arbitrary, the parameter k must take the discrete values k_n, $n = 1, \ldots$, where

$$\sin kl = 0,$$

i.e.,

$$k_n = \frac{n\pi}{l}, \quad n = 1, 2, \ldots. \qquad (12.6.5)$$

This observation is crucial; it is fundamental to deciding which patterns are selected. Proceeding precisely as before we find that the critical wave number k_c is given by

$$\frac{n^2 \pi^2}{l^2} = \frac{1}{2}\left(\frac{a_{11}}{D_1} + \frac{a_{22}}{D_2}\right), \qquad (12.6.6)$$

which may or may not be satisfied for a given l. However if one is allowed to vary l then it is possible to choose a mode characterised by n so that (12.6.6) is satisfied. In other words a critical pattern of a certain form is dependent on the size of the domain.

More generally we see that spatial patterns evolve if we can select integer values of n so that $H(k_n^2) < 0$. This can only happen if we can choose n so that

$$k_1^2 < \frac{n^2\pi^2}{l^2} < k_2^2 \qquad (12.6.7)$$

where k_i^2, $i = 1, 2\ldots$, satisfy $H(k_i^2) = 0$. That is,

$$\frac{(D_1 a_{22} + D_2 a_{11}) - \sqrt{(D_1 a_{22} + D_2 a_{11})^2 - 4D_1 D_2 A}}{2 D_1 D_2} < \frac{n^2\pi^2}{l^2}$$
$$< \frac{(D_1 a_{22} + D_2 a_{11}) + \sqrt{(D_1 a_{22} + D_2 a_{11})^2 - 4D_1 D_2 A}}{2 D_1 D_2}, \qquad (12.6.8)$$

where $A = (a_{11} a_{22} - a_{12} a_{21}) > 0$. The number of integers n for which (12.6.8) is satisfied determines the modes of pattern selection. To see this suppose the domain size l is such that (12.6.8) is satisfied only for $n = 1$. The only unstable mode is $\cos \frac{\pi x}{l}$ and morphogen concentration

$$c_1(x, t) \sim c_{1,0} + \alpha \exp\left[\lambda\left(\frac{\pi^2}{l^2}\right) t\right] \cos \frac{\pi x}{l}, \qquad (12.6.9)$$

where $\lambda(\frac{\pi^2}{l^2})$ is the positive root of (12.5.6). This unstable mode is the dominant one which emerges as t increases. If we say that **black** corresponds to a concentration above the steady state $c_{1,0}$ and **white** corresponds to a concentration below $c_{1,0}$ then we have the pattern shown in Figure 12.6.1.

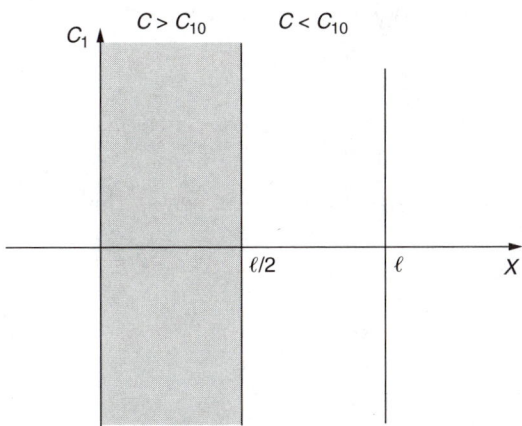

FIGURE 12.6.1: Morphogen pattern for $n = 1$.

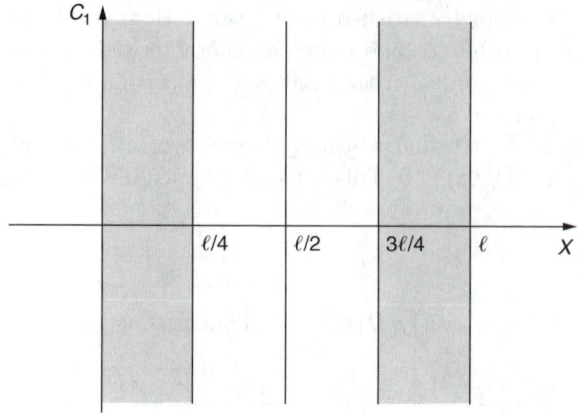

FIGURE 12.6.2: Morphogen pattern for $n = 2$.

Similarly if $n = 2$ is the only value of n for which the inequality (12.6.8) holds then (12.6.9) becomes

$$c_1(x,t) \sim c_{1,0} + \alpha \exp\left[\lambda\left(\frac{4\pi^2}{l^2}\right)t\right]\cos\frac{2\pi x}{l}, \qquad (12.6.10)$$

leading to the pattern shown in Figure 12.6.2.

Which mode or combination of modes and hence patterns are selected depends on initial conditions (12.6.1).

Now consider the two-dimensional domain Ω defined by $0 \leq x \leq l$, $0 \leq y \leq h$, with rectangular boundary $\partial\Omega$ on which no-flux boundary conditions are imposed.

Once again the theory developed above in the one-dimensional case is followed with only minor modifications. Most importantly we seek solutions of the linearised problem (12.5.11) of the form

$$\begin{bmatrix} d_1 \\ d_2 \end{bmatrix} = \begin{bmatrix} \alpha_1 \\ \alpha_2 \end{bmatrix} e^{\sigma t} \cos k_1 x \cos k_2 y, \qquad (12.6.11)$$

where the wave numbers k_1 and k_2 are chosen so that d_1 and d_2 satisfy the boundary conditions:

$$\frac{\partial d_i}{\partial x} = 0, \quad x = 0, l,$$

$$\frac{\partial d_i}{\partial y} = 0, \quad y = 0, h, \qquad (12.6.12)$$

$i = 1, 2$.

By the method of separation of variables, or otherwise, we find that

$$k_1 = \frac{m\pi}{l}, \qquad k_2 = \frac{n\pi}{h}, \tag{12.6.13}$$

for integers $m, n = 1, 2 \ldots$.

Now proceed precisely as before to see that if we define

$$K^2 = k_1^2 + k_2^2, \tag{12.6.14}$$

then we again arrive at equations (12.5.16) and (12.5.22) with k^2 simply replaced by K^2. The critical wave number K_c^2 is given by

$$K_c^2 = \frac{m^2\pi^2}{l^2} + \frac{n^2\pi^2}{h^2} = \frac{1}{2}\left(\frac{a_{11}}{D_1} + \frac{a_{22}}{D_2}\right) \tag{12.6.15}$$

and modes characterised by m and n exist if they satisfy the inequalities

$$\frac{(D_1 a_{22} + D_2 a_{11}) - \sqrt{(D_1 a_{22} + D_2 a_{11})^2 - 4 D_1 D_2 A}}{2 D_1 D_2} < \frac{m^2\pi^2}{l^2} + \frac{n^2\pi^2}{h^2}$$
$$< \frac{(D_1 a_{22} + D_2 a_{11}) + \sqrt{(D_1 a_{22} + D_2 a_{11})^2 - 4 D_1 D_2 A}}{2 D_1 D_2}. \tag{12.6.16}$$

To illustrate possible modes, suppose the domain size is sufficiently large so that (12.6.16) holds for $m = 3$, $n = 2$. Then the pattern shown in Figure 12.6.3 is possible where the shaded areas indicate regions in which the morphogen concentration is above the steady state.

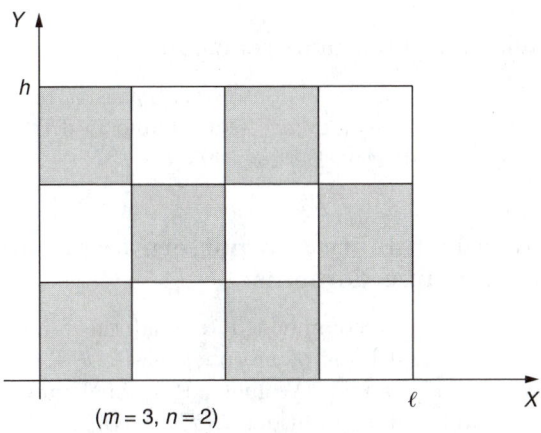

$(m = 3, n = 2)$

FIGURE 12.6.3: Morphogen pattern for $m = 3$, $n = 2$.

The fundamental assumption of pattern formation via Turing diffusion driven instabilities is that the linearly unstable modes that grow exponentially in time will eventually be bounded by the non-linear kinetic terms in (12.5.4). This is indeed the case. To prove mathematically that this happens one has to show that in the positive quadrant of morphogen space there is a compact region about the uniform steady state $(c_{1,0}, c_{2,0})$ to which c_1 and c_2 are always confined. Such a region is called a confined set or a contracting set. The determination of such sets is often difficult and is beyond the scope of this book. The interested reader is encouraged to consult the literature cited in the notes at the end of this chapter.

12.7 Notes

Diffusion through membranes

Most of the material here is based on the results of S.I. Rubinow and we recommend his book *Introduction to Mathematical Biology*, John Wiley & Sons, New York, 1975 for futher developments and a host of biological applications. See also J.D. Murray, *Mathematical Biology*, Springer-Verlag, Berlin, 1993.

Global behaviour of nerve impulse transmissions

For a fuller background to the material here, we refer to the survey article of S.P. Hastings, Some mathematical problems from neurobiology, *Am. Math. Monthly*, **82**, 881–895, 1975 and the detailed results in J. Smoller, *Shock Waves and Reaction-Diffusion Equations*, Springer-Verlag, Berlin, 1983.

Global behaviour in chemical reactions

An excellent treatment of the Belousov-Zhabotinskii reaction and many other diffusion problems in biology are to be found in J.D. Murray, *Mathematical Biology*, Springer-Verlag, Berlin, 1993.

Diffusion driven instability and pattern formation; Finite pattern forming domains

The interested reader is recommended to read the fundamental papers: A.M. Turing, The chemical basis of morphogenesis, *Phil. Trans. Roy. Soc. Lond*, **B237**, 37–72, 1952; and L. Wolpert, Positional information and the spatial pattern of cellular differentiation, *J. Theor. Biol.*, **25**, 1–47, 1969. See also the treatment in J.D. Murray, *Mathematical Biology*, Springer-Verlag, Berlin, 1993.

Exercises

12.1 A cell of solute concentration c and characteristic length δ is placed in a large bathing solution of solute with fixed concentration c_0. If the region $0 \le x \le \delta$ represents the cell while the regions $x < 0$, $x > \delta$ represent the cell exterior, show that the solution concentration of the cell is given by

$$c(x,t) = c_0 \left[1 - \frac{4}{\pi} \sum_{n=1}^{\infty} \frac{1}{2n-1} \sin\left(\frac{(2n-1)\pi x}{\delta} \right) \exp[-(2n-1)^2 \pi^2 Dt]/\delta^2 \right],$$

where D is the diffusion constant.

12.2 In one dimension, let there be a slab of solute of uniform concentration c_0 and of thickness $2a$. At $t = 0$

$$c(x,0) = \begin{cases} c_0, & -a < x < a, \\ 0, & |x| > a. \end{cases}$$

Find the concentration $c(x,t)$ for all x and t.

12.3 A stationary spherical cell of radius a is metabolising a nutrient that is at uniform concentration c_0 in the surrounding medium initially. Assume that the cell instantaneously metabolises any nutrient molecules that enter it, so that the nutrient concentration at the cell wall is zero at all times. Solve the diffusion equation in spherical polar co-ordinates and find the nutrient concentration $c = c(r,t)$ in the surrounding medium as a function of the radial postion r and the time t.

12.4 By considering the energy function

$$E(t) = \int_t^{t+T} \left(\frac{du}{dx} \right)^2 dx,$$

show that the ordinary differential equation

$$\frac{du}{dt} = F(u),$$

where $F(u) = \frac{df(u)}{du}$, cannot have a periodic solution of period T.

12.5 Consider the reaction-diffusion system

$$\frac{\partial u}{\partial t} = D_1 \frac{\partial^2 u}{\partial x^2} + f(u,v),$$

$$\frac{\partial v}{\partial t} = D_2 \frac{\partial^2 v}{\partial x^2} + g(u,v),$$

defined for $x, t \in (0,1) \times (0, \infty)$, and where u, v satisfy the following initial and boundary conditions

$$\frac{\partial u}{\partial x} = \frac{\partial v}{\partial x} = 0 \quad \text{for } x = 0, 1, \ t \geq 0,$$

$$u(x,0) = u_0(x), \quad v(x,0) = v_0(x) \quad \text{for } 0 \leq x \leq 1,$$

$$\frac{du_0}{dx} = \frac{dv_0}{dx} = 0 \quad \text{for } x = 0, 1.$$

Define the energy function

$$E(t) = \frac{1}{2} \int_0^1 \left[\left(\frac{\partial u}{\partial x} \right)^2 + \left(\frac{\partial v}{\partial x} \right)^2 \right] dx$$

and show that

$$\frac{dE}{dt} \leq -d \int_0^1 \left[\left(\frac{\partial^2 u}{\partial x^2} \right)^2 + \left(\frac{\partial^2 v}{\partial x^2} \right)^2 \right] dx + \frac{m}{2} \int_0^1 \left[\left(\frac{\partial u}{\partial x} \right)^2 + \left(\frac{\partial v}{\partial x} \right)^2 \right] dx$$

$$\leq (m - 2\pi^2 d)E,$$

where $d = min(d_1, d_2)$ and

$$m = max_{u,v} \left[\left| \frac{\partial f}{\partial u} \right| + \left| \frac{\partial f}{\partial v} \right| + \left| \frac{\partial g}{\partial u} \right| + \left| \frac{\partial g}{\partial v} \right| \right].$$

Here, $max_{u,v}$ means the maximum over the solution values of u and v. Deduce that if $m < 2\pi^2 d$ then $E(t) \to 0$ as $t \to \infty$. Interpret this result.

12.6 From the general theory of diffusion driven instability and using the notation of Section 12.5 derive the inequalities

$$D_2 > D_1 \quad \text{and} \quad D_1 \tau_1 < D_2 \tau_2,$$

where

$$\tau_i = |a_{ii}|^{-1}, \quad i = 1, 2.$$

12.7 Consider the reaction-diffusion system

$$\frac{\partial u}{\partial t} = \alpha uv - \beta u^2 + \delta^2 \Delta u,$$

$$\frac{\partial v}{\partial t} = v - uv + v^2 + \Delta v,$$

where u and v represent concentrations of morphogens and α, β and δ are positive constants. Find the non-zero equilibrium point and determine conditions on α and β for a Turing instability to occur. Calculate the values of δ for which the Turing instability can take place.

12.8 A model capable of diffusion driven instability is

$$\frac{\partial u}{\partial t} = \gamma(a - u + u^2 v) + \frac{\partial^2 u}{\partial x^2},$$

$$\frac{\partial v}{\partial t} = \gamma(b - u^2 v) + d\frac{\partial^2 v}{\partial x^2}, \quad (x, t) \in (-\infty, \infty) \times (0, \infty).$$

Determine the homogeneous steady state and show that it is stable provided

$$(b - a) - (a + b)^3 < 0.$$

By linearising the system about the homogeneous steady state show that diffusion instability can occur if

$$H(k^2) \equiv dk^4 = k^2 \left[\gamma(a + b)^2 - d\gamma\frac{(b - a)}{(a + b)} \right] + \gamma^2(a + b)^2 < 0.$$

12.9 Derive diffusion driven instability and pattern forming properties of the two dimensional reaction-diffusion system

$$\frac{\partial u}{\partial t} = a - bu - \frac{u^2}{v} + \Delta u,$$

$$\frac{\partial v}{\partial t} = u^2 - v + d\Delta v,$$

where a, b are positive parameters and d is a positive diffusion coefficient. The system is defined on the rectangular region $0 \leq x \leq A, 0 \leq y \leq B$ and the morphogens u and v satisfy the boundary conditions

$$\frac{\partial u}{\partial x} = 0, \quad \text{on } x = 0, A \quad \text{and} \quad v = 0 \quad \text{on } y = 0, B.$$

Chapter 13

Bifurcation and Chaos

13.1 Bifurcation

Equilibrium points have been considered in Chapter 5. The position of an equilibrium point may depend upon a parameter. For instance, the numbers infected when an epidemic reaches equilibrium might be different for different rates of infection. How the variation of a parameter can affect equilibrium is the topic to be discussed here. Such a parameter will be called a **control parameter** to distinguish it from any parameters which remain fixed.

We begin with an example to indicate the sort of thing that can happen.

Example 13.1.1

Consider the non-linear conservative system in which $\mu \geq 0$ and

$$\dot{x} = y, \qquad \dot{y} = \mu \sin x - x.$$

The equilibrium points are given by $y = 0$ and solutions of

$$\mu \sin x - x = 0. \tag{13.1.1}$$

One solution of (13.1.1) is $x = 0$. To see if there are any other solutions we examine the derivative with respect to x of the left-hand side of (13.1.1), namely $\mu \cos x - 1$. The derivative is always negative if $0 \leq \mu < 1$. Therefore, for this range of μ, the left-hand side of (13.1.1) decreases steadily as x increases from 0. Since it is zero when $x = 0$ it cannot vanish subsequently and there is no other solution of (13.1.1) for $0 \leq \mu < 1$.

If $\mu > 1$, the derivative is positive at $x = 0$. As x increases the derivative decreases, passes through zero and then stays negative up to $x = \pi$; for simplicity we limit x to $(-\pi, \pi)$. Therefore, as x increases, the left-hand side of (13.1.1) first increases from 0 to some positive maximum where $\mu \cos x = 1$ and then decreases steadily to $x = \pi$. At $x = \pi$ it is negative and so there is a zero $x = x_1 \neq 0$ of (13.1.1) when $\mu > 1$. By the preceding argument $\mu \cos x_1 < 1$. Since (13.1.1) is unaffected by changing the sign of x there is also a solution $x = -x_1$.

Thus, we have found one equilibrium point for $0 \leq \mu < 1$ and three for $\mu > 1$.

To examine the nature of the equilibrium we proceed as in Section 5.3. Near the equilibrium point $(0,0)$ put $x = \xi$, $y = \eta$ with ξ and η small. Then

$$\dot{\xi} = \eta, \qquad \dot{\eta} = (\mu - 1)\xi.$$

According to Section 5.4 the equation

$$\lambda^2 = \mu - 1$$

should be solved now. If $\mu < 1$, the two values of λ are imaginary complex conjugates and the equilibrium point is a centre. When $\mu > 1$, the two values of λ are real but of opposite sign; the equilibrium point is a saddle-point.

When $\mu > 1$ the equilibrium point $(x_1, 0)$ can occur. Near it put $x = x_1 + \xi$, $y = \eta$ to obtain

$$\dot{\xi} = \eta, \qquad \dot{\eta} = (\mu \cos x_1 - 1)\xi.$$

In this case

$$\lambda^2 = \mu \cos x_1 - 1.$$

Since $\mu \cos x_1 - 1$ is negative, as shown above, the equilibrium point is a centre. The same is true for the equilibrium point $(-x_1, 0)$. $\qquad$ ☐

Several features of this example should be noted. The position of the equilibrium point $x = 0$ does not change as the control parameter μ is altered but its character switches as μ passes through 1. It could be regarded as going from a stable regime to an unstable one—termed *interchange of stabilities* by Poincaré. At the switch extra equilibrium points are born; their position varies with the control parameter but not their character. The changing of the character of an equilibrium point and/or the creation of extra ones by alteration of a control parameter is known as **bifurcation**. The value of μ where bifurcation occurs may be called a **bifurcation point**.

An aid to keeping track of what is going on as the control parameter varies is the bifurcation diagram. This is a plot of the positions of the equilibrium points against the control parameter (see Figure 13.1.1). The letters c and s signify which curves represent centres and saddle-points, respectively. Adding information about λ would render the diagram too complicated; so sometimes there is a separate plot of λ against μ. It can be somewhat awkward when λ is complex. To avoid a three-dimensional picture λ can be plotted on the complex λ-plane with values of μ attached to the points.

On account of its appearance a bifurcation like that of Figure 13.1.1 is sometimes known as a *pitchfork bifurcation*.

If the equation for $\dot{y}$ in Example 13.1.1 is changed to

$$\dot{y} = \mu - x^2 \tag{13.1.2}$$

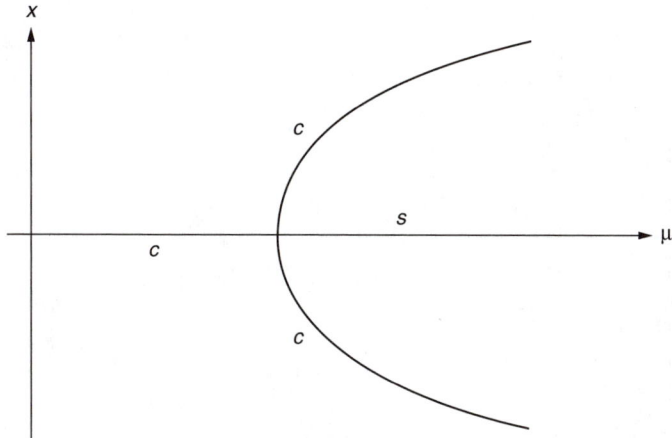

FIGURE 13.1.1: Bifurcation diagram for Example 13.1.1.

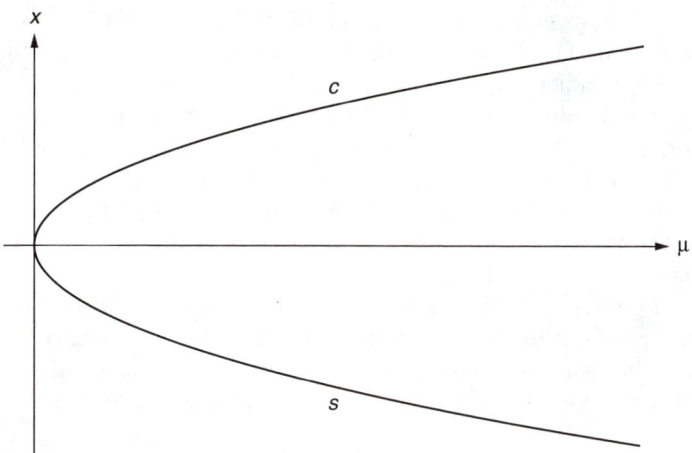

FIGURE 13.1.2: Bifurcation diagram for (13.1.2).

there are no equilibrium points for $\mu < 0$. For $\mu > 0$ there are two $(\sqrt{\mu}, 0)$ and $(-\sqrt{\mu}, 0)$. The first of these corresponds to a centre whereas the second corresponds to a saddle-point. The resulting bifurcation diagram is shown in Figure 13.1.2.

Bifurcation is not restricted to the non-linear conservative systems that have been studied so far. For example, the system

$$\dot{x} = 5x + (3 - \mu)y + 5x^3,$$
$$\dot{y} = x + 5y + x^3$$

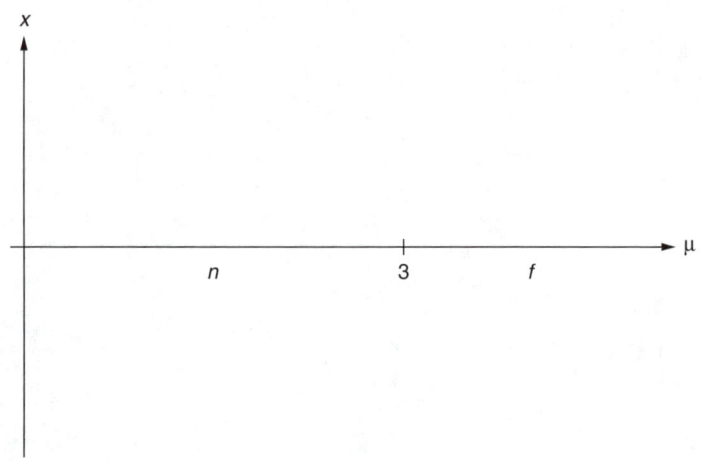

FIGURE 13.1.3:   Bifurcation from node to focus.

with $\mu \geq 0$ has an equilibrium point at $(0,0)$. It is a node for $\mu < 3$ and a focus for $\mu > 3$. The bifurcation diagram is displayed in Figure 13.1.3, the letters n and f indicating node and focus, respectively.

The reader should be warned that often the term bifurcation is employed much more strictly than has been used here. It is kept to denote points where the stability of the system changes. A diagram like Figure 13.1.3 would not be regarded as bifurcation in the strict sense because the stability is unaltered although the behaviour of the system changes radically from non-oscillatory to oscillatory.

It should be stressed that the identification of the type of equilibrium point has been carried out by the approximation of Section 5.3. Thus, the equations are valid only near the equilibrium point. They represent what is occurring locally. How the local scene fits into the global picture can be very difficult to resolve.

Of course, there may be more than one control parameter as in

$$\dot{x} = y, \qquad \dot{y} = \mu_1 x + \mu_2 x^2 + \mu_3 x^3 + x^5,$$

which has three control parameters μ_1, μ_2 and μ_3. Discussion of bifurcation is generally extremely tricky even for the equilibrium point in which $x = 0$, let alone the determination of other equilibrium points. Pictorial representation is far from straightforward since three dimensions are occupied already by the control parameters. Accordingly, we shall concentrate on cases in which there is a single control parameter.

Often what we have called equilibrium points are known as *fixed points* in bifurcation theory, for reasons that will become apparent later. Sometimes it will be convenient to use one term in the rest of this chapter and sometimes the other.

13.2 Bifurcation of a limit cycle

Bifurcation is not confined to the types of equilibrium points in Section 5.4 although they are common in systems of two differential equations. Limit cycles can be involved even with two differential equations. As the number of differential equations increases to three and upwards the possible types of common bifurcations proliferate rapidly. Also discussion of them becomes increasingly complicated, going beyond the scope of this book. So we shall stay mostly with two differential equations.

An example where limit cycles are involved in bifurcation is provided by the differential equations

$$\dot{x} = -y + x(\mu - x^2 - y^2), \tag{13.2.1}$$
$$\dot{y} = x + y(\mu - x^2 - y^2). \tag{13.2.2}$$

Make the transformation to polar coordinates by putting $x = r\cos\phi$, $y = r\sin\phi$ (see Figure 13.2.1). Note that, in this transformation, r is not allowed to be negative. It represents the length of the radius vector from the origin to

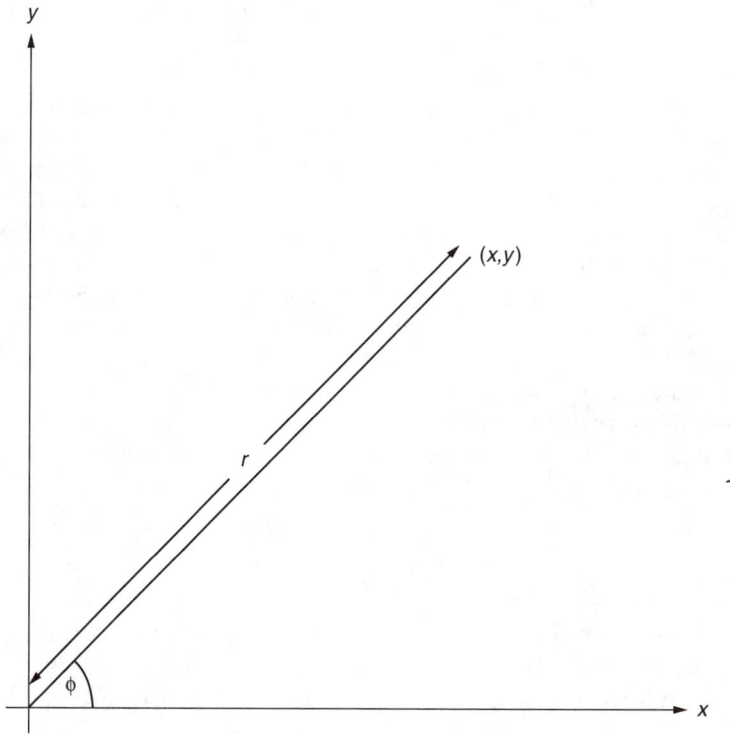

FIGURE 13.2.1: Polar coordinates.

the point (x, y). With the substitution (13.2.1) and (13.2.2) become

$$\dot{r}\cos\phi - r\dot{\phi}\sin\phi = -r\sin\phi + (\mu - r^2)r\cos\phi,$$
$$\dot{r}\sin\phi + r\dot{\phi}\cos\phi = r\cos\phi + (\mu - r^2)r\sin\phi.$$

These equations can be solved for $\dot{r}$ and $\dot{\phi}$ with the result

$$\dot{r} = r(\mu - r^2), \tag{13.2.3}$$
$$\dot{\phi} = 1. \tag{13.2.4}$$

Equation (13.2.4) can be solved immediately to give

$$\phi = \phi_0 + t \tag{13.2.5}$$

where ϕ_0 is the value of ϕ at $t = 0$. Thus, the radius vector in Figure 13.2.1 sweeps steadily counter-clockwise as t increases. The end of the radius vector describes the trajectory, which starts from (r_0, ϕ_0) where r_0 is the initial value of r.

When $\mu < 0$, (13.2.3) shows that $\dot{r}$ is negative for any positive r. Therefore, r decreases steadily and the trajectory spirals round the origin in a counter-clockwise sense and moves inwards as time progresses. Eventually, the trajectory ends up at the equilibrium point $r = 0$. The origin is a stable focus.

When $\mu > 0$, the right-hand side of (13.2.3) is positive for $r^2 < \mu$. Hence r increases and the spiral round the origin moves outwards. Now the origin is an unstable focus. The outward movement of r does not continue unabated because $r = \sqrt{\mu}$ is another equilibrium point ($r = -\sqrt{\mu}$ is not a possibility because r is not allowed to be negative). The trajectory with $r_0 = \sqrt{\mu}$ is the circle $r = \sqrt{\mu}$, i.e., a limit cycle. Consequently, a trajectory started with $r_0^2 < \mu$ spirals counter-clockwise and outwards, steadily approaching the limit cycle. If $r^2 > \mu$, $\dot{r} < 0$ and a trajectory spirals inwards to the limit cycle. Accordingly, the limit cycle is stable. The bifurcation diagram is shown in Figure 13.2.2.

The creation of a limit cycle in this way by variation of a control parameter is known as **Hopf bifurcation**.

Although we have not needed it the solution of the differential equation (13.2.3) is available. It is

$$r^2 = \frac{\mu r_0^2}{r_0^2 + (\mu - r_0^2)e^{-2\mu t}} \tag{13.2.6}$$

where the positive square root of the right-hand side gives r. You should check that it leads to the same conclusions as have been obtained above.

Notice that the linear approximation of Section 5.3 leads to

$$\dot{r} = \mu r$$

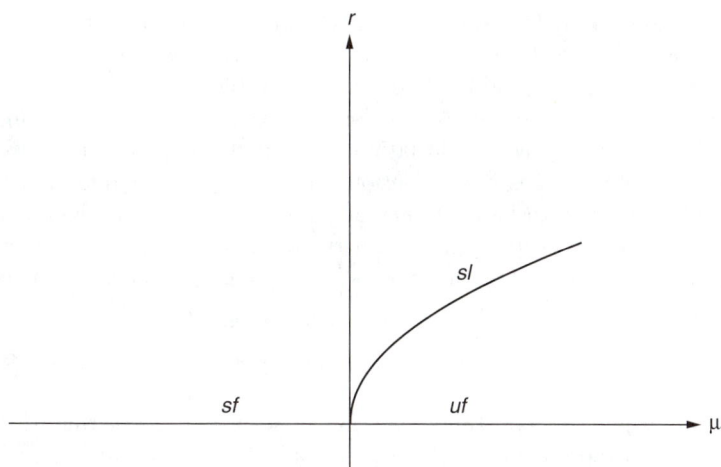

FIGURE 13.2.2: The Hopf bifurcation.

instead of (13.2.3). This does predict correctly that a stable focus switches to an unstable focus as μ passes through 0. However, it does not tell us what happens to the trajectories as they spiral away from the unstable focus. As far as the local behaviour is concerned they could be going off to infinity. It is only with the full equation (13.2.3) that we discover that they end up on a limit cycle eventually. This illustrates yet again the care that is necessary in extrapolating local behaviour to the global picture in non-linear systems.

13.3 Discrete bifurcation

Often observations on a system have to be taken at intervals that are too wide apart for it to be sensible to use a model based on differential equations. For example, if the crop of grain on a farm is measured just at the end of the year, the frequency would be insufficient for a differential equation on our normal time scale. Nevertheless, phenomena similar to those described above may occur.

As a simple model, suppose that, at the end of the first year, the crop is C_1 and the farmer sets aside a proportion of this as seed for the next year. Assume perfect germination and no loss of seed; the crop at the end of the second year C_2 will be proportional to C_1, say

$$C_2 = \mu C_1.$$

If the farmer carries out the same process at the end of each year, $C_3 = \mu C_2$ and generally $C_{n+1} = \mu C_n$. Obviously, $C_{n+1} = \mu^n C_1$, which shows that the crop increases steadily from year to year when $\mu > 1$. In contrast, when $\mu < 1$, the crop decays towards zero. There is a bifurcation as μ passes through 1 from a stable (but undesirable) state to an unstable state.

Of course, the assumption that μ does not vary from year to year is not likely to be valid in a practical situation. This might be overcome by using an average value but there is a more serious objection to the model. It takes no account of losses due to bad weather or parasites. Moreover, there could be a limit to the size of crop which the nutrients in the soil could support. The simplest way of allowing for such effects is to change the relation between C_n and C_{n+1} to

$$C_{n+1} = \mu C_n - \nu C_n^2 \tag{13.3.1}$$

with μ and ν positive constants. Making ν positive ensures that the extra term in (13.3.1) represents losses. Furthermore, because it contains C_n^2, it will inhibit the growth if the crop becomes too large.

Now, make the substitution

$$C_n = \mu x_n / \nu$$

in (13.3.1). There results

$$x_{n+1} = \mu x_n (1 - x_n). \tag{13.3.2}$$

Since negative crops are not permitted x_n must be restricted to lie between 0 and 1. Note that μ cannot exceed 4. If μ is greater than 4 it is possible to generate values of x_n outside the permitted range. For example, $x_n = 1/2$ would make x_{n+1} too large.

Equation (13.3.2) constitutes an *iteration scheme* in which each value is determined from its predecessor once x_1 has been fixed. **IF** x_n tends to a limit x_0 as $n \to \infty$ then x_0 satisfies

$$x_0 = \mu x_0 (1 - x_0). \tag{13.3.3}$$

The **IF** is in bold capitals because we do not yet know whether or not x_n tends to a limit.

The solutions of (13.3.3) are known as the **fixed points** of the iteration scheme. They are

$$x_0 = 0, \qquad x_0 = 1 - 1/\mu.$$

Since x_0 is forced to lie between 0 and 1 the second possibility does not exist unless $\mu > 1$. Consequently, there are two fixed points when $\mu > 1$ but only one when $\mu < 1$. Note that, when $x_n \to 1 - 1/\mu$, $C_n \to (\mu - 1)/\nu$.

Observe that, if $x_1 = 1$, $x_2 = 0$ and all subsequent x_n are zero. So the iteration goes straight to the fixed point $x_0 = 0$ whenever $x_1 = 1$. Accordingly, it will be sufficient to limit x_1 to $x_1 < 1$ subsequently.

To find out what happens near a fixed point in general subtract (13.3.3) from (13.3.2) to obtain

$$x_{n+1} - x_0 = \mu(x_n - x_0)(1 - x_0 - x_n).$$

Put $x_n = x_0 + \epsilon_n$ and take the modulus of both sides. Then

$$|\epsilon_{n+1}| = \mu|\epsilon_n||1 - 2x_0 - \epsilon_n|. \tag{13.3.4}$$

For the fixed point $x_0 = 0$ the deviation ϵ_n cannot be negative and (13.3.4) reduces to

$$\epsilon_{n+1} = \mu\epsilon_n(1 - \epsilon_n).$$

When $\mu < 1$ the factor $\mu(1 - \epsilon_n) < 1$ and so $\epsilon_{n+1} < \epsilon_n$. Thus, ϵ_n decreases steadily and tends to zero as $n \to \infty$. Hence $x_n \to 0$ and the iteration always ends up at zero no matter what starting values are given to x_1. The fixed point can be regarded as stable.

When $\mu > 1$ the iteration can never reach the fixed point $x_0 = 0$ (unless it starts there). For, if it could, ϵ_n would become eventually so small that $\mu(1 - \epsilon_n) > 1$. But then $\epsilon_{n+1} > \epsilon_n$ which stops the reduction in ϵ_n. In this case the fixed point $x_0 = 0$ is unstable. There is a bifurcation at $\mu = 1$.

A second fixed point is available when $\mu > 1$. For it (13.3.4) goes over to

$$|\epsilon_{n+1}| = \mu|\epsilon_n|| - 1 + 2/\mu - \epsilon_n|. \tag{13.3.5}$$

If $|\epsilon_n|$ is small enough for

$$|2 - \mu| + \mu|\epsilon_n| < 1 \tag{13.3.6}$$

then $|\epsilon_{n+1}| < |\epsilon_n|$ and ϵ_{n+1} also complies with (13.3.6). Hence $\epsilon_n \to 0$ and the iteration ends up at $x_0 = 1 - 1/\mu$ provided that (13.3.6) is satisfied. A suitable ϵ_n can be found certainly for $1 < \mu < 3$ but there is no possible choice for $\mu \geq 3$. Indeed, for small ϵ_n, (13.3.5) shows that, to the first order, $|\epsilon_{n+1}| > |\epsilon_n|$ and x_n does not approach the second fixed point.

To summarise:

the fixed point $x_0 = 0$ is stable for $\mu < 1$ and unstable for $\mu > 1$:
the fixed point $x_0 = 1 - 1/\mu$ is stable for $1 < \mu < 3$ and unstable
for $\mu > 3$.

That leaves us with the puzzle of what happens when $\mu > 3$, for then neither fixed point is stable. The possibility of x_n going off to infinity is prevented by x_n being confined to $(0,1)$; so something new must be taking place. A clue is offered by a numerical calculation. In Table 13.3.1 are shown the first 20 values of x_n when $x_1 = 0.250$ and $\mu = 3.1$. It is clear that alternative entries are pretty much the same from x_{10} onwards. In other words, the iteration

TABLE 13.3.1:

Iteration for $\mu = 3.1$.

n	x_n
1	0.250
2	0.581
3	0.754
4	0.574
5	0.758
6	0.569
7	0.760
8	0.565
9	0.762
10	0.562
11	0.763
12	0.561
13	0.764
14	0.560
15	0.764
16	0.559
17	0.764
18	0.559
19	0.764
20	0.558

is oscillating between two values without settling down to one or the other. **Period-doubling** is said to occur.

What it means, in terms of our original model, is that the crop is high one year, low the next, then high the following year, then low again and so forth.

Let us see if analysis supports the numerical evidence. For brevity, write the iteration scheme as

$$x_{n+1} = f(x_n). \tag{13.3.7}$$

Then

$$x_{n+2} = f(x_{n+1}) = f(f(x_n)) \tag{13.3.8}$$

on applying (13.3.7). Now, if alternative values of x_n approach the same limit x_0, (13.3.8) implies that

$$x_0 = f(f(x_0)). \tag{13.3.9}$$

After insertion of f from (13.3.2), (13.3.9) can be rearranged as

$$x_0\left(1 - x_0 - \frac{1}{\mu}\right)\left\{(1 - x_0)^2 + \left(\frac{1}{\mu} - 1\right)(1 - x_0) + \frac{1}{\mu^2}\right\} = 0. \tag{13.3.10}$$

That two of the solutions are the fixed points found already is not surprising because, for them, every x_n tends to the same limit and, a fortiori, so do alternate x_n. The remaining solutions of (13.3.10) are given by

$$2(1 - x_0) = 1 - \frac{1}{\mu} \pm \left\{ \left(1 - \frac{3}{\mu}\right)\left(1 + \frac{1}{\mu}\right) \right\}^{1/2} \qquad (13.3.11)$$

Period-doubling is confirmed provided that x_n tends to a limit. Moreover, one of the values in (13.3.11) is above $1 - 1/\mu$ and the other below, consistent with Table 13.3.1. Actually, for $\mu = 3.1$, the two values of x_0 in (13.3.11) are 0.5580 and 0.7646 to four significant figures. These values agree well with those at the end of Table 13.3.1.

To check on convergence a linear analysis of (13.3.8) is performed. Put $x_n = x_0 + \epsilon_n$ with ϵ_n small and x_0 as in (13.3.11). Then

$$\begin{aligned} x_0 + \epsilon_{n+2} &= f(f(x_0 + \epsilon_n)) = f(f(x_0) + \epsilon_n f'(x_0)) \\ &= f(f(x_0)) + \epsilon_n f'(x_0) f'(f(x_0)) \end{aligned}$$

whence

$$\epsilon_{n+2} = \epsilon_n f'(x_0) f'(f(x_0)) \qquad (13.3.12)$$

on account of (13.3.9).

With f as in (13.3.2), $f'(x) = \mu(1 - 2x)$ and so (13.3.11) gives

$$f'(x_0) f'(f(x_0)) = 4 + 2\mu - \mu^2.$$

The right-hand side is 1 when $\mu = 3$ and decreases as μ increases until it reaches -1 at $\mu = 1 + \sqrt{6} = 3.44949$. Hence $\epsilon_n \to 0$ for $3 < \mu < 1 + \sqrt{6}$ but not elsewhere. Accordingly, period-doubling occurs in the range $3 < \mu < 1 + \sqrt{6}$ but something else happens for $\mu > 1 + \sqrt{6}$. There is another bifurcation to be considered.

At the bifurcation each of the fixed points splits into two. Now the iteration oscillates between four values when n is large enough. For example, when $\mu = 3.5$, the four fixed points are (0.8750, 0.3828, 0.8269, 0.5009). In other words there is period-4 behaviour.

However, period-4 behaviour lasts only up to $\mu = 3.5441$. Then there is another bifurcation where period-doubling occurs resulting in period-8 behaviour. The process continues with the intervals between bifurcation points getting shorter and shorter (see Table 13.3.2).

Finally, the process comes to a halt when μ reaches the value μ_∞ just above 3.5699. An increase of μ over μ_∞ produces an entirely new regime which is comprehended best from the bifurcation diagram as μ goes from 3, where period-doubling commences, to 4, its maximum permitted value (Figure 13.3.1). It can be seen that the iteration wanders all over the place without seeming to make any attempt to converge. Here we have an example of **chaos**.

TABLE 13.3.2: Bifurcation points for period-doubling.

μ_1	3	period-2
μ_2	3.4495	period-4
μ_3	3.5441	period-8
μ_4	3.5644	period-16
μ_5	3.5687	period-32
	....	
μ_∞	3.569946	

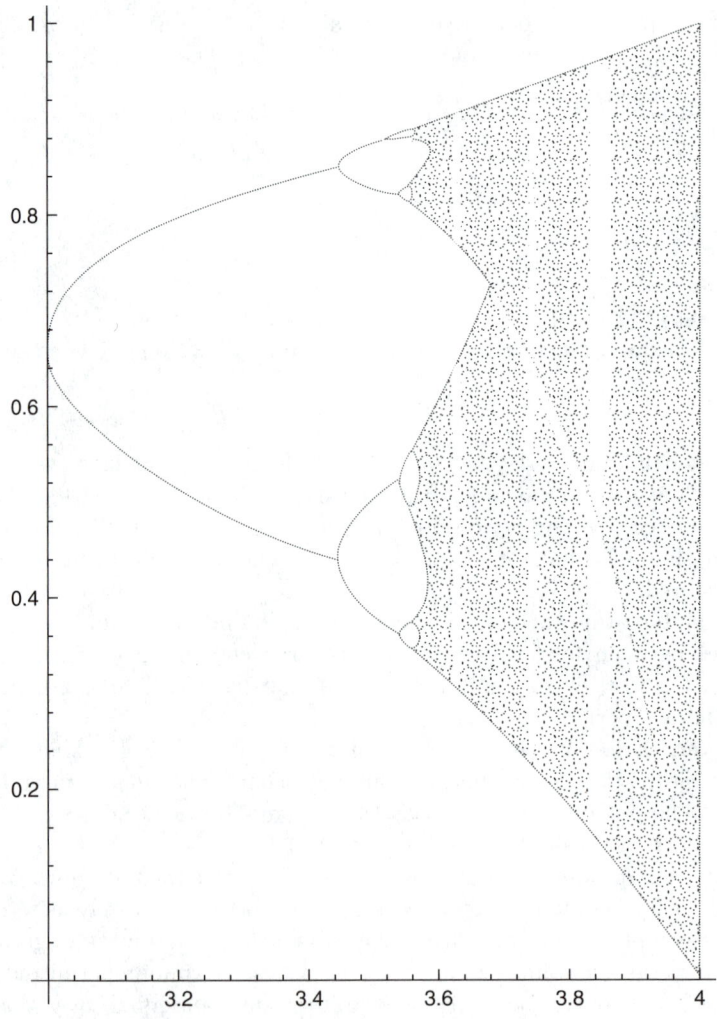

FIGURE 13.3.1: Bifurcation diagram for μ between 3 and 4.

Perhaps it should be emphasised that the bifurcation diagrams for iteration schemes are slightly different from those in earlier sections in that the only fixed points shown are stable. The reason is that the diagrams are generated usually by a computer starting from some arbitrary point and finding the points of convergence. Unless the initial point happens to coincide with an unstable fixed point its presence will never be detected and it will be absent from the bifurcation diagram.

While this simple model of crops shows extremely complex behaviour it is not likely that real crops exhibit the same phenomena because many other factors than those considered may be relevant. Nevertheless, the important point is that a simple mathematical relation, involving only a quadratic, can display a rich set of features as the control parameter μ varies. For some values of μ the chaotic behaviour would entail wild variations in the crop from year to year, which might be interpreted as due to the vagaries of the weather or some other factor when they are, in fact, part and parcel of the model.

One of the difficulties in verifying experimentally whether a system conforms to the model is the closeness of successive bifurcation points. Their difference is so small that it could be hidden easily in experimental error. Notwithstanding, a set of experiments on the flour beetle indicates that populations can exhibit period-doubling and chaotic behaviour with variation of a control parameter. In this case, the control parameter is the rate of cannibalism of pupae by adults.

The methods following (13.3.7) can be used to discuss other iteration schemes such as

$$x_{n+1} = \mu \sin(\pi x_n),$$

which also displays period-doubling and chaotic behaviour. For these and more complicated iterations, such as that for the flour beetle, it will not be possible to progress very far without extensive computation.

NOTE: The differential equation of logistics

$$\frac{dp}{dt} = N_0 p - ap^2 \tag{13.3.13}$$

was discussed in Section 1.3. Its solution tends to N_0/a as $t \to \infty$ so long as p is not initially zero. In solving a differential equation numerically the derivative is often approximated so that an estimate of the solution is obtained at intervals of time, say $t = 0, \tau, 2\tau, \dots$. For example, the fact that

$$\frac{dp}{dt} = \lim_{h \to 0} \{p(t+h) - p(t)\}/h$$

suggests the approximation

$$\frac{dp}{dt} = \{p((n+1)\tau) - p(n\tau)\}/\tau.$$

Writing $p(nr) = p_n$, we have the approximation

$$p_{n+1} = (N_0\tau + 1)p_n - a\tau p_n^2$$

for (13.3.13). This has the same form as (13.3.1) with $\mu = N_0\tau+1$. Since $C_n \to (\mu - 1)/\nu$ as $n \to \infty$ when $1 < \mu < 3$ it follows that $p_n \to N_0/a$ provided that $0 < N_0\tau < 2$. Once $N_0\tau$ exceeds 2 the iteration does not tend to the solution of (13.3.13) but can exhibit period doubling and chaos, an illustration of the importance of choosing the time interval correctly in the numerical approximation. Of course, p_n tending to the right limit when $0 < N_0\tau < 2$ does not mean that intermediate values of the iteration are good approximations to the solution of (13.3.13) without further investigation.

13.4   Chaos

The chaotic behaviour in Figure 13.3.1 looks as though it is the result of a random process. Nevertheless, the wild wandering is entirely deterministic; each point is determined directly from its predecessor via the iteration scheme. No considerations of probability are involved. One property that is not evident from the figure can be seen in Table 13.4.1, which shows the iteration for three different starting values when $\mu = 3.99$. The way that successive values swing

TABLE 13.4.1:   Iteration for $\mu = 3.99$.

n	x_n	x_n	x_n
1	0.5000	0.5005	0.5010
2	0.9975	0.9975	0.9975
3	0.0100	0.0100	0.0100
4	0.0393	0.0393	0.0394
5	0.1507	0.1507	0.1509
6	0.5106	0.5107	0.5112
7	0.9971	0.9970	0.9970
8	0.0117	0.0118	0.0119
9	0.0462	0.0464	0.0471
10	0.1759	0.1767	0.1790
11	0.5784	0.5804	0.5864
12	0.9730	0.9717	0.9677
13	0.1050	0.1097	0.1246
14	0.3750	0.3900	0.4352
15	0.9351	0.9488	0.9807
16	0.2421	0.1937	0.0753
17	0.7321	0.6232	0.2780
18	0.7825	0.9369	0.8008
19	0.6791	0.2358	0.6364
20	0.8694	0.7189	0.9232

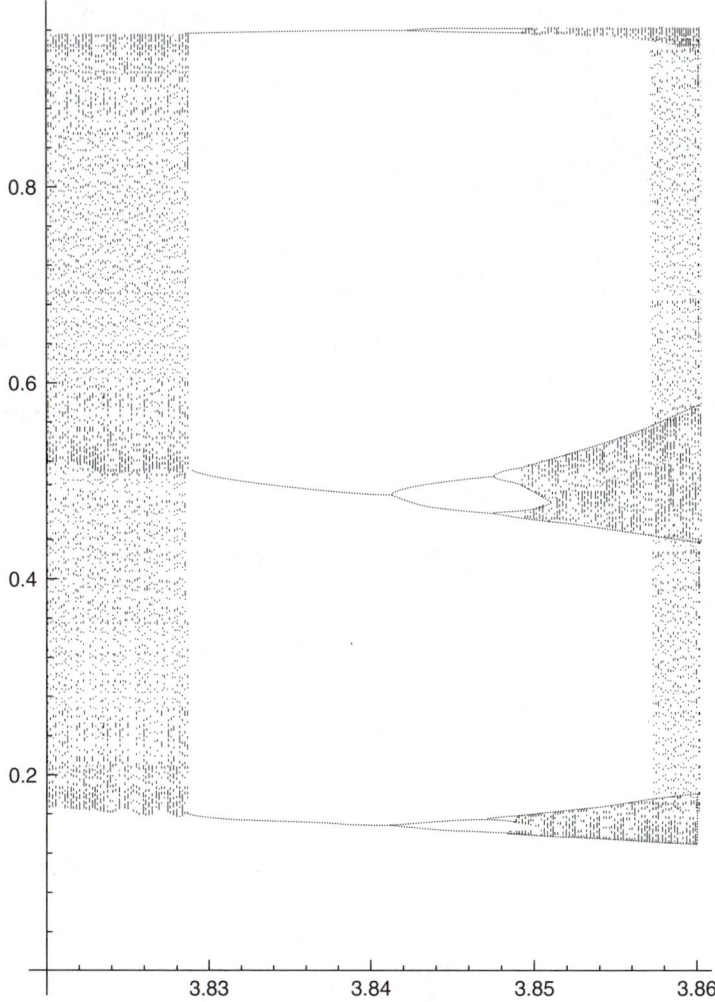

FIGURE 13.4.1:   Bifurcation diagram for μ between 3.82 and 3.86.

about is obvious but there is another feature. The initial values are close together and the first few iterates tally. But soon the iterates in the three columns begin to deviate and by the time x_{16} is reached they bear little or no resemblance to one another. *A small change in the initial value can produce a very large difference in the iterates.*

It must not be supposed that the onset of chaos precludes the possibility of fixed points and bifurcation. An expanded version of the bifurcation diagram with μ lying between 3.82 and 3.86 is displayed in Figure 13.4.1. Not only is there a gap between two regions of chaos with three fixed points but also there is bifurcation with period-doubling as well. The three fixed points when $\mu = 3.835$ are $(0.95863, 0.15207, 0.49451)$ (programs suitable for computation are in the appendix to this chapter). At $\mu = 3.848$ there are twelve fixed points $(0.96199, 0.14070, 0.46524, 0.95735, 0.15712, 0.50960, 0.96165, 0.14193, 0.46862, 0.95821, 0.15408, 0.50155)$; they are just visible on Figure 13.4.1. Similar gaps have been found to occur with other iteration schemes.

Another feature that may carry over concerns the values of μ at bifurcations before the onset of chaos (see Table 13.3.2). Define

$$\delta_n = \frac{\mu_{n+1} - \mu_n}{\mu_{n+2} - \mu_{n+1}}. \tag{13.4.1}$$

A numerical investigation reveals that, as $n \to \infty$, $\delta_n \to \delta = 4.6692016$; δ is known as *Feigenbaum's number*.

Feigenbaum's number can be useful in two ways. If μ_1, μ_2 and μ_3 are known, (13.4.1) suggests that

$$\mu_4 = \mu_3 + (\mu_3 - \mu_2)/\delta. \tag{13.4.2}$$

There is, naturally, no guarantee that μ_4 exists for an arbitrary iteration scheme but, if it does, (13.4.2) offers a clue as to where to search to find the bifurcation point.

More generally, if δ_n can be replaced by δ in (13.4.1) for $n \geq N$,

$$\mu_{n+2} - \mu_{n+1}(1 + 1/\delta) + \mu_n/\delta = 0. \tag{13.4.3}$$

Try $\mu_n = a^n$ and then

$$a^2 - a(1 + 1/\delta) + 1/\delta = 0$$

whence $a = 1$ or $1/\delta$. Hence the general solution of (13.4.3) is

$$\mu_n = A + B/\delta^n.$$

The constants A and B can be fixed by putting $n = N$ and $n = N + 1$. The result is that

$$\mu_n = \mu_{N+1} + \frac{(\mu_{N+1} - \mu_N)(1 - \delta^{N+1-n})}{\delta - 1} \tag{13.4.4}$$

for $n \geq N$.

As a test of (13.4.4) take $N = 1$ with μ_1 and μ_2 as in Table 13.3.2. Then (13.4.4) predicts that μ_3 will be 3.5458, $\mu_4 = 3.5664$ and $\mu_5 = 3.5708$. While these differ a little from the values in Table 13.3.2 they are sufficiently close to be a good guide in a search for bifurcation points where period-doubling occurs.

Since $\delta > 1$ the prediction of (13.4.4) for the onset of chaos is

$$\mu_\infty = \mu_{N+1} + \frac{\mu_{N+1} - \mu_N}{\delta - 1}. \tag{13.4.5}$$

When $N = 1$, (13.4.5) predicts $\mu_\infty = 3.5720$ which is surprisingly close to the correct value.

It is remarkable that such good predictions are obtained from $N = 1$ and one would expect to do much better with larger values of N. For other iteration schemes exhibiting period-doubling leading to chaos in the same way one would hope that (13.4.4) and (13.4.5) would prove equally valuable. Unfortunately, it is true only for functions with a single maximum near where a good approximation is a quadratic like (13.3.2). For functions of the type

$$f(x) = 1 - a|x|^m \qquad (13.4.6)$$

there is period-doubling and δ exists for each m but the value of δ changes with m. Some values are

$m =$	2	4	6	8
$\delta =$	5.12	9.32	13.37	17.40

where δ has been rounded to two decimal places. The more general iteration

$$x_{n+1} = f(x_n, y_n),$$
$$y_{n+1} = g(x_n, y_n)$$

has been found to possess bifurcations with period-doubling when

$$\frac{\partial f}{\partial x}\frac{\partial g}{\partial y} - \frac{\partial f}{\partial y}\frac{\partial g}{\partial x} = 1$$

but now $\delta = 0.72$. Period-doubling between chaotic regions as in Figure 13.4.1 is governed by yet other values of δ.

To sum up, the basic features that have been observed in the iteration are:

(i) There are successive bifurcations at which period-doubling occurs, the values of μ becoming ever more compressed.

(ii) Chaotic regions appear beyond μ_∞.

(iii) In a chaotic region the iteration is highly sensitive to changes in initial value.

(iv) Period-3, period-5, ... and multiples thereof can appear between chaotic regions.

(v) Positions of bifurcation points can be estimated if an appropriate Feigenbaum δ is available.

As already explained, chaos is not a random process. What distinguishes a chaotic solution to a deterministic equation is that it is highly sensitive to the initial conditions and wanders all over the place in a manner that seems to be random. Solutions that alter sharply after small changes in initial conditions but do not exhibit random-like behaviour are not classed as chaotic.

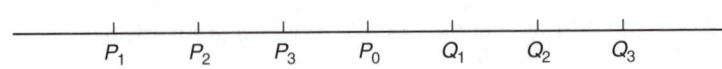

FIGURE 13.5.1:   The Poincaré section.

13.5   Stability

In the discussion of the Hopf bifurcation in Section 13.2 it was shown that the limit cycle was stable. There is another method of demonstrating this property; it uses the **Poincaré section** of a limit cycle.

The Poincaré section is formed by the following procedure. Draw a line segment in the phase plane that intersects the limit cycle. Any convenient line can be chosen so long as it intersects the limit cycle. Let P_0 be a point of intersection with the limit cycle. Start a trajectory from a point near to, but not on, the limit cycle. As we follow the trajectory it will cross the line of the Poincaré section at a point, say P_1 (Figure 13.5.1). The next time around, the crossing will be at P_2, say, and the next one at P_3. If the sequence of points P_1, P_2, P_3, ... approaches P_0 as in Figure 13.5.1 the trajectory is tending to the limit cycle. If the sequence leaves P_0 as for Q_1, Q_2, Q_3, ... in Figure 13.5.1 the trajectory is going away from the limit cycle. Thus, a stable limit cycle can be identified by the sequences on both sides of P_0 tending to P_0. A sequence always stays on one side of P_0 because distinct trajectories in the phase plane cannot intersect in finite time. Two trajectories are said to be distinct when the starting point of one does not lie on the other.

Example 13.5.1

The discussion of the Hopf bifurcation in Section 13.2 was based on the polar coordinates r and ϕ. Explicit formulae for them were obtained, namely

$$r^2 = \frac{\mu r_0^2}{r_0^2 + (\mu - r_0^2)e^{-2\mu t}},$$
$$\phi = \phi_0 + t$$

(see (13.2.5) and (13.2.6)) for a trajectory starting at (r_0, ϕ_0).

Take the line $\phi = \phi_0$ for the Poincaré section. Then r will give the distance of the point of intersection from the origin. The first return of the trajectory to the Poincaré line occurs when $t = 2\pi$ and so

$$r_1^2 = \frac{\mu r_0^2}{r_0^2 + (\mu - r_0^2)e^{-4\pi\mu}}.$$

At the nth return $t = 2n\pi$ and

$$r_n^2 = \frac{\mu r_0^2}{r_0^2 + (\mu - r_0^2)e^{-4\pi\mu n}}.$$

We deduce that

$$r_{n+1}^2 = \frac{\mu r_n^2 e^{4\pi\mu}}{\mu + r_n^2(e^{4\pi\mu} - 1)}, \tag{13.5.1}$$

which is an iteration scheme to determine r_n from its initial value. It is easy to check that, when $\mu > 0$, the iteration possesses fixed points at $r = 0$ and $r = \sqrt{\mu}$, the second corresponding to the limit cycle, in harmony with what is known already.

Equation (13.5.1) can be rewritten as

$$r_{n+1}^2 = \mu + \frac{\mu(r_n^2 - \mu)}{\mu + r_n^2(e^{4\pi\mu} - 1)} \tag{13.5.2}$$

$$= r_n^2 + \frac{(\mu - r_n^2)r_n^2(e^{4\pi\mu} - 1)}{\mu + r_n^2(e^{4\pi\mu} - 1)}. \tag{13.5.3}$$

These equations are helpful in noting certain features. From (13.5.2) can be inferred $r_{n+1}^2 \gtrless \mu$ according as $r_n^2 \gtrless \mu$. Thus the sequence of points on the Poincaré section does stay on one side of $P_0(r^2 = \mu)$ in conformity with the general theory. Also (13.5.3) shows that $r_{n+1}^2 \gtrless r_n^2$ according as $\mu \gtrless r_n^2$. Hence the sequence is steadily increasing (or decreasing) and is bounded above (or below). In either case, the sequence converges and the only possibility is convergence to P_0. Consequently, the limit cycle is stable, in agreement with the conclusion of Section 13.2. $\qquad\square$

The Poincaré section has two characteristics. Firstly, it transforms a two-dimensional problem into one on a line, thereby saving a dimension. Secondly, instead of working through differential equations with continuous variations in time, it employs values at discrete time intervals, each interval being roughly the time to go round the limit cycle once.

Since a trajectory that starts at P_n is completely specified, so is P_{n+1}, i.e., P_{n+1} can be inferred from P_n. To put this on a mathematical footing select any convenient reference point P_r on the Poincaré line and identify a point on the line by its distance s from P_r, with sign depending on which side of P_r it lies. If P_n is at s_n the relation between P_{n+1} and P_n can be expressed as

$$s_{n+1} = f(s_n). \tag{13.5.4}$$

The point P_0 is a fixed point of (13.5.4) because the trajectory that starts at P_0 is a limit cycle and, therefore, must cross the Poincaré line exactly at P_0 on every circuit. Accordingly, if P_0 is designated by s_0,

$$s_0 = f(s_0). \tag{13.5.5}$$

From (13.5.4) and (13.5.5)

$$s_{n+1} - s_0 = f(s_n) - f(s_0).$$

For s_n close to s_0, a Taylor expansion to the first order gives

$$s_{n+1} - s_0 = (s_n - s_0)\left[\frac{df(s)}{ds}\right]_{s=s_0}.$$

Writing

$$M = \left[\frac{df(s)}{ds}\right]_{s=s_0} \tag{13.5.6}$$

we have

$$s_{n+1} - s_0 = M^n(s_1 - s_0).$$

It follows that, if $|M| < 1$, $s_{n+1} \to s_0$, i.e., the points on the Poincaré line tend to P_0 and the limit cycle is stable. In contrast, when $|M| > 1$, s_{n+1} separates further and further from s_0 and the limit cycle is unstable. The case $|M| = 1$ needs further consideration. Actually, M cannot be negative. If M were negative s_{n+1} and s_n would be on opposite sides of P_0. For that to occur the trajectory would have to cross the limit cycle and that is not allowed.

For the scheme of (13.5.3) you can verify that $M = e^{-4\pi\mu} < 1$ another confirmation that the limit cycle involved is stable.

The Poincaré section offers another method of determining the character of a limit cycle by using an iteration scheme. The main difficulty in its application is finding the function f in (13.5.4). This is equivalent to solving the original set of differential equations, a task which may be extremely difficult or even impossible in practice. Without explicit knowledge of f or some adequate approximation to it the applicability of the Poincaré section is restricted.

We have seen in Section 13.3 that iteration schemes can display chaotic behaviour. That raises the question: Could an iteration scheme for the Poincaré line generate chaos and, if so, would the associated trajectory in the phase plane be chaotic? The question cannot be answered without a decision on what constitutes a chaotic trajectory, which is a solution of differential equations. Confining our attention to systems in which the trajectories stay in a bounded region (so that there is no possibility of the solution becoming infinite) we expect the conditions for chaos to be

(a) distinct trajectories do not intersect,

(b) the trajectories are bounded,

(c) trajectories, which are initially close, diverge widely and rapidly.

Not all of these conditions can be met by trajectories in the phase plane so that *chaos does not arise for the phase plane or Poincaré line.*

The situation is quite different in *phase space* with three or more dimensions. Here the trajectories can intertwine without intersection and stay in a bounded region so that chaos is possible. A chaotic trajectory cannot return to its initial point; if it did, it would correspond to a periodic solution of the differential system. It might, however, return to near the initial point but then, on account of (c), its subsequent path would bear little or no resemblance to the path from the initial point. That chaos can occur in phase space is illustrated in the next section.

13.6   The Poincaré plane

Rather than attempt a treatment of general phase space we shall consider only three dimensions. Then a first order system would take the form

$$\dot{x} = f(x, y, z),$$
$$\dot{y} = g(x, y, z),$$
$$\dot{z} = h(x, y, z),$$

where f, g and h do not involve the time derivatives of x, y and z or the time itself. Control parameters may be present also. Such a system is said to be **autonomous**.

When the functions involve the time as well, as in forced oscillations, the system is called **non-autonomous**. There is a trick for turning a non-autonomous system into the autonomous type in one higher dimension. Suppose

$$\dot{x} = f(x, y, t),$$
$$\dot{y} = g(x, y, t). \tag{13.6.1}$$

Put $z = t$ so that $\dot{z} = 1$. The system becomes

$$\dot{x} = f(x, y, z),$$
$$\dot{y} = g(x, y, z),$$
$$\dot{z} = 1, \tag{13.6.2}$$

which is autonomous but in three dimensions instead of the original two. This device enables the treatment of non-autonomous systems on the same footing as autonomous systems. The cost is the necessity to work with an extra dimension in the phase space. Furthermore, the new system has no fixed points because the right-hand side of the equation for $\dot{z}$ never vanishes. However, that difficulty can be overcome by generalising the Poincaré line of two dimensions to higher dimensions.

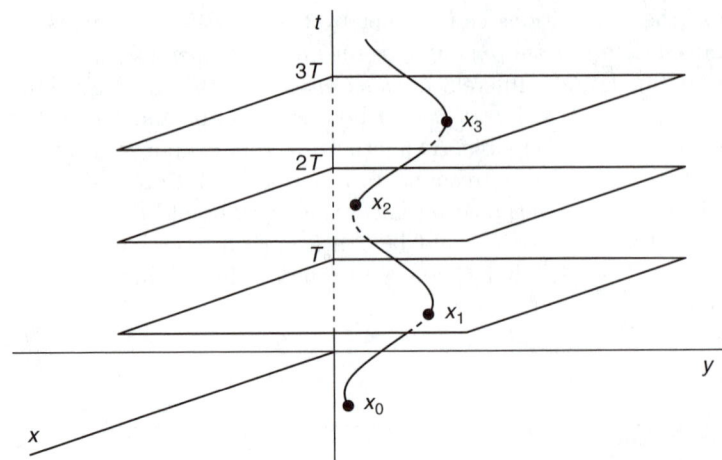

FIGURE 13.6.1:　A three-dimensional trajectory.

The simplest generalisation is when the functions in (13.6.1) are periodic, i.e., there is a constant T such that

$$f(x, y, t + T) = f(x, y, t), \qquad g(x, y, t + T) = g(x, y, t)$$

for all x, y and t. If, now, a trajectory of (13.6.2) is drawn in the three-dimensional space from $t = 0$ it will intersect the planes $t = T, 2T, \ldots$. Denote (x, y) by $\mathbf{x}$ and let the points of intersection starting from $\mathbf{x}_0$ be $\mathbf{x}_1$, $\mathbf{x}_2, \ldots$ (see Figure 13.6.1). Plot the points $\mathbf{x}_0$, $\mathbf{x}_1$, $\mathbf{x}_2$, $\ldots$ on the (x, y)-plane (Figure 13.6.2). The resulting diagram is known as the **Poincaré plane**. To put it another way, the Poincaré plane is constructed by plotting the points $\mathbf{x}(0)$, $\mathbf{x}(T)$, $\mathbf{x}(2T)$, $\ldots$ of the trajectory on the (x, y)-plane.

Of course, construction of the Poincaré plane is equivalent to solving the system of differential equations. So it does not simplify the determination of a solution. Fortunately, there are cases where a reliable approximation can be devised as will be seen in the next section. It is evident from Figures 13.6.1 and 13.6.2 that a complete trajectory will produce a scatter of points on the Poincaré plane. As a result it can be awkward to understand what is going on. Therefore, it is generally wise to delay plotting until any transients have died away so that the long-term behaviour is more easily visible. Once the transients are unimportant the picture on the Poincaré plane will consist of a single point if the system tends to periodic motion with period T. On the other hand, if the period is $2T$, the picture will show two points and, generally, for period nT, there will be n points. When T corresponds to the period of a periodic forcing term, two or more points on the Poincaré plane indicate that, in the long-term, the system will oscillate in a subharmonic of the periodic force.

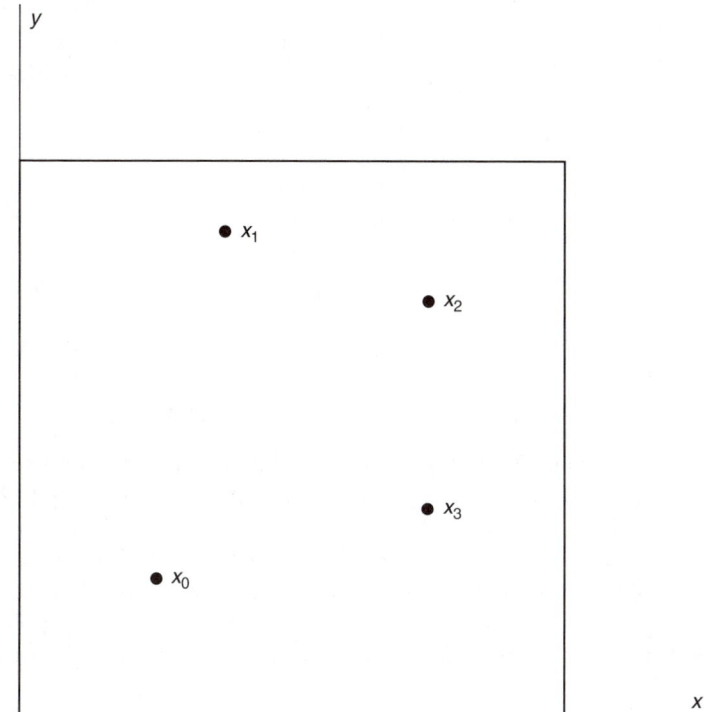

FIGURE 13.6.2:   The Poincaré plane.

On a trajectory abbreviate $x(nT)$ and $y(nT)$ to x_n and y_n, respectively. Since the point (x_n, y_n) determines (x_{n+1}, y_{n+1}) there is a relation

$$x_{n+1} = F(x_n, y_n),$$
$$y_{n+1} = G(x_n, y_n), \qquad (13.6.3)$$

i.e., the points on the Poincaré plane satisfy an iteration scheme. The long-term behaviour is dictated by the fixed points of this scheme; they are the solutions of

$$x_0 = F(x_0, y_0),$$
$$y_0 = G(x_0, y_0). \qquad (13.6.4)$$

Every solution of (13.6.4) corresponds to a limit cycle in the three-dimensional phase space.

If (x_n, y_n) is close to (x_0, y_0) the subtraction of (13.6.4) from (13.6.3) followed by a Taylor expansion leads to

$$x_{n+1} - x_0 = A(x_n - x_0) + B(y_n - y_0),$$
$$y_{n+1} - y_0 = C(x_n - x_0) + D(y_n - y_0) \qquad (13.6.5)$$

where

$$A = \frac{\partial F(x,y)}{\partial x}, \quad B = \frac{\partial}{\partial y} F(x,y), \quad C = \frac{\partial}{\partial x} G(x,y), \quad D = \frac{\partial}{\partial y} G(x,y),$$

$$(13.6.6)$$

all the derivatives being evaluated at (x_0, y_0).

To solve (13.6.5) try both $x_n - x_0$ and $y_n - y_0$ being proportional to λ^n. Then λ must satisfy

$$\lambda^2 - \lambda(A + D) + AD - BC = 0. \qquad (13.6.7)$$

If the solutions of (13.6.7) are λ_1, λ_2 there are constants A_1, A_2, B_1 and B_2 such that

$$x_n - x_0 = A_1 \lambda_1^n + A_2 \lambda_2^n, \qquad y_n - y_0 = B_1 \lambda_1^n + B_2 \lambda_2^n$$

unless $\lambda_1 = \lambda_2$ when λ_2^n is replaced by $n\lambda_1^n$. In any case, for (x_n, y_n) to tend to (x_0, y_0), whatever the initial conditions it is necessary that $|\lambda_1| < 1$ and $|\lambda_2| < 1$. If $|\lambda_1| > 1$ and $|\lambda_2| > 1$, (x_n, y_n) will tend to separate from (x_0, y_0). When $|\lambda_1| < 1$ and $|\lambda_2| > 1$, (x_n, y_n) will approach or deviate from (x_0, y_0) according as A_2 or B_2 is or is not zero; this will depend on the initial conditions. Limit cycles that are approached from some nearby points and separated from others are often known as *saddle limit cycles*. In summary we have

$$|\lambda_1| < 1 \text{ and } |\lambda_2| < 1 \qquad \text{stable limit cycle,}$$
$$|\lambda_1| > 1 \text{ and } |\lambda_2| > 1 \qquad \text{unstable limit cycle,}$$
$$|\lambda_1| < 1 \text{ and } |\lambda_2| > 1 \qquad \text{saddle limit cycle.}$$

Obviously, if varying a control parameter causes $|\lambda_1|$ or $|\lambda_2|$ to pass through unity, a bifurcation occurs.

To illustrate these points we consider the motion of a particle that is given a constant velocity for an interval of time, the constant depending upon the position of the particle at the beginning of the interval. The resulting behaviour is amazingly complex.

Example 13.6.1
For every positive integer n, (x, y) satisfies

in $n < t < n + 1/3$, $\qquad \dot{x} = 0$, $\qquad\qquad\qquad \dot{y} = 3(1 - ax^2)$,
in $n + 1/3 < t < n + 2/3$, $\quad \dot{x} = 3(b-1)x(n+1/3)$, $\qquad \dot{y} = 0$,
in $n + 2/3 < t < n + 1$, $\quad \dot{x} = 3y(n+2/3) - 3x(n+2/3)$, $\quad \dot{y} = -\dot{x}$.

The quantities a and b are non-negative constants. Integration of the differential equations gives

$$x(n + 1/3) = x(n), \qquad\qquad y(n + 1/3) = y(n) + 1 - ax(n)^2,$$
$$x(n + 2/3) = bx(n + 1/3), \qquad y(n + 2/3) = y(n + 1/3),$$
$$x(n + 1) = y(n + 2/3), \qquad\quad y(n + 1) = x(n + 2/3).$$

Combining these equations and writing x_n for $x(n)$ we have the iteration scheme

$$x_{n+1} = y_n + 1 - ax_n^2,$$
$$y_{n+1} = bx_n. \qquad (13.6.8)$$

In the discussion of the iteration it will be assumed that $0 < b < 1$ and $a \geq 0$. Also b will be kept fixed while a is regarded as the control parameter. The values of x at fixed points satisfy

$$ax^2 + (1 - b)x - 1 = 0$$

with solutions

$$2ax_0^{(1)} = b - 1 + \{4a + (1 - b)^2\}^{1/2}, \qquad (13.6.9)$$
$$2ax_0^{(2)} = b - 1 - \{4a + (1 - b)^2\}^{1/2}. \qquad (13.6.10)$$

At $x_0^{(1)}$, (13.6.6) gives

$$A = -2ax_0^{(1)}, \quad B = 1, \quad C = b, \quad D = 0$$

so that (13.6.7) becomes

$$\lambda^2 + 2ax_0^{(1)}\lambda - b = 0. \qquad (13.6.11)$$

Evidently λ_1 and λ_2 have opposite signs; take $\lambda_1 > 0$ and $\lambda_2 < 0$. When $a = 0$, $ax_0^{(1)} = 0$ so that $\lambda_1 = \sqrt{b}$, $\lambda_2 = -\sqrt{b}$. Since $b < 1$ there is stability at $a = 0$. In

$$\lambda_1 - b/\lambda_1 = -2ax_0^{(1)} \qquad (13.6.12)$$

the right-hand side decreases as a increases. Therefore λ_1 decreases and hence so does λ_2. Consequently, stability continues until $\lambda_2 = -1$ and $\lambda_1 = b$. This occurs, according to (13.6.12) and (13.6.9), when

$$4a = 3(1 - b)^2. \qquad (13.6.13)$$

Thus (13.6.13) specifies when bifurcation will take place.

For the fixed point $x_0^{(2)}$ replace $x_0^{(1)}$ by $x_0^{(2)}$ in (13.6.11) and (13.6.12). At $a = 0$, $2ax_0^{(2)} = 2(b - 1)$ and then the left-hand side of (13.6.11) is negative when $\lambda = 1$ so that $\lambda_1 > 1$. Since $-ax_0^{(2)}$ increases with a so does λ_1 on account of (13.6.12). Hence $x_0^{(2)}$ is never stable; since $|\lambda_2| < b$ it corresponds to a saddle limit cycle.

Hence, for a less than the value in (13.6.13), the iteration has two fixed points of which $x_0^{(1)}$ is stable and $x_0^{(2)}$ is not. For $a > 3(1 - b)^2/4$, period-doubling

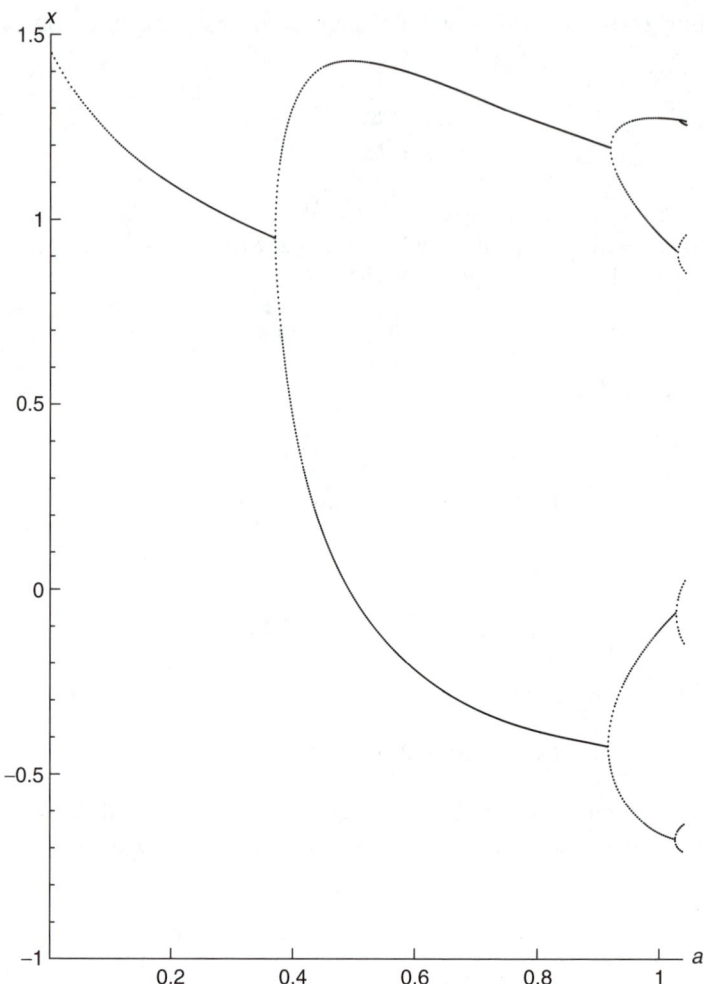

FIGURE 13.6.3: Bifurcation diagram for x when $b = 0.3$ and a ranges from 0 to 1.04.

occurs. A bifurcation diagram for x when $b = 0.3$ is shown in Figure 13.6.3 (fixed points that are not stable do not appear on such diagrams as explained in Section 13.3). Here a goes from 0 to 1.04 and it is clear that there are further bifurcations with accompanying period-doubling. Moreover, the distance between successive bifurcations grows steadily shorter just as in Section 13.3.

As a increases beyond 1.04 matters become much more complicated. Figure 13.6.4 shows the continuation of the bifurcation diagram as a goes from 1.04 to 1.06. More period-doubling takes place and regions of chaotic

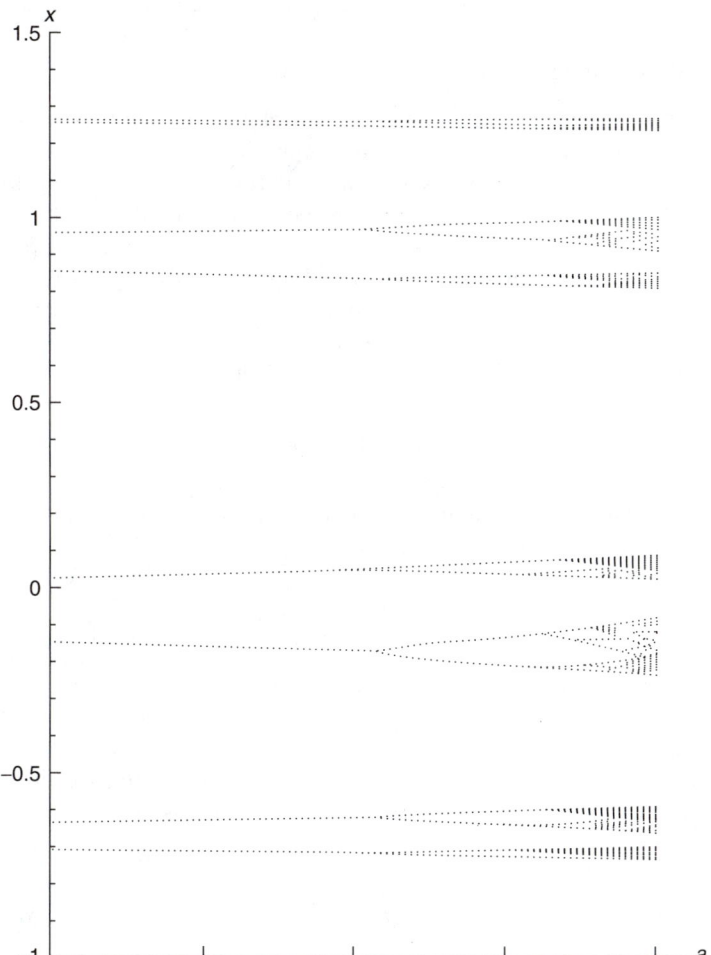

FIGURE 13.6.4: Bifurcation diagram for x when $b = 0.3$ and a ranges from 1.04 to 1.06.

behaviour make their appearance. Beyond 1.06 the situation is even more complex but discussion of the details would take us too far afield. ☐

The example has demonstrated that period-doubling and chaos can be present when three variables are involved. It shows also that even a particle moving at constant velocity at intervals can exhibit behaviour that is far from simple. In this case we were able to find the exact formula connecting points on the Poincaré plane but in many circumstances this is not feasible. So we turn now to a method of approximation that allows progress when conditions are suitable.

13.7 Averaging

The method of averaging is designed to relate a non-autonomous system to an autonomous one by means of an approximation. It is appropriate for problems where a small perturbation is made to a system that can be solved exactly. Although the perturbation may be small that does not mean that the solutions of the perturbed system stay closed to those of the unperturbed. Therefore, just taking the unperturbed system as the first approximation may lead quickly to erroneous results; an approximation with a behaviour similar to that of the perturbed system is needed.

The type of system to be considered is, in vector form,

$$\dot{\mathbf{x}} = \epsilon \mathbf{f}(\mathbf{x}, t, \epsilon) \tag{13.7.1}$$

where the scalar ϵ is very small compared with unity and $\epsilon \geq 0$. As regards $\mathbf{f}$ we require it to be periodic in t so that

$$\mathbf{f}(\mathbf{x}, t + T, \epsilon) = \mathbf{f}(\mathbf{x}, t, \epsilon) \tag{13.7.2}$$

for all t under consideration. Moreover, $\mathbf{f}$ has to possess at least two continuous derivatives in its variables and to be bounded for bounded values of its arguments.

The related autonomous system is obtained by averaging $\mathbf{f}$ over a period. To be specific define $\mathbf{w}$ to be the solution of

$$\dot{\mathbf{w}} = \frac{\epsilon}{T} \int_0^T \mathbf{f}(\mathbf{w}, t, 0) dt. \tag{13.7.3}$$

In the integration on the right-hand side of (13.7.3) $\mathbf{w}$ is treated as a constant.

Under the above assumptions it is possible to prove the following:

(a) If $\mathbf{x}(t)$ is a solution of (13.7.1) such that $\mathbf{x}(0) = \mathbf{x}_0$ and $\mathbf{w}(t)$ is a solution of (13.7.3) such that $\mathbf{w}(0) = \mathbf{w}_0$, then, if $|\mathbf{x}_0 - \mathbf{w}_0| = O(\epsilon)$, $|\mathbf{x}(t) - \mathbf{w}(t)| = O(\epsilon)$ for values of t up to $O(1/\epsilon)$.

(b) If $\mathbf{w} \equiv (u, v)$ and (13.7.3) has a fixed point $\mathbf{w}_p$, which corresponds to a stable, unstable, or saddle limit cycle, then the Poincaré plane of (13.7.1) has a fixed point at $\mathbf{w}_p + O(\epsilon)$ with the same stability.

Property (a) shows that solutions of (13.7.1) and (13.7.3), which start close together, stay close for a long time since ϵ is small. Thus (13.7.3) does provide a good approximation to the trajectories of (13.7.1). In addition, property (b) indicates that a good idea of the limit cycles of (13.7.1) can be obtained from (13.7.3).

Example 13.7.1

Find an approximate solution of

$$\dot{x} = -\epsilon x \sin^2 t \qquad (13.7.4)$$

by the method of averaging.

Actually, the exact solution of (13.7.4) can be found by separation of variables. It gives, if $x(0) = x_0$,

$$x(t) = x_0 \exp\left\{-\epsilon\left(\tfrac{1}{2}t - \tfrac{1}{4}\sin 2t\right)\right\}. \qquad (13.7.5)$$

The averaged equation corresponding to (13.7.4) is, from (13.7.3),

$$\dot{w} = -\epsilon w \frac{1}{2\pi} \int_0^{2\pi} \sin^2 t \, dt = -\frac{1}{2}\epsilon w \qquad (13.7.6)$$

since w is regarded as a constant in the integration. The solution of (13.7.6) such that $w = w_0$ at $t = 0$ is

$$w(t) = w_0 \exp\left(-\tfrac{1}{2}\epsilon t\right). \qquad (13.7.7)$$

The difference between (13.7.5) and (13.7.7) is

$$\dot{x}(t) - w(t) = \exp\left(-\tfrac{1}{2}\epsilon t\right)\left[x_0 - w_0 + x_0\left\{\exp\left(\tfrac{1}{4}\epsilon \sin 2t\right) - 1\right\}\right]. \qquad (13.7.8)$$

This shows immediately that, if $x_0 - w_0 = O(\epsilon)$, then $x(t) - w(t) = O(\epsilon)$ for $t \geq 0$. Thus, property (a) is confirmed and for a larger range of t than specified. However, if (13.7.4) were replaced by

$$\dot{x} = \epsilon x \sin^2 t,$$

the only modification to (13.7.8) would be a change in the sign of ϵ. Then, as t increased beyond $1/\epsilon$, the first factor of the right-hand side of (13.7.8) could become large enough to offset the $O(\epsilon)$ of the second factor. Hence the upper limit on t in (a) cannot be removed.

The only fixed point of (13.7.6) is $w = 0$. According to property (b) there is a nearby fixed point associated with (13.7.4) and obviously it is $x = 0$, as may be checked from (13.7.5). ⬚

Example 13.7.2

Consider the perturbed oscillator

$$\ddot{x} + x = E \cos \omega t - D\dot{x} - Bx^3$$

where E, D and B are to be regarded as small positive constants.

Converting the differential equation into a first order system by the substitution $y = \dot{x}$ we have

$$\dot{x} = y, \tag{13.7.9}$$
$$\dot{y} = -x - Dy - Bx^3 + E\cos\omega t. \tag{13.7.10}$$

The system does not have the structure of (13.7.1) so that the method of averaging cannot be applied directly. To transform it to the requisite form introduce polar coordinates. Write

$$x(t) = r(t)\cos\{\omega t - \theta(t)\}, \tag{13.7.11}$$
$$y(t) = -\omega r(t)\sin\{\omega t - \theta(t)\}. \tag{13.7.12}$$

Substitution in (13.7.9) and (13.7.10) leads to

$$\dot{r}\cos(\omega t - \theta) + r\dot{\theta}\sin(\omega t - \theta) = 0,$$
$$-\omega\dot{r}\sin(\omega t - \theta) + \omega r\dot{\theta}\cos(\omega t - \theta) = F$$

where

$$F = D\omega r\sin(\omega t - \theta) + (\omega^2 - 1)r\cos(\omega t - \theta) - Br^3\cos^3(\omega t - \theta) + E\cos\omega t.$$

Solving these equations for $\dot{r}$ and $\dot{\theta}$ we have

$$\omega\dot{r} = -F\sin(\omega t - \theta), \tag{13.7.13}$$
$$\omega r\dot{\theta} = F\cos(\omega t - \theta). \tag{13.7.14}$$

If, now, F were replaced by $\epsilon(F/\epsilon)$ this system would be of the desired form with the various small coefficients in F divided by ϵ. However, it is slightly more convenient to proceed with the original form without explicitly writing in the multiplication and division by ϵ.

The averaged equations are obtained from formulae like

$$\frac{\omega}{2\pi}\int_0^{2\pi/\omega}\cos^2(\omega t - \theta)dt = \frac{1}{2},$$
$$\frac{\omega}{2\pi}\int_0^{2\pi/\omega}\cos(\omega t - \theta)\cos\omega t\, dt = \frac{1}{2}\cos\theta.$$

The result is

$$\omega\dot{r} = -\tfrac{1}{2}D\omega r + \tfrac{1}{2}E\sin\theta, \tag{13.7.15}$$
$$\omega r\dot{\theta} = \tfrac{1}{2}r(\omega^2 - 1) - \tfrac{3}{8}Br^3 + \tfrac{1}{2}E\cos\theta. \tag{13.7.16}$$

A fixed point (r_0, θ_0) of (13.7.15) and (13.7.16) satisfies

$$E\sin\theta_0 = D\omega r_0, \tag{13.7.17}$$
$$E\cos\theta_0 = \tfrac{3}{4}Br_0^3 + r_0(1 - \omega^2). \tag{13.7.18}$$

According to (13.7.17), $\sin\theta_0$ can never be negative so that θ_0 can be restricted to the interval $(0,\pi)$. With that convention θ_0 can be determined from (13.7.18) once r_0 is known. The elimination of θ_0 gives

$$r_0^2 \left(\tfrac{3}{4}Br_0^2 + 1 - \omega^2\right)^2 + D^2\omega^2 r_0^2 = E^2 \qquad (13.7.19)$$

or, with $R = r_0^2$,

$$9B^2R^3 + 24BR^2(1-\omega^2) + 16R\{D^2\omega^2 + (1-\omega^2)^2\} - 16E^2 = 0, \qquad (13.7.20)$$

which is a cubic equation to fix r_0. It is transparent from (13.7.19) that no root of the cubic can be negative. Since the left-hand side of (13.7.20) is negative when $R = 0$ there is at least one positive root. When $\omega^2 \leq 1$ there is at most one positive root by Descartes's rule of signs. Hence there is precisely one positive root when $\omega^2 \leq 1$. For $\omega^2 > 1$ there may be three positive roots or one.

Keep B, D and E fixed and treat ω as a control parameter. Typical values of r_0 obtained from (13.7.20) are shown in Figure 13.7.1. The graph was drawn for $B = 0.2$, $D = 0.2$, $E = 1.25$, values chosen to make what happens clearly visible rather than ensuring that they are small enough to comply with the requirements of the theory of averaging. For most of the range of ω the cubic has a single root but for ω in the neighbourhood of 1.5 there are three roots. The corresponding values of θ_0 are displayed in Figure 13.7.2; over much of the range θ_0 is not far from 0 or π.

Since r is the length of the radius vector the value of r_0 specifies the radius of a limit cycle. Therefore, we expect the original system to have a limit cycle whose distance from the origin is approximately r_0. In an interval of ω there are three possible limit cycles; which of these does the system adopt?

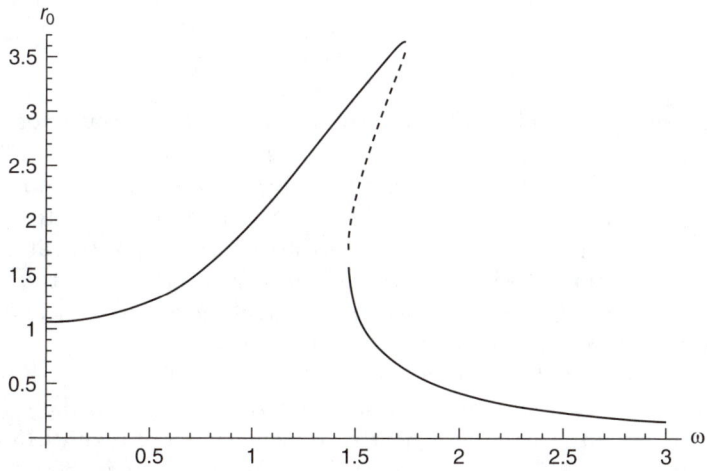

FIGURE 13.7.1: Behaviour of r_0 as ω varies.

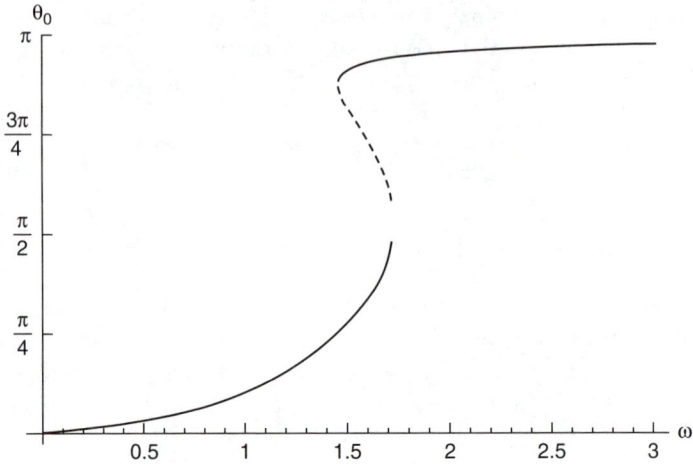

FIGURE 13.7.2: Variation of phase with ω.

To answer the question consider a nearby trajectory where $r = r_0 + \rho$, $\theta = \theta_0 + \eta$. Then, for a first approximation, (13.7.15) and (13.7.16) become

$$\omega\dot{\rho} = -\tfrac{1}{2}D\omega\rho + \tfrac{1}{2}E\eta\cos\theta_0,$$
$$\omega r_0\dot{\eta} = \tfrac{1}{2}(\omega^2 - 1)\rho - \tfrac{9}{8}Br_0^2\rho - \tfrac{1}{2}E\eta\sin\theta_0.$$

According to Section 5.4 the behaviour of this system is governed by the roots λ_1, λ_2 of

$$\lambda^2 + \frac{1}{2}\lambda\{D\omega + (E/r_0)\sin\theta_0\} + \frac{D}{4r_0}E\omega\sin\theta_0$$
$$- \frac{E}{4r_0}\left(\omega^2 - 1 - \frac{9}{4}Br_0^2\right) = 0. \tag{13.7.21}$$

Evidently, one of λ_1, λ_2 is always negative and the other is negative or positive according as the term in (13.7.21) independent of λ is positive or negative. Substitution for θ_0 from (13.7.17) and (13.7.18) reveals that this term is $1/64$ times the derivative of the left-hand side of (13.7.20) with respect to R. When the cubic has a single positive root the derivative must be positive at the root. When there are three positive roots the derivative is positive at the smallest and largest roots but negative at the intermediate one. Thus, λ_1 and λ_2 are both negative except at the intermediate r_0 of three.

Since λ_1 and λ_2 both negative corresponds to stability it follows that the limit cycle associated with r_0 is stable unless r_0 is the intermediate value of three when it is unstable. Consequently, the solid curves of Figures 13.7.1 and 13.7.2 indicate stable limit cycles while the dashed curves indicate an absence of stability.

Now, we see that as ω increases the system can occupy a stable limit cycle related to the upper solid curve of Figure 13.7.1 until ω reaches the point where the solid and dashed curves meet. If ω is increased further the system cannot transfer to the limit cycle, which is not stable; so it must jump to the lower solid curve and continue along it. Decreasing ω now induces a similar phenomenon but there is a jump upwards when the lower solid curve meets the dashed curve. Translating this back to the perturbed oscillator we expect it to exhibit similar jumps as ω is varied. The existence of jumps has been verified experimentally. □

13.8 Appendix: programs

In this appendix various routines for calculations on iteration schemes by means of MATHEMATICA are given.

The scheme of (13.3.2) is based on

```
logis[x_]:= mu x (1-x)
```

Early values, such as those of Table 13.3.1, can be obtained by instructions like

```
mu = 3.1; NestList [logis,.25,20]
```

If later values are desired, say those after a 1000 iterations, it is better to use Nest, which is much faster than NestList, at the beginning, e.g.,

```
NestList [logis,Nest[logis,.25,1000],20]
```

Finding the limit points of an iteration requires a test of the equality of values being produced. The default test is provided by

```
optionlp = {test-> SameQ, rofftest-> SameQ};
```

The default is stringent in that two numbers have to be precisely the same to be passed as equal. As a result two numbers that differ only by round-off error will be treated as unequal. Since many iterations are likely to introduce round-off error, some milder tests, in decreasing order of stringency, are

```
roff12[x_,y_]:=Abs[x-y]<10^(-12)
roff6[x_,y_]:=Abs[x-y]<10^(-6)
roff5[x_,y_]:=Abs[x-y]<10^(-5)
```

They can be used as options to modify the default when calling the routine to find the limit points.

The instructions for determining the limit points are

```
limpoint[f_,start_,options___]:=limpoint[f,start,
    Infinity]

limpoint[f_,start_,maxit_?NumberQ,options___]:=Module[
    {check,roffcheck,single,double,count,ans,ans2},
{check,roffcheck}={test,rofftest}/.{options}
    /.optionlp;
single=f[start];
double=f[f[start]];
count=1;
If[maxit=!=Infinity,While[(!check[single,double])&&
    (count<maxit),single=f[single];
    double=f[f[double]];++count],
    While[!check[single,double],
        single=f[single];double=f[f[double]];++count]];
If[count==maxit,ans=single,ans={double};
    ans2=f[double];
    While[!roffcheck[ans2,double],AppendTo[ans,ans2];
        ans2=f[ans2]]];
ans
]
```

Apart from the test options the arguments of `limpoint` are `f`, the function to be iterated, `start`, the first value and `maxit`, the limit to the number of iterations allowed. If `maxit` is not specified the iterations can go on forever if convergence is not secured; this is undesirable when the value of mu corresponds to chaos because `limpoint` will never finish then. If the iteration fails to converge by the time `maxit` is reached the output of `limpoint` consists of a single plain number such as 0.154081. When convergence occurs the output is enclosed in { }; for example, {0.764567, 0.558014} in the iteration of (13.3.2) with $\mu = 3.1$ (compare Table 13.3.1).

A typical instruction for determining the limit points of Sections 13.3 and 13.4 is

```
muperiod[mu_,maxit_]:=limpoint[mu # (1-#)&,0.5,maxit,
    rofftest->roff12]
```

which can be adapted easily to other iteration schemes by altering the function. The effects of round-off should be borne in mind. If `muperiod` produces two numbers that look pretty much the same it is worth repeating it with a milder test of equality to check if one number is spurious, differing only in round-off error.

The final program is `bifdiag` for drawing a bifurcation diagram. It is specified by

```
quickperiod[mu_,maxit_]:=limpoint[mu # (1-#)&,0.5,
   maxit,test->roff6,rofftest->roff5]

bifdiag[lowmu_,topmu_,numbmu_,options___]:=Module[{mu,
   lis1,period},
lis1=Table[period=quickperiod[mu,500];Map[{mu,#}&,
   If[Head[period]===List,Take[period,
      Min[Length[period],128]],
      NestList[N[mu # (1-#)]&,period,128]]],
   {mu,lowmu,topmu,(topmu-lowmu)/numbmu}];
ListPlot[Flatten[lis1,1],options,
   PlotStyle->PointSize[0.002],AspectRatio->10/7]
]
```

It covers from `lowmu` to `topmu` in the number of steps set by `numbmu`. Because the resolution of screens and printers is limited only the weakest tests of equality are employed. The options in `bifdiag` are any acceptable to `ListPlot`; in particular, you may wish to adjust `PointSize` and `AspectRatio` to your own context. Again, modification of the function should permit other iteration schemes to be discussed.

Exercises

13.1 Show that

$$\dot{x} = x(y-1),$$
$$\dot{y} = \mu - y(x+1)$$

has an equilibrium point, which is a stable node, for $\mu < 1$ that becomes a saddle-point as μ passes through the bifurcation point $\mu = 1$. Also show that there is an additional equilibrium point when $\mu > 1$, which is a stable node.

13.2 Show that

$$\dot{x} = \mu x(2y-1),$$
$$\dot{y} = \mu - y(2x+1)$$

has an equilibrium point, which is a stable node, for $1/2 < \mu < 1$ and a stable focus for $\mu > 1$. Prove that $\mu = 1/2$ is a bifurcation point.

13.3 Show that

$$\ddot{x} + \mu\dot{x} + 2x + x^2\dot{x} + x^3 = 0$$

has a Hopf bifurcation at $\mu = 0$.

13.4 Show that the iteration scheme

$$x_{n+1} = 1 - \mu x_n(1 - \dot{x}_n)$$

has a stable fixed point $x_0 = 1$ for $\mu < 1$ and that $\mu = 1$ is a bifurcation point where the fixed point $x_0 = 1/\mu$ appears. Show that period-doubling occurs as soon as μ exceeds 3.

13.5 Study the fixed points and period-doubling of

$$x_{n+1} = 1 - \mu x_n^2.$$

13.6 Demonstrate that period-doubling can occur for

$$x_{n+1} = \mu \sin \pi x_n$$

and try to draw a bifurcation diagram for μ between 0 and 0.07.

13.7 Show that, if the iteration

$$x_{n+1} = \exp(-7.5x_n^2) - 0.9$$

is started with $x_1 = 0$, x_n tends to 0.067 approximately but that, if $x_1 = 0.7$, x_n tends to -0.898 approximately.

13.8 Draw a bifurcation diagram for

$$x_{n+1} = \exp(-4x_n^2) + \mu$$

with μ ranging from -1 to 1.

13.9 Estimate the Feigenbaum number for

$$x_{n+1} = 1 - \mu x_n^4.$$

13.10 By changing to polar coordinates and using the Poincaré line show that

$$\dot{x} = -y + \tfrac{1}{2}\mu x - \tfrac{1}{2}x(x^2 + y^2)^2,$$
$$\dot{y} = x + \tfrac{1}{2}\mu y - \tfrac{1}{2}y(x^2 + y^2)^2$$

has a stable limit cycle when $\mu > 0$.

13.11 On a Poincaré plane successive points are related by

$$x_{n+1} = \tfrac{1}{2} y_n,$$
$$y_{n+1} = -x_n + \tfrac{1}{2} \mu y_n - y_n^3.$$

Show that there is a bifurcation at $\mu = 3$. Show that the limit cycle in $0 < \mu < 3$ is stable and becomes of saddle type when μ exceeds 3. Verify that the two other limit cycles for μ just above 3 are stable.

13.12 Demonstrate that the solutions of

$$\dot{x} = -\epsilon x \cos^2 t$$

and the averaged equation stay close together if they are initially close.

13.13 A trajectory in the phase plane of

$$\ddot{x} + \epsilon x = -\epsilon x \cos^2 t$$

starts from the same point as a trajectory of the averaged equations. Show that the trajectories stay close together.

13.14 Transform the first order system derived from

$$\ddot{x} + \epsilon(x^2 - 1)\dot{x} + x = \omega \epsilon E \cos \omega t$$

to polar coordinates and then average. If $E = 1/2$ and $\mu = (\omega^2 - 1)/\omega\epsilon$ show that the averaged equations have three fixed points, two of which are unstable, when $\mu = 0$. Draw a graph of r_0 against μ for μ going from 0 to 1.

Chapter 14

Growth of Tumours

14.1 Introduction

Tumours can arise from the cells of nearly all types of body tissue and this diversity of origin is largely responsible for the wide variety of the structural appearance of tumours. In this chapter we endeavour to describe the main features of tumour growth and one of the possible ways in which to model the processes involved. During the course of discussion we shall draw attention to some of the recent ideas and developments. In the earliest stages of development tumour growth seems to be regulated by direct diffusion of nutrients and wastes from and to surrounding tissue. When a tumour is very small, every cell receives nourishment by simple diffusion and the growth rate is exponential in time. However, this stage cannot be sustained because as a nutrient is consumed its concentration must decrease towards the centre of the tumour. Eventually the concentration of a vital nutrient at the centre will fall below a critical level insufficient to sustain cell life. Then a central necrotic core develops. The rate of growth of the tumour then falls away and it becomes more difficult to obtain nourishment and to dispose of wastes solely by diffusion.

Unfortunately this is not the end of the process. Indeed a majority of tumours exhibit the phenomenon of angiogenesis marking the transition from the relatively harmless and localised avascular state described above to the more dangerous vascular state wherein the tumour develops the ability to proliferate, invade surrounding tissue and metastasize to distant parts of the body.

In its early stages of growth the tumour achieves only a few millimetres in diameter consisting of an outer shell several cell layers thick, which grow and divide. See Figure 14.1.1. As we move into the interior of the tumour the proliferation of cells decreases markedly until we reach a region of quiescent non-dividing cells and further inward until we reach the central core of necrotic debris in various stages of disintegration.

We have discussed various processes of diffusion before, but here we are faced with a completely new situation, namely, that the boundary of the tumour is moving and is unknown except in the initial stages of growth. Indeed the basic problem to be addressed is to formulate a mathematical model that enables us to focus attention on the movement of the tumour outer cell layer, to track its movement and to examine its stability.

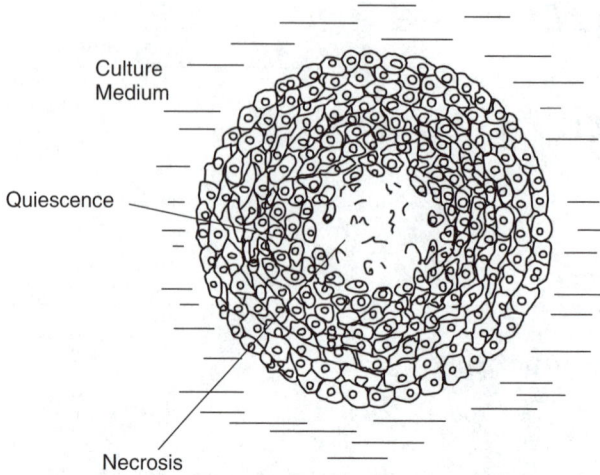

Culture Medium

Quiescence

Necrosis

FIGURE 14.1.1: An avascular tumour cell colony.

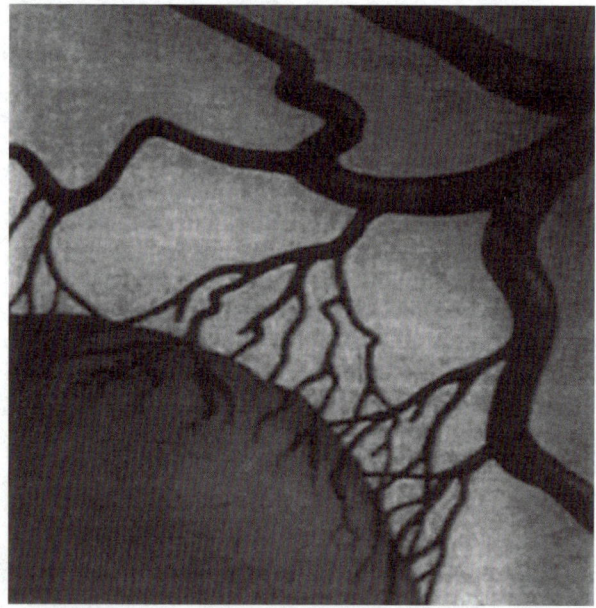

FIGURE 14.1.2: A vascularised tumour as the result of angiogenesis.

The question of stability is extremely important; it is central to determining whether a tumour is non-malignant or like cancer cells which create malignant tumours. Most tumours involve some vascularisation due to angiogenesis. That is the process by which tumours induce blood vessels from the host tissue to sprout capillaries which migrate towards and ultimately penetrate the tumour, providing it with a circulating blood supply. See Figure 14.1.2. To be able to control vascularisation in cancer cells is of paramount importance. It is well documented that cancer cells produce a variety of chemical growth factors called **tumour angiogenesis factors** (or TAF) that stimulate the formation of new capillaries. That is, as the tumour approaches its diffusion-limited size, the TAF initiates angiogenesis. The malignancy becomes vascularised and perfusion replaces diffusion as the dominant mechanism for the supply of nutrients and the removal of wastes. Once the tumour connects with the circulatory system all constraints imposed by diffusion are eliminated and subsequent growth is explosive. If one could block the chemical messages for vascularisation sent from the tumour to surrounding tissue, it might be possible to maintain the tumour indefinitely in its dormant or non-malignant state or even to kill the tumour completely by cutting off its blood supply. Currently there are drugs undergoing clinical trials that are designed specifically to starve malignant tumours. These are called anti-angiogenesis drugs and include angiostatin and endostatin.

Mathematical modelling of tumour growth and the processes of angiogenesis are being actively pursued and hopefully will help bio-medical scientists and clinicians to develop strategies with which to better understand and combat this life threatening disease.

The mathematical model we develop here is concerned with avascular tumour growth under the following simplifying assumptions.

(a) The cell colony and surrounding medium are essentially in a diffusive equilibrium state at all times. The tumour has a three-layer structure comprising an outer layer of live proliferating cells enveloping a thin inner layer of quiescent non-proliferating cells, which, in turn, envelops a large core of necrotic debris.

(b) Cells proliferate as long as the available concentration of nutrient supply denoted by $\sigma(x, y, z, t)$ remains above a critical level σ_1. Cells die when σ falls below a critical level σ_2. In the quiescent region $\sigma_2 < \sigma < \sigma_1$. The thickness h of the layer of live proliferating cells depends on σ_1 and the value of σ at the outer surface of the tumour. Experimental evidence suggests that

$$h = \begin{cases} \nu\sqrt{\sigma - \sigma_1}, & \text{for } \sigma > \sigma_1, \\ 0 & \text{for } \sigma < \sigma_1, \end{cases} \tag{14.1.1}$$

where ν is a positive constant.

(c) If dA is an element of surface area of the tumour, then the incremental volume of live cells $dV = h\,dA$ creates new cell volume at the rate $\beta h\,dA$ where β is a constant. Nutrient is consumed by this volume at the rate $\gamma h\,dA$ where γ is another constant.

(d) Proliferating cells become quiescent when the nutrient supply σ lies in the region $\sigma_2 < \sigma < \sigma_1$ and the rate of gain of quiescent mass per unit volume is constant.

(e) Necrotic debris disintegrates continually into simpler compounds. The rate of loss of necrotic mass per unit volume is constant.

(f) A surface tension force T proportional to the mean curvature K of the boundary keeps the tumour a compact and continuous mass.

(g) The birth or death of cells produces internal pressure differentials which cause the motion of cellular material. This is assumed to be governed by

$$q = -\nabla P \qquad (14.1.2)$$

where $\mathbf{q}(x, y, z, t)$ is the cell velocity and $P(x, y, z, t)$ is proportional to internal pressure. Indeed the colony is assumed to behave like an incompressible fluid composed of cells and cellular debris.

14.2   A mathematical model of tumour growth

In order to develop a mathematical model of tumour growth, we endeavour to combine diffusion processes with the above assumptions, to arrive at a set of equations which allow us to relate the dynamics of the surface of the tumour with variations in the nutrient concentration σ and the internal pressure P.

Suppose the outer surface is represented by the unknown functional equation

$$\Gamma(x, y, z, t) = 0. \qquad (14.2.1)$$

Similarly the outer surface of the necrotic core is represented by the unknown functional equations $\Gamma_N(x, y, z, t) = 0$.

Apply the law of conservation of mass to the elemental volume shown in Figure 14.2.1. This says that since h is small, the mass/volume flow out of the surface dA of the elemental volume dV, namely $(\mathbf{q}_+ \cdot \hat{n} - \mathbf{q}_- \cdot \hat{n})dA$, equals the rate of mass/volume production within this small volume, namely $\beta h dA$. Thus

$$\mathbf{q}_+ \cdot \hat{n} = \mathbf{q}_- \cdot \hat{n} + \beta h, \quad \text{on } \Gamma = 0. \qquad (14.2.2)$$

Similarly the rate of nutrient diffusion into dV (with diffusion coefficient k) through the outer surface is $k\hat{n} \cdot \nabla \sigma dA$, which is equal to the rate at which nutrient is consumed in this small volume, namely $\gamma h dA$. Thus

$$k\hat{n} \cdot \nabla \sigma = \gamma h, \quad \text{on } \Gamma = 0. \qquad (14.2.3)$$

Notice that since $\sigma \leq \sigma_1$ in the quiescent region and the necrotic core, there is no diffusive transport from the interior.

Suppose the proliferation rate of new cells is so large that its product with small quantities, such as outer shell thickness, is of order one, i.e.,

$$\beta h = \beta v \sqrt{(\sigma - \sigma_1)} = \lambda \sqrt{(\sigma - \sigma_1)}, \quad \lambda = O(1)$$

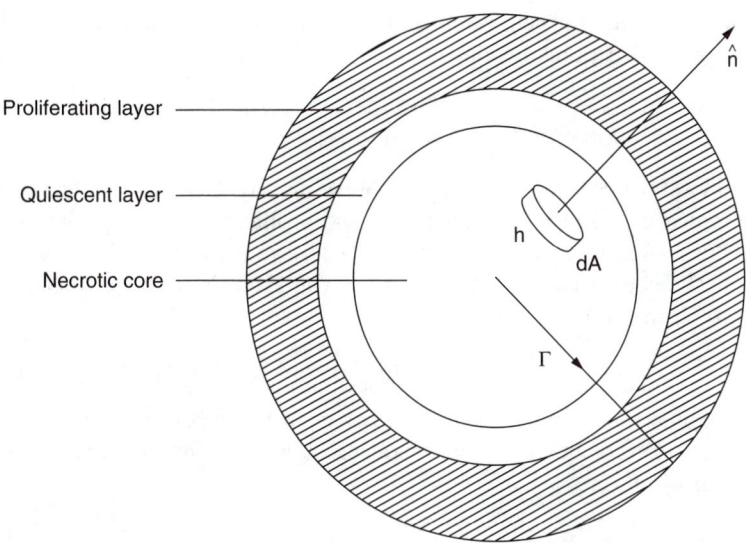

FIGURE 14.2.1: Model of a tumour.

and

$$\gamma h = \gamma \nu \sqrt{(\sigma - \sigma_1)} = \mu \sqrt{(\sigma - \sigma_1)}, \quad \mu = O(1).$$

Next we suppose that the pressure P and tangential velocity components are continuous across each of the surfaces $\Gamma = \Gamma_N = 0$.

For example, at the outer surface $\Gamma = 0$

$$P_+ = P_- = P,$$
$$\mathbf{q}_+ \times \hat{n} = \mathbf{q}_- \times \hat{n}. \tag{14.2.4}$$

From assumption (f) the pressure on the surface of the tumour must equal the surface tension T and thus in turn is proportional to the mean curvature K, i.e.,

$$P = \alpha K \quad \text{on } \Gamma = 0, \tag{14.2.5}$$

where α is a constant.

If a typical point on the outer surface of the tumour is represented by the vector $\mathbf{r}$, then the motion of $\Gamma = 0$ is represented by

$$\frac{d\mathbf{r}}{dt} = \mathbf{q}_+ \tag{14.2.6}$$

where $\Gamma(x, y, z, t) = 0$ is assumed known.

If we let $\mathbf{q}$ denote cell velocity within the tumour and $S(x,y,z,t)$ the cell loss rate at a point inside the tumour then conservation of mass can be written as

$$\nabla \cdot \mathbf{q} = -S. \qquad (14.2.7)$$

The cell loss rate S is modelled in the following way. Cell loss due to apoptosis, programmed cell death, is restricted to the proliferating and quiescent region and occurs at the constant rate S_1. Cell loss due to necrosis is assumed to occur at the constant rate S_2. In terms of the **Heaviside** step function H we can write S in the concise form

$$S(x,y,z,t) = S_1 H(|\mathbf{r}| - |\mathbf{r_N}|) + S_2 H(|\mathbf{r_N}| - |\mathbf{r}|) \qquad (14.2.8)$$

where $\mathbf{r_N}$ is a point on the surface of the necrotic region $\Gamma_N = 0$.

The equation for nutrient concentration σ, which is assumed to be in diffusive equilibrium, is

$$\nabla^2 \sigma = 0, \qquad (14.2.9)$$

outside and within the tumour colony.

There are several problems that can be investigated with this model, including the effect of a nearby source of nutrient or the presence of another tumour cell colony. It is also of importance to examine the effect of the presence of an impermeable wall (e.g., artery). Here we shall consider the surrounding medium to be large in comparison with the tumour size and that there is a constant supply of the nutrient, i.e.,

$$\sigma \to \sigma_\infty \quad \text{as } |\mathbf{r}| \to \infty. \qquad (14.2.10)$$

It is convenient to bring our mathematical model together in the following collection of equations and boundary conditions:

$$\nabla^2 P = S \text{ inside } \Gamma = 0,$$
$$\nabla^2 \sigma = 0 \text{ in the tumour and the surrounding medium}, \qquad (14.2.11)$$

where the first of equations (14.2.11) comes from combining (14.1.2) and (14.2.7).

On the boundary $\Gamma = 0$ of the tumour we have

$$P = \alpha K, \qquad (14.2.12)$$
$$\mathbf{q_+} \cdot \hat{n} = -\hat{n} \cdot \nabla P + \lambda \sqrt{(\sigma - \sigma_1)}, \qquad (14.2.13)$$
$$\mathbf{q_+} \times \hat{n} = -\nabla P \times \hat{n}, \qquad (14.2.14)$$
$$\hat{n} \cdot \nabla \sigma = \mu \sqrt{(\sigma - \sigma_1)}. \qquad (14.2.15)$$

The boundary surface is defined by

$$\frac{d\mathbf{r}}{dt} = \mathbf{q_+} \qquad (14.2.16)$$

and the initial configuration is given by

$$\mathbf{r} = \mathbf{a}, \tag{14.2.17}$$

at $t = 0$ where $\mathbf{a}$ is assumed known.

Furthermore P and σ are continuous together with their normal derivatives across the surface

$$\Gamma_N = 0.$$

Finally

$$\sigma = \sigma_2 \quad \text{on } \Gamma_N = 0. \tag{14.2.18}$$

The set of equations (14.2.11)–(14.2.18) constitutes a **moving boundary problem** which is usually very difficult to solve both analytically and computationally. However, for some prescribed geometric configurations we can solve the equations exactly. This is the subject of the following section.

14.3 A spherical tumour

Suppose that the tumour is initially a sphere of radius a and grows with time, keeping its spherical shape. In this situation an exact solution may be obtained. Because the tumour maintains its spherical shape the equation of the outer surface of proliferating cells can be represented by

$$r = R(t), \tag{14.3.1}$$

where $R(t)$ denotes the radius of the tumour at time t. Clearly $R(0) = a$. Equations (14.2.11) expressed in spherical polar coordinates reduce, because of spherical symmetry, to

$$\frac{1}{r^2} \frac{\partial}{\partial r}\left(r^2 \frac{\partial P}{\partial r}\right) = S, \quad r \leq R(t),$$

$$\frac{1}{r^2} \frac{\partial}{\partial r}\left(r^2 \frac{\partial \sigma}{\partial r}\right) = 0. \tag{14.3.2}$$

In order to proceed with the solution we have to construct appropriate solutions in each of the proliferating cell regions and the necrotic core, respectively.

Region: $R_N < r \leq R$:

Here we have to solve the equations

$$\frac{1}{r^2} \frac{\partial}{\partial r}\left(r^2 \frac{\partial P}{\partial r}\right) = S_1, \quad R_N < r \leq R(t),$$

$$\frac{1}{r^2} \frac{\partial}{\partial r}\left(r^2 \frac{\partial \sigma}{\partial r}\right) = 0. \tag{14.3.3}$$

From these we find that

$$P = S_1 \frac{r^2}{6} + \frac{A_1}{r} + B_1 \tag{14.3.4}$$

and

$$\sigma = \frac{C_1}{r} + D_1, \tag{14.3.5}$$

where A_1, B_1, C_1 and D_1 are constants to be determined by the boundary conditions. Using the condition (14.2.12) we obtain from (14.3.3)

$$S_1 \frac{R^2}{6} + \frac{A_1}{R} + B_1 = \frac{\alpha}{R} \tag{14.3.6}$$

and from (14.2.10) and (14.3.5) we find

$$D_1 = \sigma_\infty. \tag{14.3.7}$$

Region: $0 < r \leq R_N$: In this region we find

$$P = S_2 \frac{r^2}{6} + \frac{A_2}{r} + B_2,$$

$$\sigma = \frac{C_2}{r} + D_2. \tag{14.3.8}$$

Here we require both P and σ to exist at $r = 0$ and so we must take $A_2 = C_2 = 0$. On the boundary $r = R_N$ we demand that

$$S_2 \frac{R_N^2}{6} + B_2 = S_1 \frac{R_N^2}{6} + \frac{A_1}{R_N} + B_1,$$

$$S_2 \frac{R_N}{3} = S_1 \frac{R_N}{3} - \frac{A_1}{R_N^2} \tag{14.3.9}$$

and also

$$C_1 = (\sigma_2 - \sigma_\infty)R_N,$$
$$D_2 = \sigma_2.$$

In principle we now have enough information to solve for all the unknown quantities $A_i, B_i, (i = 1, 2)$, provided we know the boundary radius R_N. Since $A_2 = 0$ the three equations (14.3.6) and (14.3.9) can be solved for A_1, B_1, B_2.

The all important question now is to determine how the boundary of the tumour, $r = R(t)$, evolves with time. To do this we need to use equations (14.2.13) and (14.2.16). First of all combine (14.2.13) and (14.2.15) to get

$$q = -\frac{\partial P}{\partial r} + \frac{\lambda}{\mu} \frac{\partial \sigma}{\partial r}$$

on $r = R(t)$ which when combined with (14.2.16) gives the evolution

$$\frac{dr}{dt} = -\frac{\partial P}{\partial r} + \frac{\lambda}{\mu}\frac{\partial \sigma}{\partial r}. \qquad (14.3.10)$$

We now carry out the task of solving for all the unknown quantities A_1, B_1, B_2. From (14.3.9) we can solve directly for A_1 to get

$$A_1 = (S_1 - S_2)\frac{R_N^3}{3}. \qquad (14.3.11)$$

Knowing A_1 we can use equation (14.3.6) to find B_1, namely,

$$B_1 = \frac{\alpha}{R} - S_1\frac{R^2}{6} - (S_1 - S_2)\frac{R_N^3}{3R}. \qquad (14.3.12)$$

With A_1, B_1 known, the first of equations (14.3.9) can be used to get B_2, i.e.,

$$B_2 = \frac{\alpha}{R} - S_1\frac{R^2}{6} - (S_1 - S_2)\frac{R_N^3}{3R} + (S_1 - S_2)\frac{R_N^2}{2},$$
$$= \frac{\alpha}{R} - S_1\frac{R^2}{6} + (S_1 - S_2)(3R - 2R_N)\frac{R_N^2}{6R}. \qquad (14.3.13)$$

Having determined all the quantities $A_i, B_i, C_i, D_i, i = 1, 2$ we summarise the pressure and nutrient distributions in each of the two regions as follows:

Region: $R_N < r \le R$:

$$P = \frac{\alpha}{R} - \frac{S_1}{6}(R^2 - r^2) + (S_1 - S_2)\frac{R_N^3}{3}\left(\frac{1}{r} - \frac{1}{R}\right),$$
$$\sigma = \frac{(\sigma_2 - \sigma_\infty)}{r}R_N + \sigma_\infty. \qquad (14.3.14)$$

Region: $0 < r \le R_N$:

$$P = \frac{\alpha}{R} + S_2\frac{r^2}{6} - S_1\frac{R^2}{6} + (S_1 - S_2)\frac{R_N^2}{6R}(3R - 2R_N),$$
$$\sigma = \sigma_2. \qquad (14.3.15)$$

To determine the evolution of the outer boundary, $R(t)$, of the tumour we substitute (14.3.14) in (14.3.10) to get

$$\frac{dR}{dt} = -S_1\frac{R}{3} + (S_1 - S_2)\frac{R_N^3}{3R^2} - \frac{\lambda}{\mu}(\sigma_2 - \sigma_\infty)\frac{R_N}{R^2}. \qquad (14.3.16)$$

This nonlinear ordinary differential equation can be solved numerically provided R_N is known. We can find R_N in terms of R by using (14.3.14) in (14.2.15) to obtain the quadratic equation

$$(\sigma_\infty - \sigma_2)^2 R_N^2 + \mu^2(\sigma_\infty - \sigma_2)R^3 R_N - \mu^2(\sigma_\infty - \sigma_1)R^4 = 0, \qquad (14.3.17)$$

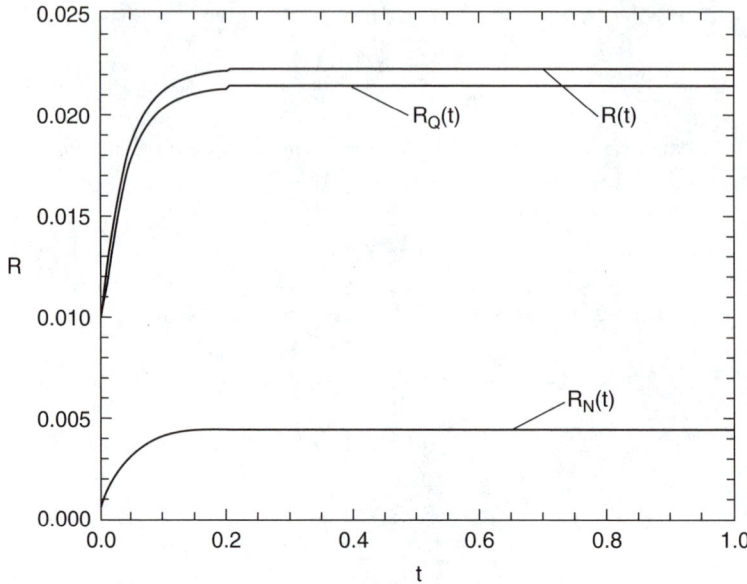

FIGURE 14.3.1:   Evolution of spherical tumour with time.

which has the positive solution

$$R_N = \frac{-\mu^2 R^3 + \mu R^2 \sqrt{\mu^2 R^2 + 4(\sigma_\infty - \sigma_1)}}{2(\sigma_\infty - \sigma_2)}. \tag{14.3.18}$$

This expression for R_N can now be substituted into the ordinary differential equation (14.3.16) for the determination of $R(t)$. Figure 14.3.1 shows the numerical solution of (14.3.16), (14.3.18) and also the outer boundary of the quiescent layer, R_Q, determined by calculating the depth of the thin layer h of proliferating cells, using (14.1.1), for the parameter values $S_1 = 60$, $S_2 = 100$ together with $\lambda = 1$, $\mu = 10$, $\nu = 0.002$ and $\sigma_1 = 0.7$, $\sigma_2 = 0.5$, $\sigma_\infty = 1$.

In the opening section of this chapter we gave a brief account of the processes of angiogenesis whereby a tumour may become vascularised and receive its own blood supply. The growth of the tumour is then freed of its diffusion limiting mechanism and tumour invasion and mestastases occurs. Our model of unvascularised tumour growth can be modified to give some insight into this important phenomenon as follows.

If the tumour has become vascularised then we may assume that both the quiescent layer and the necrotic core are absent. In other words we may assume $R_Q = R_N = 0$. However assumption b) and equation (14.1.1) no longer apply. In other words, cell proliferation may take place throughout the tumour. If we assume cell proliferation due to mitosis occurs at the constant rate S_3 then

(14.2.8) may be replaced by

$$S(x, y, z, t) = (S_1 - S_3)H(|\mathbf{r}|).$$ (14.3.19)

This time the pressure and nutrient concentration within the tumour is given by

$$P = \frac{\alpha}{R} + \frac{(S_3 - S_1)}{6}(R^2 - r^2),$$
$$\sigma = \sigma_\infty.$$ (14.3.20)

Notice that we retain the cell loss due to apoptosis by including the rate S_1. The rate of growth of the tumour is now governed by the simple first order differential equation

$$\frac{dR}{dt} = \frac{(S_3 - S_1)}{3}R,$$ (14.3.21)

subject to the initial condition $R(0) = a$.
Thus

$$R(t) = a \exp\left(S_3 - S_1\right)t/3.$$ (14.3.22)

From this simple result we deduce that if the rate of cell production exceeds the rate of apoptosis ($S_3 > S_1$) then the tumour grows exponentially fast, whereas if apoptosis exceeds cell mitosis ($S_3 < S_1$) the tumour decays exponentially with time.

14.4 Stability

Surface tension, as we have already mentioned, plays an important role in maintaining the compactness of the tumour. Any perturbation that allows surface tension to be overcome by pressure forces could be an important feature in determining whether the tumour becomes vascularised and consequently malignant.

In other words, the question of whether the spherical tumour is stable with respect to small deviations from its spherical shape is of fundamental importance. The models described above can be examined for stability under such perturbations and, in the case of instability, lead to some quite dramatic deviations from the initial spherical shape. In some situations the tumour may even break into two or more pieces.

A detailed analysis of stability is beyond the scope of this book. Nevertheless we can indicate how such an analysis can be performed. To do this we motivate the ideas by considering a further modification of our general model. Specifically we consider a tumour with a very small proliferating outer layer enclosing a large necrotic core. In fact this model is the one originally developed by H.P. Greenspan (see the notes at the end of this chapter). In other words we take the cell loss rate to be

$$S(x, y, z, t) = S_2. \tag{14.4.1}$$

Carrying out the by now familiar analysis we find

$$P = \frac{S_2}{6}(r^2 - R^2) + \frac{\alpha}{R}, \quad r \le R(t) \tag{14.4.2}$$

and

$$\sigma = \frac{D}{r} + \sigma_\infty, \quad r \ge R(t), \tag{14.4.3}$$

where the constant D is determined from the boundary conditon (14.2.15) to be

$$D = \frac{1}{2}\mu R^2\{\mu R - \sqrt{[\mu^2 R^2 + 4(\sigma_\infty - \sigma_2)]}\}. \tag{14.4.4}$$

Notice that since there is no quiescent layer we have set $\sigma_1 = \sigma_2$. The rate of growth of the radius of the tumour is determined from (14.2.13) and (14.2.16) as the solution to the ordinary differential equation

$$\frac{dR}{dt} = -\frac{1}{2}S_2 R + \lambda\sqrt{(\sigma_\infty - \sigma_2) + \frac{1}{2}\mu R[\mu R - (\mu^2 R^2 + 4(\sigma_\infty - \sigma_2))^{1/2}]}. \tag{14.4.5}$$

Without solving this equation we can estimate the ultimate size of a *stable* spherical tumour by setting $\frac{dR}{dt} = 0$ to get the limiting size

$$R_\infty = \frac{3\lambda}{S_2}\left(\frac{\sigma_\infty - \sigma_2}{3\lambda\mu/S_2 + 1}\right)^{1/2}. \tag{14.4.6}$$

To indicate how a stability analysis should proceed let $\bar{P}(r, t)$, $\bar{\sigma}(r, t)$ and $\bar{R}(r, t)$ be the quantities determined by (14.4.2), (14.4.3) and the solution to (14.4.5). Suppose these quantities are perturbed by amounts $\epsilon\tilde{P}(r, \theta, t)$, $\epsilon\tilde{\sigma}(r, \theta, t)$ and $\epsilon\tilde{\xi}(\theta, t)$ independent of the azimuthal angle ϕ and where ϵ is a small parameter. Under these perturbations the total pressure and nutrient concentration are represented by

$$P(r, \theta, t) = P(r, t) + \epsilon\tilde{P}(r, \theta, t),$$
$$\sigma(r, \theta, t) = \sigma(r, t) + \epsilon\tilde{\sigma}(r, \theta, t) \tag{14.4.7}$$

and the surface of the tumour is given at any time by

$$\Gamma(r, \theta, t) = r - R(t) - \epsilon\xi(\theta, t) = 0. \tag{14.4.8}$$

On substituting (14.4.7) and (14.4.8) into equations (14.2.11)–(14.2.17) and equating coefficients of ϵ to zero, we have the following problem to solve:

$$\nabla^2 \tilde{P} = 0, \quad r \le R(t),$$
$$\nabla^2 \tilde{\sigma} = 0, \quad r \ge R(t), \tag{14.4.9}$$

with $\tilde{\sigma} \to 0$ as $r \to \infty$, while on $r = R(t)$,

$$\frac{\partial\xi}{\partial t} = -\left(\frac{\partial^2 \tilde{P}}{\partial r^2}\xi + \frac{\partial \tilde{P}}{\partial r}\right) + \frac{\lambda}{2\sqrt{\tilde{\sigma}}}\left(\xi\frac{\partial\tilde{\sigma}}{\partial r} + \tilde{\sigma}\right),$$

$$\frac{\partial^2 \tilde{\sigma}}{\partial r^2}\xi + \frac{\partial\tilde{\sigma}}{\partial r} = \frac{\mu}{2\sqrt{\tilde{\sigma}}}\left(\xi\frac{\partial\tilde{\sigma}}{\partial r} + \tilde{\sigma}\right),$$

$$\frac{\partial\tilde{P}}{\partial r}\xi + \tilde{P} = -\frac{\alpha}{2R^2}\left(\frac{\partial}{\partial\eta}(1 - \eta^2)\frac{\partial\xi}{\partial\eta} + 2\xi\right), \tag{14.4.10}$$

where $\eta = \cos\theta$. The solutions of (14.4.10) are time-dependent multiples of harmonic functions from which the equation for $\xi(t)$ can be solved to show, for a range of parameter values, that if the harmonics are of sufficiently high order then $\xi(t)$ has exponential growth. The tumour is then unstable and radically departs from its original spherical shape no matter how small ϵ may be.

14.5 Notes

The models of tumour growth discussed in this chapter were inspired by the model due to H.P. Greenspan, On the growth and stability of cell cultures and solid tumors, *J. Theor. Biol.*, **56**, 229–242, 1976 and is the outcome of a previous article by the same author, Models for the growth of a solid tumor by diffusion, *Stud. Appl. Math.*, **51**, 317–340, 1972.

In the former paper, Greenspan carries out a detailed stability analysis along the lines briefly described in Section 14.4 and also shows that tumour cell colonies which share the same nutrient supply repel each other and move apart. The model is quite versatile and can be modified to describe several important problems; for example, that of describing the movement of tumours in the presence of solid boundaries or the effect on growth with changes in the nutrient supply. Similar problems can be posed for the models developed in this chapter.

Since the work of Greenspan there has been a considerable development in our understanding of growth of solid avascular tumours both in *in vitro* and in *in vivo* studies. This has led to other mathematical models being developed.

Among such models are those that model more closely the nutrient supply and diffusion processes. See H.M. Byrne and M.A.J. Chaplain, Free boundary value problems associated with the growth and development of multicellular spheroids, *Eur. J. Appl. Math.*, **8**, 639–658, 1997. The techniques used to study these more recent models are essentially based on the ideas developed in this chapter.

We have mentioned in the introduction to this chapter the important role of angiogenesis, the process whereby the tumour is able to acquire its own blood supply from host tissue and nearby blood vessels. Many of the fundamental discoveries associated with tumour angiogenesis were made by J. Folkman and his colleagues (see the biography *Dr. Folkman's War* by Robert Cooke, Random House, New York, 2000). Mathematical models of angiogenesis of varying degrees of complexity have and continue to be developed; see for example M.A.J. Chaplain and A.R.A. Anderson, Modelling the growth and form of capillary networks in *On Growth and Form*, M.A.J. Chaplain, G.D. Singh, and J.C. McLachlan, Eds. John Wiley & Sons, 1999 and H.A. Levine, S. Pamuk, B.D. Sleeman and M. Nilsen-Hamilton, Mathematical modeling of capillary formation and development in tumor angiogenesis: Penetration into the stroma, *Bull. Math. Biol.*, **63**, 801–863, 2001.

Further Reading

Alberts, B., D. Bray, J. Lewis, M. Raff, K. Roberts, and J. D. Watson, *Molecular Biology of the Cell*, 3rd ed., Garland, New York and London, 1994.

Folkman, J., The vascularisation of tumors. *Sci. Am.*, **234**, 58–64, 1976.

King, R. J. B., *Cancer Biology*, 2nd ed., Harlow, London and NY: Prentice Hall, 2000.

Paweletz, N. and M. Knierim, Tumor related angiogenesis, *Crit. Rev. Oncol. Hematol.*, **9**, 197–242, 1989.

Sherratt, J. A. and M. A. J. Chaplain, A new mathematical model for avascular tumour growth. *J. Math. Biol.*, **43**, 291–312, 2001.

Sleeman, B. D., Solid tumour growth: a case study in mathematical biology, *Nonlin. Math. Appl.*, Ed. P. J. Aston, C.U.P. 237–256, 1996.

Exercises

14.1 Write a computer programme to solve the differential equation (14.3.16) for the evolution of the outer boundary, $R(t)$, of a tumour subject to the initial condition $R(0) = 1$ and where R_N is given by (14.3.18). Experiment by choosing various values of $S_i, i = 1, 2$ and $\sigma_j, j = 1, 2$ and σ_∞ to investigate the effects of proliferation, quiescence and necrosis.

14.2 Verify the steady state radius given by (14.4.6).

14.3 A tumour colony is cultured in a circular dish of radius A. Assume the nutrient concentration σ has the constant value σ_A at the edge of the dish. If the tumour colony is initially a circle of radius a and is assumed to grow radially, formulate a mathematical model of colony growth under the same assumptions as for the growth of a spherical tumour colony.

Hint: formulate the model using polar coordinates.

14.4 In problem 14.3 show that the outer boundary of the colony satisfies the differential equation

$$\frac{dR}{dt} = -S_1 \frac{R}{2} - (S_2 - S_1)\frac{R_N^2}{2R} + \frac{\lambda}{\mu}\frac{(\sigma_A - \sigma_2)}{R\ln(A/R_N)},$$

where $R(t)$ and $R_N(t)$ are related by

$$\frac{(\sigma_A - \sigma_2)^2}{R^2}$$
$$= \mu^2[(\sigma_A - \sigma_2)\ln(R/A)\ln(A/R_N) + (\sigma_A - \sigma_1)(\ln(A/R_N))^2]$$

14.5 Investigate the growth of a fully vascularised tumour colony growing in a circular dish.

14.6 A vascularised tumour colony in a laboratory experiment grows along a straight narrow tube. Ignoring any curvature at the growing boundary, determine the rate of growth of the tumour colony.

Chapter 15

Epidemics

15.1 The Kermack-McKendrick model

The problem of epidemics is to assess how a group of individuals with a communicable infection spreads the disease to a population able to catch it. The model constructed depends upon the assumption made about the disease and the behaviour of the population. One possibility has been discussed (Section 5.2) and the aim is now to amplify the study, as well as to take account of more characteristics of the process of infection.

In the Kermack-McKendrick model, the population is presumed to be constant in size and to be divided into three classes. There are I infected individuals who can pass on the disease to others and s susceptibles who have yet to contract the disease and become infectious. The remaining group contains r members who have been infected but cannot transmit the disease for some reason, e.g., they have been isolated from the rest of the population. The governing equations will be taken as

$$\dot{s} = -asI, \qquad (15.1.1)$$
$$\dot{I} = asI - bI, \qquad (15.1.2)$$
$$\dot{r} = bI \qquad (15.1.3)$$

with a and b positive constants.

The basis for (15.1.1) is that the susceptibles become infected at a rate that is proportional to the number of contacts between individuals of s and I, assuming that contact depends only on the numbers of each group, i.e., there is uniform mixing of the population. The assumption in (15.1.3) is that the rate at which individuals become unable to transmit the disease is proportional to the number infected. It represents some kind of average of the process in which particular individuals take different lengths of time to reach a state in which they neither contract nor pass on the infection.

If now we add on the hypothesis of the constancy of the population, we have

$$s(t) + I(t) + r(t) = N \qquad (15.1.4)$$

where N is the population's invariable size. A derivative of (15.1.4) then adds (15.1.2) to (15.1.1) and (15.1.3).

At the beginning, $t = 0$ and $r = 0$ so that

$$s(0) + I(0) = N. \qquad (15.1.5)$$

There must be some infected and some available for infection at the start, so $I(0) > 0$ and $s(0) > 0$.

Only solutions in which s, I and r are non-negative are of concern. Therefore, (15.1.1) implies that $\dot{s} < 0$ when both infected and susceptibles are present. Hence s decreases steadily and

$$s(t) < s(t_1) < s(0) \qquad (15.1.6)$$

for $t > t_1 > 0$. The constant diminution of s and the fact that s must be non-negative means that, as $t \to \infty$, s must tend to a limit (which may be zero), i.e., $s(\infty) = \lim_{t\to\infty} s(t)$ exists.

From (15.1.2), $\dot{I} < 0$ if $as < b$. In view of (15.1.6) it follows that, if $as(0) < b$, $\dot{I} < 0$ for all t and the infection is eventually wiped out. Thus there is a critical level, or **threshold value,** of b/a which the initial population of susceptibles must exceed if the epidemic is to spread. The threshold will be low if $b \ll a$, i.e., the rate at which immunity is conferred is small enough for the disease to prosper.

According to (15.1.3), r increases monotonically while (15.1.4) forces $r(t) \le N$. Hence $r(\infty) = \lim_{t\to\infty} r(t)$ exists. It follows from (15.1.4) and what has been proved about s that $I(\infty) = \lim_{t\to\infty} I(t)$ also exists. The quantity $\{I(\infty) - r(\infty)\}/N$ is a measure of the extent to which the infection swept through the population and so now we attempt to estimate the limits.

From (15.1.1) and (15.1.3)

$$\frac{ds}{dr} = -\frac{as}{b}$$

so that

$$s = s(0)e^{-ar/b}. \qquad (15.1.7)$$

Since $r \le N$, we are forced to have $s \ge s(0)e^{-aN/b}$ and therefore $s(\infty) > 0$, i.e., there are always susceptibles available. Thus *some individuals never suffer from the disease*; there are still susceptibles available when the disease stops spreading.

The trajectories in the phase plane for s and I can be drawn. They are solutions of, according to (15.1.1) and (15.1.2),

$$\frac{dI}{d\dot{s}} = -1 + \frac{b}{as}$$

whence, on account of (15.1.5),

$$I = N - s + (b/a)\ln\{s/s(0)\}.$$

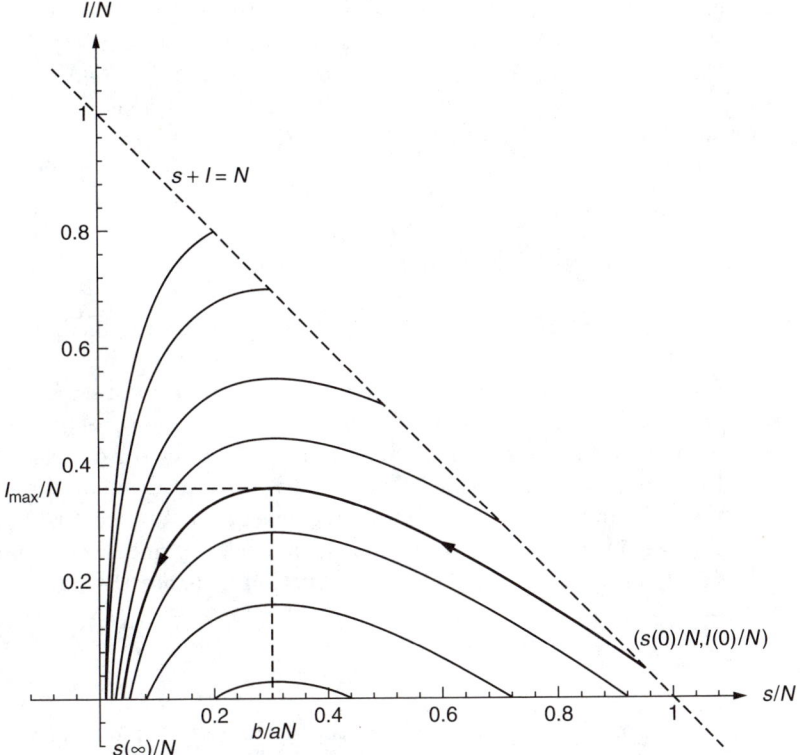

FIGURE 15.1.1: The trajectories for an epidemic.

They are plotted in Figure 15.1.1, the direction of the arrows being dictated by the steady decrease of s. Evidently, $s(t) - s(\infty) > 0$ and $I(t) \to 0$ as $t \to \infty$ (there are no critical points off $I = 0$). The consequence of this, (15.1.7) and (15.1.4), is that

$$s(\infty) = s(0) \exp[-a\{N - s(\infty)\}/b], \qquad (15.1.8)$$

a transcendental equation to determine $s(\infty)$. The equation is satisfied by only one positive value of $s(\infty)$ less than b/a. Once $s(\infty)$ is known, $r(\infty)$ can be calculated from $r(\infty) = N - s(\infty)$ and the spread of the infection measured by $r(\infty)/N$ or $1 - s(\infty)/N$.

15.2   Vaccination

In the preceding section, it has been implicitly assumed that the individuals who are naturally immune are small enough in number to be neglected and we shall continue to adopt this hypothesis. However, if a calculation by the previous model suggests that the infection will spread to unacceptable levels, we may decide to introduce some immunity by means of vaccination.

For a simple model assume that vaccination removes an individual instantaneously from s or r without joining the group I and that a device is available that prevents the infected from being vaccinated. There is then a group v of vaccinated and we have

$$\dot{s} = -asI - \alpha(t), \tag{15.2.1}$$
$$\dot{v} = \alpha(t) \tag{15.2.2}$$

where (15.2.1) replaces (15.1.1) but (15.1.2) and (15.1.3) are retained, α being the vaccination rate.

A decision on the function to be selected as the vaccination rate is not at all easy. Any vaccination programme involves cost through the employment of people, equipment and supplies. Against this must be set the damage that is caused to society by the infection. It may be desirable to vaccinate in order to limit the total number who are infected in a given time interval, or to keep the peak of those infected at any particular instant below some level, or both. To ensure that no more than N_1 of the population contract the disease for $0 \le t \le T$ we want

$$r(T) + I(T) \le N_1 \tag{15.2.3}$$

and, to force the peak of infection below N_2 in the same time interval, we need

$$\max_{0 \le t \le T} I(t) \le N_2. \tag{15.2.4}$$

To control the disease in this way, $\alpha(t)$ would have to be chosen so that (15.2.3) and (15.2.4) were satisfied while the cost of vaccination was kept to a minimum. This is a problem in **dynamic programming** for which there are available techniques, but they are beyond the scope of this book.

Before pursuing the model too far for a particular disease, it would be necessary to verify that it complied with the two hypotheses on vaccination, for vaccination may not grant immediate immunity and there may be no acceptable way of determining those already infected.

15.3   An incubation model

In the Kermack-McKendrick model of Section 15.1, an individual is capable of passing on the infection as soon as he or she has succumbed to it. For some diseases there is an incubation period during which an individual has become infected but cannot communicate the disease to others. The incorporation of this feature into the model of an epidemic will be considered in this section. At the same time, the possibility that an individual may be infectious for not more than a finite time will be allowed for.

Let $s(t)$ be the number of those who are susceptible but who have not yet been exposed to the disease and let $E(t)$ be the number who have been exposed to but cannot yet transmit the infection. The groups I and r are defined as before.

Continue to assume that there is homogeneous mixing of the population so that the rate of exposure is proportional to the number of contacts between individuals of s and I. Then

$$\dot{s} = -asI. \tag{15.3.1}$$

Thus (15.1.6) is still valid and $s(\infty)$ exists. Also a solution of (15.3.1) satisfies

$$s(t) = s(0) \exp\left\{-a \int_0^t I(u)du\right\}. \tag{15.3.2}$$

Suppose that the epidemic is started by a number of infectious individuals entering a population which has been unexposed previously to the disease. Assume that we know how many of these infectious individuals will remain infectious as time varies. Until the population produces members who are infectious

$$I(t) = I_0(t) \tag{15.3.3}$$

where $I_0(t)$ is a known function.

Assume now that an individual who contracts the infection at time t does not become infectious until time $t+T$ so that T specifies the incubation period. Then, for $0 \le t \le T$,

$$E(t) = s(0) - s(t) \tag{15.3.4}$$

and, for $t \ge T$,

$$E(t) = s(t-T) - s(t). \tag{15.3.5}$$

Keeping the population constant entails

$$s(t) + E(t) + I(t) + r(t) = N. \tag{15.3.6}$$

While $t \le T$, incubation prevents any new infectious individuals from appearing. Therefore, (15.3.3) is valid and $s(t)$ can be found from (15.3.2). Then $E(t)$ follows from (15.3.4) and $r(t)$ from (15.3.6). Thus, all quantities are known during the initial incubation period.

When $t \geq T$, some of those who have been incubating the disease will have become infectious. In fact, all those who were infected up to the time $t - T$ will be infectious. Consequently, (15.3.3) must be modified and replaced by

$$I(t) = I_0(t) + s(0) - s(t - T). \tag{15.3.7}$$

Substitution of (15.3.7) into (15.3.1) leads to a differential-difference equation for $s(t)$ but its solution will not be required for our purpose.

Now introduce the assumption that an individual remains infectious only for the time interval σ. A necessary consequence is that $I_0(t) = 0$ for $t > \sigma$ because all those who started the epidemic were infectious from the beginning. Since none of those infected in the initial incubation period cease being infectious for $t \leq T + \sigma$, the equation (15.3.7) is valid for $T \leq t \leq T + \sigma$. However, when $t \geq T + \sigma$, some of the infected are no longer infectious; they are the ones who were infected up to the time $t - T - \sigma$. Therefore, their numbers must be removed from the infectious and (15.3.7) must be replaced by

$$I(t) = I_0(t) + s(t - T - \sigma) - s(t - T). \tag{15.3.8}$$

Actually, the term $I_0(t)$ could be dropped because it is zero for $t > \sigma$ and, a fortiori, for $t \geq T + \sigma$.

To determine $s(\infty)$ the integral in (15.3.2) has to be calculated. Now, for $t > T + \sigma$,

$$\int_0^t I(u)du = \int_0^T I(u)du + \int_T^{T+\sigma} I(u)du + \int_{T+\sigma}^t I(u)du.$$

Insert (15.3.3), (15.3.7) and (15.3.8), respectively, into the three integrals. Then, after a change of variable of integration where s is involved,

$$\int_0^t I(u)du = \int_0^\sigma I_0(u)du + \sigma s(0) - \int_{t-T-\sigma}^{t-T} s(u)du$$

since $I_0(u)$ vanishes for $u > \sigma$. Let $t \to \infty$ so that both limits in the integral of $s(u)$ become large. But it has been pointed out that $s(u) \to s(\infty)$ as $u \to \infty$ so that $s(u)$ can be replaced effectively by $s(\infty)$ in the integral, i.e.,

$$\int_{t-T-\sigma}^{t-T} s(u)du \sim s(\infty) \int_{t-T-\sigma}^{t-T} du = \sigma s(\infty)$$

as $t \to \infty$. Hence

$$\int_0^\infty I(u)du = \int_0^\sigma I_0(u)du + \sigma\{s(0) - s(\infty)\}.$$

Consequently, we deduce from (15.3.2) that

$$s(\infty) = s(0) \exp\left[-a \int_0^\sigma I_0(v)dv - a\sigma\{s(0) - s(\infty)\}\right], \tag{15.3.9}$$

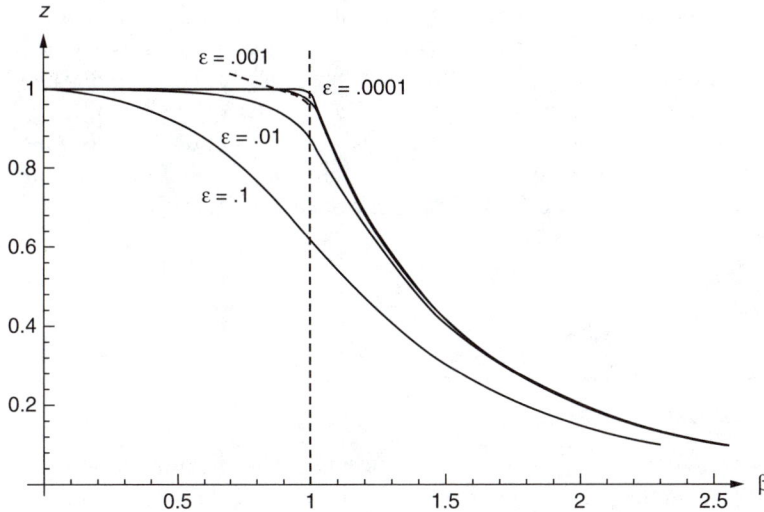

FIGURE 15.3.1: Variation of $z = s(\infty)/s(0)$ with β.

which constitutes an equation to determine $s(\infty)$.

Equation (15.3.9) can be expressed in the form

$$z = \exp\{\beta(z - 1 - \epsilon)\}, \qquad (15.3.10)$$

where $z = s(\infty)/s(0), \beta = a\sigma s(0)$ and

$$\epsilon = (a/\beta) \int_0^\sigma I_0(v)dv.$$

The left-hand side of (15.3.10) is 0 at $z = 0$ and 1 at $z = 1$, whereas the right-hand side is positive and less than 1 (since $\epsilon > 0$), respectively, at the two points. Hence there is a positive root with $z \leq 1$, i.e., $s(\infty)/s(0) \leq 1$ as there should be. The root does not differ by much from $z = 1$ when $\beta\epsilon$ is fairly small. Some graphs of the value of z for various values of β and ϵ are shown in Figure 15.3.1. They are knee-shaped with a distinct kick near $\beta = 1$ when ϵ is very small. Interpreting ϵ as a measure of the infectiousness of the initial infectives, and β as the number of susceptibles likely to be infected by each infective, we see that, when the initial infectives are only a very small proportion of the population, the infection tends to die out without altering the susceptibles much if $\beta < 1$, but to originate a substantial epidemic if $\beta > 1$. This is a kind of threshold effect. The effect becomes less pronounced for higher values of ϵ, indicating that when there are sufficient initial infectives there is a tendency for a significant epidemic to occur.

Various refinements to the foregoing model have been proposed, some of which are in the exercises. One possibility is that the parameter a should be made a function of time in order to permit seasonal and cyclic changes. For example, measles has relatively high and low incidences in alternate years, whereas chickenpox recurs annually. It is more probable that the low incidence of measles is due to periodic alteration of a than to a threshold effect.

The model can be generalised to the case of two populations as when parasites transmit disease (e.g., mosquitoes infecting humans with malaria) or when an infection is transferred from one sex to another, but details will be omitted.

15.4   Spreading in space

A basic hypothesis in the preceding sections is that the population is mixing thoroughly, so that there is no distinction between the individuals in one place and those in another. When this is not so, the disease may spread faster in some parts than in others and it is necessary to allow the variables to depend on space as well as time. For simplicity, we shall assume that only one space variable is involved, though most of the ideas carry over to more general cases.

Take x as the single space variable. Then, in the Kermack-McKendrick model, s and I will be functions of t and x giving the distributions of susceptibles and infected, respectively, at time t at the position x. Moreover, it is necessary to allow for the effect of the infected at one position on the susceptibles at another place. Therefore, replace (15.1.1) by

$$\frac{\partial s}{\partial t} = -\left\{ \int_{-\infty}^{\infty} F(x, x')I(t, x')dx' \right\} s(t, x) \tag{15.4.1}$$

where the non-negative function $F(x, x')$ is a measure of the influence of the infected at the position x' on the susceptibles at x. The integration permits the possibility of all those infected at time t affecting the suceptibles at x. The corresponding replacements for (15.1.2) and (15.1.3) are

$$\frac{\partial I}{\partial t} = \left\{ \int_{-\infty}^{\infty} F(x, x')I(t, x')dx' \right\} s(t, x) - g(t, x)I(t, x) \tag{15.4.2}$$

and

$$\partial r/\partial t = g(t, x)I(t, x), \tag{15.4.3}$$

respectively.

The presence of F in the system (15.4.1)–(15.4.3) makes the system highly flexible and offers the ability to cover many different situations. Unfortunately, this very flexibility makes the system awkward from an experimental point of view. Unless the structure of F is very simple it is unlikely that its functional behaviour can be determined by experimental observations alone. The best approach is to make plausible assumptions about F based on any information available and compare the consequent predictions with experiment. For instance, it is known that certain epidemics have a propensity for spreading like a wave travelling in space. One possible supposition on F that reproduces such behaviour will be discussed now.

A plausible hypothesis for some diseases is to imagine that the transmission of infection depends only on the distance between individuals. Furthermore, it seems reasonable to suppose that the transmission will be relatively poor unless the individuals are close to one another. This suggests trying

$$F(x, x') = \begin{cases} f(|x - x'|) & (|x - x'| < \delta) \\ 0 & (|x - x'| \geq \delta) \end{cases}$$

where δ is a small quantity. Then

$$\int_{-\infty}^{\infty} F(x, x') I(t, x') dx' = \int_{x-\delta}^{x+\delta} f(|x - x'|) I(t, x') dx'$$

$$= \int_{-\delta}^{\delta} f(|u|) I(t, x + u) du$$

on putting $x' = x + u$. On account of the smallness of δ introduce the approximation

$$I(t, x + u) = I(t, x) + u \frac{\partial}{\partial x} I(t, x) + \frac{1}{2} u^2 \frac{\partial^2}{\partial x^2} I(t, x).$$

As a result

$$\int_{-\infty}^{\infty} F(x, x') I(t, x') dx' = I(t, x) \int_{-\delta}^{\delta} f(|u|) du + \frac{1}{2} \frac{\partial^2 I}{\partial x^2} \int_{-\delta}^{\delta} u^2 f(|u|) du$$

since $uf(|u|)$ is an odd function of u. With this approximation equations (15.4.1) and (15.4.2) convert to

$$\frac{\partial s}{\partial t} = -s \left(\theta I + \phi \frac{\partial^2 I}{\partial x^2} \right), \qquad (15.4.4)$$

$$\frac{\partial I}{\partial t} = -\frac{\partial s}{\partial t} - gI \qquad (15.4.5)$$

where θ and ϕ are positive constants defined by

$$\theta = \int_{-\delta}^{\delta} f(|u|) du, \qquad \phi = \frac{1}{2} \int_{-\delta}^{\delta} u^2 f(|u|) du.$$

The partial differential equations (15.4.4) and (15.4.5) constitute the **diffusion approximation** to the problem of the spatial spread of an epidemic.

Let us examine whether the diffusion approximation is capable of sustaining a wave-like epidemic. If it is, then there must be solutions of (15.4.4) and (15.4.5) which are progressing waves. According to Section 10.7 a progressing wave is a function of $x - ct$ where c is the speed of the wave. Consequently, solutions of the form

$$s(t, x) = s_1(x - ct), \qquad I(t, x) = I_1(x - ct)$$

are sought in which the constant c is real and positive. Whether c can be chosen freely or is subject to limitations will have to be investigated.

Write $\zeta = x - ct$ and substitute the assumed progressive waveforms for s and I into the partial differential equations. There results

$$cs_1' = (\theta I_1 + \phi I_1'')s_1, \tag{15.4.6}$$
$$cI_1' = -cs_1' + gI_1 \tag{15.4.7}$$

where the primes signify derivatives with respect to ζ.

Divide (15.4.6) by s_1 and integrate with respect to ζ starting from $\zeta = A$. Then

$$c \ln\{s_1(\zeta)/s_1(A)\} = \int_A^\zeta \{\theta I_1(u) + \phi I_1''(u)\}du$$

$$= \theta \int_A^\zeta I_1(u)du + \phi\{I_1'(\zeta) - I_1'(A)\}. \tag{15.4.8}$$

Integration of (15.4.7) gives, when g is a constant,

$$c\{I_1(\zeta) + s_1(\zeta) - I_1(A) - s_1(A)\} = g \int_A^\zeta I_1(u)du. \tag{15.4.9}$$

Elimination of the integral from (15.4.8) and (15.4.9) leads to

$$\phi I_1'(\zeta) = \phi I_1'(A) + c \ln\{s_1(\zeta)/s_1(A)\}$$
$$-\{\theta c/g\}\{I_1(\zeta) + s_1(\zeta) - I_1(A) - s_1(A)\}, \tag{15.4.10}$$

which is a differential equation of the first order for I_1. Insertion of (15.4.10) into (15.4.7) gives

$$s_1'(\zeta) = (g/c)I_1(\zeta) - I_1'(A) - (c/\phi)\ln\{s_1(\zeta)/s_1(A)\}$$
$$+ \{\theta c/g\phi\}\{I_1(\zeta) + s_1(\zeta) - I_1(A) - s_1(A)\}. \tag{15.4.11}$$

In this way (15.4.6) and (15.4.7) are transformed into the first-order system (15.4.10) and (15.4.11)

Since the spread of the epidemic is dictated by the motion of the wave we do not expect any individuals to be infected at the position x until the wave arrives. The number of susceptibles at that position before the advent of the wave is fixed by the initial state of the susceptibles. Since $t \to -\infty$ corresponds to the time before the wave gets going these conditions are met by requiring: $s_1(\infty)$ to be some positive constant and $I_1(\infty) = 0, I_1'(\infty) = 0$ assuming that the wave starts smoothly. Letting $A \to \infty$ in (15.4.10) and (15.4.11) we have

$$I_1'(\zeta) = (c/\phi) \ln\{s_1(\zeta)/s_1(\infty)\}$$
$$- \{\theta c/\phi g)(I_1(\zeta) + s_1(\zeta) - s_1(\infty)\}, \tag{15.4.12}$$
$$s_1'(\zeta) = (g/c)I_1(\zeta) - (c/\phi) \ln\{s_1(\zeta)/s_1(\infty)\}$$
$$+ \{\theta c/g\phi\}\{I_1(\zeta) + s_1(\zeta) - s_1(\infty)\}. \tag{15.4.13}$$

The discussion of (15.4.12) and (15.4.13) is simplified somewhat by switching to the variables $X(\zeta) = s_1(\zeta)/s_1(\infty)$ and $Y(\zeta) = I_1(\zeta)/s_1(\infty)$. This has the advantage that only values of $X(\zeta)$ between 0 and 1 need be considered. In terms of the new variables, (15.4.12) and (15.4.13) become

$$X'(\zeta)/\alpha = -\ln X(\zeta) + \beta\{X(\zeta) - 1\} + (\beta + \gamma)Y(\zeta), \tag{15.4.14}$$
$$Y'(\zeta)/\alpha = \ln X(\zeta) - \beta\{X(\zeta) - 1\} - \beta Y(\zeta) \tag{15.4.15}$$

where

$$\alpha = c/\phi s_1(\infty), \qquad \beta = \theta s_1(\infty)/g, \qquad \gamma = \phi g s_1(\infty)/c^2. \tag{15.4.16}$$

Note that the constants α, β and γ are all positive.

Once the wave has passed a particular position there should be no infected there and the number of susceptibles should have settled down to a steady level (which may be zero). This will be the state at a point as $t \to \infty$ and so the conditions $Y(-\infty) = 0$ and $X'(-\infty) = 0$ are imposed. It follows from (15.4.14) that the final number of susceptibles, obtained from $X(-\infty)$, is determined by

$$\ln X(-\infty) = \beta\{X(-\infty) - 1\}. \tag{15.4.17}$$

The exponential of (15.4.17) is the same as (15.3.10) with $\epsilon = 0$, which enables us to say that, when $\beta \leq 1$, the only solution of (15.4.17) in the permitted range of X is $X(-\infty) = 1$. But, in that case, the susceptibles have been unaffected and there has been no epidemic. Therefore, in order for an epidemic to occur, it is necessary that $\beta > 1$, i.e.,

$$s_1(\infty) > g/\theta \tag{15.4.18}$$

from (15.4.16). Thus (15.4.18) is a necessary condition for the existence of a wave and sets a lower bound on the initial density of susceptibles in order for the population to support a progressing epidemic. Hence there is a *threshold* below which the susceptibles are inadequate for a wave-like epidemic.

When (15.4.18) is satisfied there is a solution of (15.4.17) with $X(-\infty) < 1$ and, furthermore, $X(-\infty) < 1/\beta$ (see Exercise 15.5). Consequently,

$$s_1(-\infty) < g/\theta. \tag{15.4.19}$$

Comparison with (15.4.18) indicates that, after the passage of the wave, the number of susceptibles is insufficient to sustain the wave again. In other words, *according to this model, once a wave epidemic has swept through a population it cannot do so again until more susceptibles have been added to the population.*

So far nothing has been said about the speed at which the epidemic travels through the population. Some information can be garnered from the phase plane for X and Y on the assumption that (15.4.18) and (15.4.19) hold. Equations (15.4.14) and (15.4.15) possess then only two critical points at $(X(-\infty), 0)$ and $(1, 0)$, respectively. Near the point $(X(-\infty), 0)$ put $X(\zeta) = X(-\infty) + \xi$, $Y(\zeta) = \eta$ and linearise the differential equations (Section 5.3) to give

$$\xi'/\alpha = \left\{ \beta - \frac{1}{X(-\infty)} \right\} \xi + (\beta + \gamma)\eta,$$

$$\eta'/\alpha = \left\{ \frac{1}{X(-\infty)} - \beta \right\} \xi - \beta\eta.$$

The equation for the exponents (see Section 5.4) is

$$\lambda^2 + \frac{\alpha\lambda}{X(-\infty)} - \alpha^2\gamma\left\{ \frac{1}{X(-\infty)} - \beta \right\} = 0, \tag{15.4.20}$$

which corresponds to a saddle-point since $\beta X(-\infty) < 1$. The dividing lines of Figure 5.4.3 have equations

$$X(-\infty)(\beta + \gamma)\eta = \left[\frac{1}{2} - \beta X(-\infty) \pm \frac{1}{2}\{1 + 4\gamma X(-\infty)(1 - \beta X(-\infty))\}^{1/2} \right]\xi$$

with the upper sign for the line which leaves the saddle-point. Evidently, the line with the upper sign has positive slope whereas that with the lower sign has negative slope. Consequently, there is just one trajectory from $(X(-\infty), 0)$ entering the first quadrant of the phase plane in $X(\zeta) > X(-\infty)$.

The linearised equations for the critical point $(1, 0)$ are the same except for $X(-\infty)$ being replaced by unity. Hence (15.4.20) continues to hold with the same replacement. Since $\beta > 1$ the critical point is either a stable node or a stable focus. A focus is unacceptable because a trajectory in its neighbourhood would spiral around the critical point $(1, 0)$ and values of $X(\zeta)$ greater than 1 would occur on it. Since $X(\zeta)$ is not permitted to exceed 1 the critical point cannot be a focus. In order that it be a stable node it is necessary that

$$4(\beta - 1)\gamma \le 1. \tag{15.4.21}$$

Expressed in terms of the original quantities via (15.4.16) this implies that

$$c \ge 2[\phi s_1(\infty)\{\theta s_1(\infty) - g\}]^{1/2}, \tag{15.4.22}$$

which sets a lower bound on the speed of a possible wave.

To confirm that there is a wave with a speed satisfying (15.4.22) we must show that there is a trajectory connecting the two critical points subject to (15.4.18), (15.4.19) and (15.4.21). Since it has been shown already that there is a trajectory leaving $(X(-\infty), 0)$ on which $X(\zeta)$ increases it will be sufficient to verify that this trajectory must end up at $(1,0)$. On $Y = 0$ we see from (15.4.14) and (15.4.15) that $Y' > 0$ and $X' < 0$ between the critical points. Hence no trajectory can leave the first quadrant by crossing $Y = 0$ between the critical points. On $X = X(-\infty)$ it is clear that $X' > 0$ and $Y' < 0$ so that no trajectory in $X > X(-\infty)$ can depart through the line $X = X(-\infty)$.

Now consider what happens on the line

$$(\beta + \gamma)Y = \left(\tfrac{1}{2} - \beta\right)(X - 1). \tag{15.4.23}$$

Since $\ln X \le X - 1$ for $X > 0$

$$X'/\alpha \ge -\tfrac{1}{2}(X - 1) > 0,$$
$$Y'/\alpha \le \left(\tfrac{1}{2}\beta + \gamma - \gamma\beta\right)(X - 1)/(\beta + \gamma).$$

Accordingly, on a trajectory

$$\frac{dY}{dX} \le \frac{2\gamma(\beta - 1) - \beta}{\beta + \gamma} \le \frac{1/2 - \beta}{\beta + \gamma}$$

by virtue of (15.4.21). The conclusion is that, since $\beta > 1$, no trajectory can pass through the line (15.4.23) from below to above. Thus the sides of the triangle bounded by $Y = 0, X = X(-\infty)$ and (15.4.23) are impassable barriers to a trajectory in the interior. It follows from the Poincaré-Bendixson theorem (Section 5.6) that a trajectory from $(X(-\infty), 0)$ must go to either a limit cycle or the other critical point. If there were a limit cycle the observation at the end of Section 5.6 indicates that there would have to be a critical point interior to it and, therefore, in the interior of the triangle. But that is contrary to what has been demonstrated already. Any trajectory inside the triangle must go more or less directly to the stable node at $(1,0)$. Therefore, the trajectory entering from $(X(-\infty), 0)$ must go to $(1,0)$, i.e., there is a unique trajectory starting from $(X(-\infty), 0)$ and ending at $(1,0)$.

The trajectory is traversed from start to finish as ζ increases. Increasing ζ corresponds to decreasing t. Therefore, as t increases, ζ falls and progress along the trajectory must be in the opposite direction. In other words, as t grows with x fixed, s_1 changes continuously from $s_1(\infty)$ to $s_1(-\infty)$. An epidemic wave can exist that alters the number of susceptibles from $s_1(\infty)$ before the arrival of the wave to $s_1(-\infty)$ after the wave has finally died away provided that (15.4.18), (15.4.19) and (15.4.22) are valid. The only restriction on the wave speed is that of (15.4.22). For every value of c which satisfies (15.4.22) there is a trajectory in the phase plane which accomplishes the transition of susceptibles from start to finish.

In summary the predictions of this model are as follows:

(i) There can be no epidemic wave unless there are sufficient initial susceptibles as specified by (15.4.18).

(ii) If there are enough susceptibles initially for a wave the number of susceptibles left behind is insufficient for another wave without regeneration.

(iii) There can be no wave with a speed less than

$$2[\phi s_1(\infty)\{\theta s_1(\infty) - g\}]^{1/2}$$

and the larger $s_1(\infty)$ the faster the wave must travel, i.e., the more susceptibles there are initially the quicker the epidemic must spread. There can be epidemic waves with speeds higher than the minimum. Altering c changes γ (but not β or $X(-\infty)$) and thereby the pattern of trajectories but there is still one from $(X(-\infty), 0)$ to $(1, 0)$.

Exercises

15.1 Show that (15.1.8) determines one and only one positive value of $s(\infty)$ less than b/a.

15.2 In another version of the incubation model, a variable time to become infectious is allowed for by saying that an individual becomes infectious at time t if exposed to the disease at the earlier time $\tau(t)$ so that $\tau(t) \leq t$. Assume that $\tau(0) = 0$ and that $\tau(t)$ increases steadily with t. Show that the equations for $I(t)$ in Section 15.3 are replaced by

$$I(t) = I_0(t) + s(0) - s\{\tau(t)\} \quad (0 \leq t \leq \sigma),$$
$$I(t) = I_0(t) + s\{\tau(t - \sigma)\} - s\{\tau(t)\} \quad (t \geq \sigma)$$

when an individual remains infectious for the time σ.

Deduce that

$$\int_0^\infty I(v)\,dv = \int_0^\sigma I_0(v)\,dv + \sigma\{s(0) - s(\tau(\infty))\}.$$

Show that (15.3.10) is still satisfied with $z = s(\tau(\infty))/s(0)$ and

$$\epsilon = \frac{1}{\beta}\left[a \int_0^\sigma I_0(v)\,dv - \ln\left\{\frac{s(\tau(\infty))}{s(\infty)}\right\}\right].$$

Prove that $\epsilon > 0$.

15.3 To allow for a non-constant time during which an individual is infectious in Exercise 15.2 let $p(x)$ be the proportion of individuals who are still infectious after a time x has elapsed from becoming infectious. Assume that $p(x)$ is positive, decreases steadily as x increases and $p(x) = 0$ if $x > \sigma$. Show that

$$I(t) = I_0(t) - \int_0^t p(t-u)\frac{d}{du}s\{\tau(u)\}du.$$

Prove that

$$\int_0^\infty I(v)dv = \int_0^\sigma I_0(v)dv + \{s(0) - s(\tau(\infty))\}\int_0^\sigma p(u)du$$

and that the other conclusions to Exercise 15.2 still hold, but with

$$\beta = as(0)\int_0^\sigma p(u)du.$$

15.4 Show that, if $\epsilon > 0$ in (15.3.10) and $\beta > 1$, there is a unique positive root with $z < 1$ and it satisfies $0 < z < 1/\beta$.

[*Hint:* $\ln x \le x - 1$ for $x > 0$.]

15.5 If $\epsilon = 0$ in (15.3.10) show that there is one positive root with $z < 1$ if, and only if, $\beta > 1$ and that then $0 < z < 1/\beta$.

15.6 Prove that (15.4.1)–(15.4.3) imply that

$$s(t, x) + I(t, x) + r(t, x) = N(x)$$

where $N(x)$ is the initial population at point x, and deduce that $\lim_{t\to\infty} r(t, x) = R(x)$ exists.

15.7 If $g(t, x) = h(x)$ in (15.4.1)–(15.4.3), prove that

$$N(x) = R(x) + s(0, x)\exp\left\{-\int_{-\infty}^\infty F(x, x')R(x')dx'/h(x')\right\}$$

where N and R are defined in Exercise 15.6.

15.8 Generalise (15.4.1)–(15.4.3) to two space dimensions x, y and show that, if

$$F(x, y, x', y') = f_1(|x - x'|)f_2(|y - y'|)$$

in the small region $|x-x'| < \delta_1, |y-y'| < \delta_2$ but is otherwise zero, the associated diffusion approximation includes a partial differential equation of the form

$$\frac{\partial s}{\partial t} = -s\left(\theta I + \phi_1\frac{\partial^2 I}{\partial x^2} + \phi_2\frac{\partial^2 I}{\partial y^2}\right).$$

15.9 Substitute for $\partial s/\partial t$ in (15.4.5) from (15.4.4). Show that the resulting partial differential equation for I has a solution

$$I(t,x) = \frac{e^{(\theta s-g)t}}{t^{1/2}}e^{-x^2/4\phi ts}$$

when s and g are constants. If a is positive and $\int_a^\infty I(t,x)dx = C(t)$ find the formula for $C(t)$. Deduce that, as $t \to 0$ with a fixed,

$$aC(t) = 2\phi t^{1/2}s\exp\{(\theta s - g)t - a^2/4\phi ts\}$$

approximately.

Answers to Exercises

Chapter 1

1.1 $r_0 - \alpha t$.

1.2 $p_0(n)^{t/T}$.

1.4 The curve drops steadily, without any point of inflexion, from an initial value of $p(0)$ to an eventual value of N_0/a. This demonstrates that, if the population could be started above N_0/a by some means, crowding would force it down to N_0/a in the end.

1.6 (a) $y = Dte^t$; (b) $y^2 + 3y = Dt$; (c) $y^2 + 1 = Dt(1+t)$.

1.7 (a) $Dt = e^{y^2/2t^2}$; (b) $\{(y/t) - 1\}e^{y/t} = C + \ln t$; (c) $y = t/(C + \ln t)$.

1.8 (a) New equation is $dy/dt = (t+y)/(t-y)$, which is of homogeneous type (with solution $\tan^{-1} z - \frac{1}{2}\ln(1+z^2) - \ln t = C, z = y/t$).

   (b) New equation is $dy/dt = (at+by)/(a't+b'y)$, which is of homogeneous type.

1.9 (a) $y = (t+C)(1-t^2)^{-1/2}$; (b) $y = Ce^{t^2/2} - e^{2t}$; (c) $w = \pm t(C+t^2)^{1/2}$.

1.10 $y = (1 + Ce^{t^2/2})^{-1}$.

1.11 (a) $y^2 = C(t+1)^2 - 2(t+1)$; (b) $t^2 = Ce^{2y} + 2y$.

1.12 $y = \dfrac{2}{t} + \dfrac{1}{Ct^8 - 2t/7}$.

1.13 (a) $y = \dfrac{2}{t} + \dfrac{1}{Ct^5 - t/4}$; (b) $y = t + \dfrac{1}{t+C}$.

1.14 (a) $\left\{\dfrac{t-1}{t+1}\right\}^{1/2}$ $y = \frac{1}{2}\ln(t+1) + C$;

   (b) $\sinh^{-1}(2y/t) = \ln t + C$;

   (c) $e^t + e^{-2y} = C$;

   (d) $\frac{1}{2}\ln\{4(t+1)^2 + (2y+3)^2\} + \tan^{-1}\left(\dfrac{y+3/2}{t+1}\right) = C$;

(e) $\frac{1}{2}\ln\left(\dfrac{1+2t-y}{1-2t+y}\right) = 2t + C$;

(f) $(t+a)^{-3}y = (t+a)^2 + C$;

(g) $\tan(2y/t) + \ln t = C$;

(h) $\frac{1}{8}(2y - 2t + 5)^2 + 2t = C$;

(i) $(2y - 1)\cos^2 2t = C(2y + 1)$;

(j) $(1-t)^{1/2}y = 2(1+t)^{1/2} + C$;

(k) $2t - 2y + \ln\{(2t + 2y + 1)^2 + 1\} = C$;

(l) $te^y = \tan y + C$.

1.16 40 min.

1.17 200 h.

1.18 98.1%.

1.19 $c = (2D/W)e^{-t/6}$; about 8 hours; about $7\frac{1}{2}$ hours.

1.20 25–100 mg/h.

1.21 $p = p_0 \exp\{-K(t - t_0)/R\}; p = RI_0 + (p_0 - RI_0)\exp\{-K(t - t_0)/R\}$.

Chapter 2

2.1 (a) $C_1 e^{3t} + C_2 e^{5t}$;

(b) $(C_1 + C_2 t)e^{4t}$;

(c) $(C_1 \cos 2t + C_2 \sin 2t)e^{-t} + t^2 + \frac{1}{5}t - \frac{12}{25}$;

(d) $C_1 e^{2t} + C_2 e^{-t} + 10te^{2t}$;

(e) $C_1 \cos \omega t + C_2 \sin \omega t + \cos \Omega t/(\omega^2 - \Omega^2)$;

(f) $C_1 \cos \omega t + C_2 \sin \omega t + (t/2\omega)\sin \omega t$;

(g) $(C_1 \cos \frac{1}{2}\sqrt{3}t + C_2 \sin \frac{1}{2}\sqrt{3}t)e^{-t/2} + \frac{3}{2} + \frac{9}{26}\cos 2t - \frac{3}{13}\sin 2t$.

2.2 (a) $C_1 e^{-2t} + C_2 e^{-t} + 2t^2 - 7t + 17/2$;

(b) $C_1 e^{(\sqrt{3}-1)t} + C_2 e^{-(\sqrt{3}+1)t} - \frac{3}{10}t^5 - \frac{3}{2}t^4 - 9t^3 - 36t^2 - 99t - 135$;

(c) $C_1 + C_2 e^{-3t} + \frac{1}{2}t^4 - \frac{2}{3}t^3 + \frac{7}{6}t^2 + \frac{2}{9}t$;

(d) $(C_1 \cos \frac{1}{2}\sqrt{3}t + C_2 \sin \frac{1}{2}\sqrt{3}t)e^{-t/2} + 2e^t(2\sin t - 3\cos t)$;

(e) $C_1 + C_2 e^{-3t} + \frac{5}{3}t - \frac{3}{2}(3\cos t + \sin t)$;

(f) $(C_1 + C_2 t)e^t + \cos t + 3t^2 e^t$;

(g) $C_1 \cos 3t + C_2 \sin 3t + (\sin 3t - \frac{1}{2}\cos 3t)t$.

2.3 (a) $\{(At+B)\cos t + (Ct+D)\sin t\}t + Et + F$;

(b) $At^2 + Bt + C + (Dt+E)t^2e^{2t} + (Ft+G)\cos 2t + (Ht+I)\sin 2t$;

(c) $(At^2+Bt+C)e^t\sin 2t + (Et^2+Ft+G)e^t\cos 2t + e^t(H\cos t+I\sin t) + Ke^t$;

(d) $(At+B)e^t + \{(Ct+D)\cos 2t + (Et+F)\sin 2t\}t$.

2.4 $(C_1 + C_2\ln t + \ln^2 t)t$.

2.5 $C_1e^{2t} + C_2e^{-t} + 5te^{2t}$.

2.6 $C_1\cos t + C_2\sin t + 3t$.

2.7 $\frac{1}{5}e^t - \frac{1}{4}t\cos 2t$.

2.8 (a) $\frac{4}{3}e^{-t}(1+t)^{5/2}$;

(b) $(C_1 - t)\cos t + (C_2 + \ln\sin t)\sin t$.

2.10 (a) $A\cos t + B\sin t + C\cos 2t + D\sin 2t$;

(b) $(A+Bt)e^{-2t} + Ce^{-3t}$;

(c) $A + Bt + (C\cos 3t + D\sin 3t)e^{-t}$;

(d) $(C_1 + C_2t + C_3t^2 + C_4t^3)e^t$.

2.12 (a) $C_1 + C_2e^t + C_3e^{-t} + \frac{1}{2}\cos t$;

(b) $C_1\cos t + C_2\sin t + C_3e^t + C_4e^{-t} - 1 - t^2$;

(c) $(t^2 - 5t)e^t + C_1 + C_2t + C_3e^t + C_4e^{-t}$.

2.13 (b) $(At+B)e^{-t^2}$;

(c) $(\cos 1 + 2\sin 1 - 1)/t^2 + (2 + \cos 1 - \sin 1)/t - (t\sin t + 2\cos t)/t^2$.

Chapter 3

3.1 (a) $x = C_1e^t + C_2e^{-5t} - \frac{6}{7}e^{2t}$,
$y = -C_1e^t + C_2e^{-5t} + \frac{8}{7}e^{2t}$;

(b) $x = (C_1 + C_2t - 8t^{5/2})e^t$,
$y = (C_1 - \frac{1}{2}C_3 + C_2t + 10t^{3/2} - 8t^{5/2})e^t$.

3.2 $x = \frac{3}{2} + Ce^{-t/2}$, $y = -\frac{1}{2} - 3Ce^{-t/2}$.

3.3 (a) $x = C_1e^{-7t} + 2C_2e^{-3t}$,
$y = 2C_1e^{-7t} + 3C_2e^{-3t}$;

(b) $x = (C_1 \cos 2t + C_2 \sin 2t)e^{-t}$,
$\quad y = (C_2 \cos 2t - C_1 \sin 2t)e^{-t}$;

(c) $x = (C_1 + C_2 t)e^t$,
$\quad y = C_2 e^t$;

(d) $x = C_1 + C_3 e^{2t}$,
$\quad y = C_2 e^{-t} - 2C_3 e^{2t}$,
$\quad z = C_1 - 2C_2 e^{-t} + C_3 e^{2t}$;

(e) $x = C_1 e^{2t} + \{C_2 + (2+t)C_3\}e^t$,
$\quad y = (C_2 + C_3 t)e^t$,
$\quad z = C_1 e^{2t} + \{C_2 + (1-t)C_3\}e^t$.

3.4 $x = (2+3t)e^t - t - 1$,
$\quad y = (6t+1)e^t - 3t - 1$.

3.5 (a) $x = e^t - \frac{1}{7}e^{-5t} - \frac{6}{7}e^{2t}$,
$\quad y = -e^t - \frac{1}{7}e^{-5t} + \frac{8}{7}e^{2t}$;

(b) $x = 2e^{-4t} + e^{3t} + t - 2$,
$\quad y = e^{-4t} - 3e^{3t} + 2$.

3.6 $x = t^3 e^{3t} + t$,
$\quad y = 1 - 3t$,
$\quad z = -6$.

3.7 $t \ (0 \le t \le \pi)$, $\frac{1}{2}\pi e^{\pi - t} - \frac{1}{2}\pi \cos t - (\frac{1}{2}\pi + 1)\sin t \ (t > \pi)$.

3.13 $-\dfrac{4}{\pi} t^{1/2} \displaystyle\sum_{m=0}^{\infty} \dfrac{\sin(2m+1)(t-\pi)}{(2m+1)\{(2m+1)^2 + 1\}}$.

Chapter 4

4.1

$$\epsilon \frac{d^2 x}{dt^2} + (3x^2 + a)\frac{dx}{dt} + x = x_a.$$

4.5

$$v^{(0)}(r,t) = \frac{(a^2 - r^2)}{4\nu}\exp(\nu\lambda t/\rho),$$

$$v^{(1)}(r,t) = \frac{(a^2 - r^2)}{4\nu}\exp(\nu\lambda t/\rho) - \frac{\lambda}{64\nu}(a^2 - r^2)(3a^2 - r^2)\exp(\nu\lambda t/\rho),$$

$$v^{(2)}(r,t) = \frac{(a^2 - r^2)}{4\nu}\exp(\nu\lambda t/\rho) - \frac{\lambda}{64\nu}(a^2 - r^2)(3a^2 - r^2)\exp(\nu\lambda t/\rho),$$

$$+ \frac{\lambda^2}{2304\nu}(a^2 - r^2)(19a^4 - 8a^2 r^2 + r^4)\exp(\nu\lambda t/\rho).$$

4.7

$$\phi'' - c\phi' + \phi(1 - \phi)(\phi - a) = \psi,$$
$$c\psi' = b\phi - \gamma\psi.$$

4.9

$$c = 0.$$

4.11

$$a^{-1} = 1 + \left[\frac{c}{2} + \sqrt{\left(\frac{c^2}{4} + 1\right)}\right]^2, \quad \theta = \frac{\pi}{2};$$

$$a^{-1} = 1 + \left(\frac{1 + \left(\frac{c}{2} + \gamma\right)\left[c + \left(\frac{c}{2} + \gamma\right)p\right]}{1 + \left(\gamma - \frac{c}{2}\right)\left(\left(\gamma - \frac{c}{2}\right)p - c\right)}\right)^{1/2}$$

$$\times \exp\left[\frac{c\beta}{2}\cos^{-1}\left(\frac{p - \frac{c^2}{2} - 1}{p + 1}\right)\right],$$

where

$$\gamma = \sqrt{\left(\frac{c^2}{4} + 1\right)}, \quad \beta = \sqrt{\left(p - \frac{c^2}{4}\right)}, \quad p = \tan\theta.$$

4.12

$$(0, 0, 0); \quad \left(X_i, \frac{f k_3 A X_i}{(k_1 A + k_2 X_i)}, \frac{k_3 A X_i}{k_5}\right),$$

where X_i, $i = 1, 2$, satisfies

$$2k_2 k_4 X^2 + [k_2 k_3 A(f - 1) + 2k_1 k_4 A]X - k_1 k_2 A^2(f + 1) = 0.$$

4.13

$$\frac{dX}{dt} = A - (B + 1)X + X^2 Y,$$

$$\frac{dY}{dt} = BX - X^2 Y.$$

4.14

$$(0, 0); \quad \left(\frac{a}{b}, 0\right); \quad \left(\frac{e}{f}, \frac{af - be}{cf}\right).$$

4.15

$$(0,0); \quad \left(\frac{a}{b},0\right); \quad \left(\frac{e}{f}+\bar{X},\ \frac{(e+f\bar{X})\left[af-b(e+f\bar{X})\right]}{cef}\right),$$

$$c \neq 0, \ af > b(e+f\bar{X}).$$

$$(0,0); \quad \left(\frac{a}{b},0\right); \quad \left(0,\frac{e}{g}\right); \quad \left(\frac{ce-ag}{cf-gb},\frac{fa-bc}{cf-gb}\right);$$

$$\frac{ce-ag}{cf-gb} > 0 \quad \text{and} \quad \frac{fa-bc}{cf-gb} > 0.$$

Chapter 5

5.1 (a) None; (b) $y_0 = 0$, singular; (c) $t_0 = 0$, singular; (d) $t_0 = 0$, singular; (e) $t_0 = 0$, singular.

5.2 (a) Unstable node, (b) saddle-point, (c) centre, (d) stable node, (e) unstable focus, (f) saddle-point, (g) stable focus, (h) centre.

5.3 Stable node.

5.4 $a < 0$, saddle-point; $b^2 > a > 0$ node (stable $b > 0$, unstable $b < 0$); $b = 0, a > 0$, centre; $b^2 < a$ focus (stable $b > 0$, unstable $b < 0$).

5.5 Focus.

5.6 The trajectories are $\frac{1}{2}cy^2 + x^2 + ax^3 + bx^4 = \text{constant}$ and $x = 0, y = 0$ is a centre.

5.8 If $b > 0$ there is a centre at $x = 0, y = 0$ and, if $b < 0$, a saddle-point. If $b = 0$ *either* $x = \text{constant}, y = 0$ for all t *or* y remains of one sign.

5.9 $\theta = \pm\pi$ are saddle-points. For $0 < \mu < 1$ there is a saddle-point at $\theta = 0$ and centres at $\theta = \pm\cos^{-1}\mu$. For $\mu > 1$, there is a centre at $\theta = 0$.

5.10 The trajectories are $2\lambda^2 y^2 = 2C\lambda^2 e^{-2\lambda x} + 1 - 2\lambda x$ where C is a constant. Closed curves for $-1 < 2\lambda^2 C < 0$.

5.14 Unstable limit cycles.

Chapter 7

7.2 Critical points are $(u_i, \frac{b}{\gamma} u_i)$, where u_i, $i = 1, 2, 3$, satisfies

$$u^3 - u^2(1 + a) + \left(a + \frac{b}{\gamma}\right) u - I = 0,$$

$$I = \frac{b}{\gamma}: \quad (u, w) = \left(1, \frac{b}{\gamma}\right), \quad \text{if } a^2 < \frac{4b}{\gamma};$$

$$(u, w) = \left(1, \frac{b}{\gamma}\right), \left(\frac{a \pm \sqrt{(a^2 - 4b/\gamma)}}{2}, \frac{b}{\gamma}\left[\frac{a \pm \sqrt{(a^2 - 4b/\gamma)}}{2}\right]\right),$$

$$\text{if } a^2 \geq \frac{4b}{\gamma};$$

$$I = \frac{ab}{\gamma}: \quad (u, w) = \left(a, \frac{ab}{\gamma}\right), \quad \text{if } \frac{b}{\gamma} > \frac{1}{4};$$

$$(u, w) = \left(a, \frac{ab}{\gamma}\right), \left(\frac{1 \pm \sqrt{(1 - 4b/\gamma)}}{2}, \frac{b}{\gamma}\left[\frac{1 \pm \sqrt{(1 - 4b/\gamma)}}{2}\right]\right), \quad \text{if } \frac{1}{4} \geq \frac{b}{\gamma}.$$

7.4

$$c\phi' = \phi'' + f(\phi) - \psi,$$
$$c\psi' = b\phi - \gamma\psi.$$

Unique rest state if $f(\phi) = \phi g(\phi)$ and $g(\phi) \neq b/\gamma$.

Chapter 9

9.1 47 years; 200 whales.

9.2 1.26 million (in fact 5 million were killed each year in the early 1870s).

9.3 If $A = ad - cg$ there are three critical points where $A > 0$: saddle-points at $(0, 0)$ and $(a/g, 0)$ whereas $(c/d, A/bd)$ is a stable node if $A < g^2 c/4d$ but a stable focus if $A > g^2 c/4d$. Predator population becomes A/bd eventually. If $A < 0$ there is a saddle-point at $(0, 0)$ and a stable node at $(a/g, 0)$; the predators are eventually wiped out.

9.4 Four critical points $(0, 0)$, $(a/g, 0)$, $(0, c'/g')$, $\left(\frac{ag' - bc'}{gg' + bd}, \frac{ad + c'g}{gg' + bd}\right)$. No.

Chapter 10

10.2 $2y + 2(y - x)^2$.

10.3 $3x$.

10.4 $4\theta + 2\rho e^{-\theta}$.

10.5 $3(y - x)(x - 1) + 3(y - x)^2 e^{-2x} + 3(y - x)e^{-x}$.

10.6 $\ln y - x^2 + y^2 - \frac{1}{2}\ln(y^2 - x^2)$.

10.7 $\frac{1}{2}x^2 - \frac{1}{4}y^2 + \frac{1}{2}x^2 y + \frac{1}{4}$.

10.8 $x(1 + 2y/x)^2 - 2xy$.

10.9 $\frac{1}{2}xy + f(y/x)$ where f is any function such that $f(1) = 0$.

10.10 (a) Elliptic everywhere; (b) hyperbolic everywhere; (c) elliptic in $|x| > 2|y|$, hyperbolic in $|x| < 2|y|$; (d) hyperbolic in first and third quadrants, elliptic in second and fourth; (e) parabolic everywhere; (f) parabolic everywhere.

10.11 $4\cos^2 \alpha \dfrac{\partial^2 u}{\partial \xi \partial \eta} = \dfrac{\partial u}{\partial \xi} + \dfrac{\partial u}{\partial \eta}$.

10.12 $2\dfrac{\partial^2 u}{\partial \xi \partial \eta} = \dfrac{1}{\xi - \eta}\left(\dfrac{\partial u}{\partial \xi} - \dfrac{\partial u}{\partial \eta}\right)$.

10.13 $3\ln y + 3x^2 y - x^2 y^4 - 6$.

10.14 $(Ax + B)^{2\nu/(2\nu - 1)}$.

10.15 (a) One solution; (b) several solutions; (c) either one or none.

Chapter 11

11.2

$$u(x, t) = 1 - \phi(x/\sqrt{4t}), \qquad \phi(s) = \frac{2}{\sqrt{\pi}} \int_0^s e^{-t^2} dt.$$

11.3

$$u(x, t) = \sum_{n=1}^{\infty} A_n e^{-n^2 \pi^2 t} \sin n\pi x, \qquad A_n = 2 \int_0^1 f(x) \sin n\pi x \, dx.$$

11.4

$$u(x,t) = \frac{8}{\pi} \sum_{n=1}^{\infty} \frac{e^{-n^2 t}}{n^3} \sin nx.$$

11.5

$$u(x,t) = 2 + 2 \sum_{n=1}^{\infty} \frac{[(-1)^{n+1} - 1]}{(2 + n^2)} e^{-n^2 t} \cos nx.$$

11.6

$$u(x,y,t) = \frac{4}{\pi^2} \sum_{m=1}^{\infty} \sum_{n=1}^{\infty} \frac{e^{-(m^2 + n^2)\pi^2 t}}{mn} (-1)^{m+n} \sin m\pi x \sin n\pi y.$$

11.8

$$u = \frac{1}{2} \exp(c\xi), \quad -\infty < \xi < 0,$$

$$= \frac{1}{2} + \frac{c}{2\sqrt{(c^2 - 4)}} \exp\left[\left(\frac{c + \sqrt{(c^2 - 4)}}{2}\right)\xi\right]$$

$$- \frac{c}{2\sqrt{(c^2 - 4)}} \exp\left[\left(\frac{c - \sqrt{(c^2 - 4)}}{2}\right)\xi\right], \quad 0 \le \xi \le \xi_1$$

$$= 1 - \frac{1}{4} \exp\left[\left(\frac{c - \sqrt{(c^2 + 4)}}{2}\right)(\xi - \xi_1)\right], \quad \xi_1 < \xi < \infty,$$

where $c > 2$ with c and ξ_1 satisfying

$$\exp\left(\frac{1}{2} c \xi_1\right) \sinh\left(\frac{\sqrt{(c^2 - 4)}}{2}\right) \xi_1 = \frac{\sqrt{(c^2 - 4)}}{4c},$$

$$\coth\left(\frac{\sqrt{(c^2 - 4)}}{2}\right) \xi_1 = 4 \frac{\sqrt{(c^2 + 4)}}{c} - 2.$$

11.14 No.

Chapter 12

12.2

$$c(x,t) = \frac{c_0}{(4\pi t)^{\frac{1}{2}}} \int_{-a}^{a} \exp\left(-\frac{(x - y)^2}{4t}\right) dy.$$

12.3

$$c(r,t) = \frac{2c_0 a}{r\pi} \sum_{n=1}^{\infty} \frac{(-1)^{n+1}}{n} \sin\left(\frac{n\pi r}{a}\right) \exp\left(-\frac{Dn^2\pi^2 t}{a^2}\right), \quad r \leq a,$$

$$= c_0 \left(1 - \frac{2a}{r\pi^{\frac{1}{2}}} \int_{(r-a)/(4\pi t)^{1/2}}^{\infty} e^{-x^2} dx\right), \quad r > a.$$

12.7

$$(u,v) = \left(\frac{\alpha}{\alpha - \beta}, \frac{\beta}{\alpha - \beta}\right), \quad \alpha > \beta, \ \alpha > 1$$

$$\delta > \sqrt{\frac{\alpha}{\beta}}(\sqrt{\alpha - \beta} + \sqrt{\alpha}).$$

Chapter 14

14.5

$$P = \frac{(S_2 - S_1)}{4}(r^2 - R^2) + \frac{\alpha}{R},$$

$$R(t) = R(0)\exp(S_1 - S_2)\frac{t}{2}.$$

14.6

$$X(t) = X(0)\exp(S_1 - S_2)t,$$

where $X(t)$ is the outer boundary of the tumour colony.

Index

NOTE: Italicized page numbers refer to illustrations